INTERMEDIATE
ELECTRONICS

INTERMEDIATE ELECTRONICS

R. J. MADDOCK
C. Eng., M.I.E.R.E.

Lecturer in Electronics,
Department in Electrical Engineering,
Southampton College of Technology

Book 1

Springer Science+Business Media, LLC

ISBN 978-1-4899-5861-7   ISBN 978-1-4899-5859-4 (eBook)
DOI 10.1007/978-1-4899-5859-4

© Springer Science+Business Media New York 1969

Originally published by Butterworth & Co. (Publishers) Ltd. in 1969.

Softcover reprint of the hardcover 1st edition 1969

Suggested U.D.C. number: 621·37/·38
Library of Congress Catalog Card Number 69-18297

PREFACE

Many textbooks adequately cover the development of small signal equivalent circuits for electronic devices. Students frequently experience difficulty in applying such circuits to the solution of the various configurations found in practice. In this volume I have demonstrated step by step procedures by which the required solutions may be obtained by the use of such equivalent circuits.

The first chapter shows clearly the necessity for equivalent circuit methods and also discusses the d.c. biasing requirements and circuits for valve and transistor amplifiers. A chapter follows covering the fundamentals of network analysis required for the remainder of the book. Valve and transistor equivalent circuits are then introduced and detailed methods of application are given with many fully worked examples. Later chapters extend these methods to a discussion on the effects of feedback, the solution of practical feedback amplifiers and to the analysis and design of a range of sinusoidal oscillators. A final chapter introduces the modifications required when considering high frequency operation. The required high frequency equivalent circuits are given together with sample calculations.

I have included worked solutions and problems with answers in all chapters. The parameters chosen for the various equivalent circuits are those most commonly available from manufacturers' published data. British Standard symbols and units are used throughout except where very recent changes may be unfamiliar to the majority of students.

Both valve and transistor circuits are discussed as I feel that for many students the valve provides a useful teaching medium, but in the later chapters, the stress is, in general, on transistor circuits. The physical operation of the devices has not been covered, as adequate literature is available elsewhere. A list of works for further reading is provided at the end of the book.

I have written this book as a result of the experience gained through teaching students at Higher National Diploma and Certificate level but I feel that it should also be useful for students at the early stages of degree courses and the final year of technician courses.

PREFACE

I would like to offer my thanks to those members of the staff and students of Southampton College of Technology who have helped in the preparation of this book and to Miss L. Lavender and Mrs. R. Huntingford for their assistance in typing the manuscript.

R. M.

CONTENTS

1

GRAPHICAL ANALYSIS OF ELECTRONIC CIRCUITS

Electronic engineering is principally concerned with the behaviour of electrical circuits or networks containing various non-linear and active devices. These include the various types of rectifier, thermionic valves, transistors, transducers such as photocells and many other devices. The difficulty in designing or analysing such circuits lies in their inherent non-linearity. The physical explanation of their behaviour will not in general be discussed in this book since this aspect is more than adequately covered by other authors.

We must first consider suitable methods of describing the behaviour of a particular device. Any chosen description will be suitable only for a limited set of conditions. This concept is not peculiar to these electronic devices; a capacitor, for example, may be shown as a pure capacitance shunted by a resistance representing the dielectric loss. This representation holds only if the peak voltage does not exceed the breakdown value for the capacitor. Also large changes in temperature or humidity may well affect this model.

D.C. CHARACTERISTICS

The first description of valves and transistors that will be considered are the so-called d.c. characteristics. These characteristics are graphs showing how the various direct voltages applied to the device

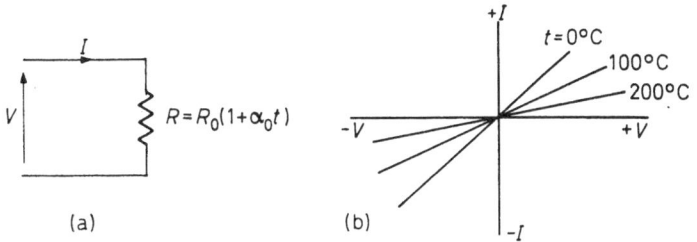

Figure 1.1. The characteristics of a resistor

1

change the direct currents flowing in it. If there are more than two variables, the values of one are plotted as a second is changed with all others held constant. A variable that is fixed is known as a parameter of the characteristics. A simple example of this would be the variation of current flowing in a resistor with variation of applied voltage. In this case the parameter could be the ambient temperature. The resulting family of characteristics are shown in *Figure 1.1*.

Similarly with thermionic and semiconductor diodes the cathode and ambient temperature respectively could be taken as the parameters, as shown in *Figure 1.2a* and *b*.

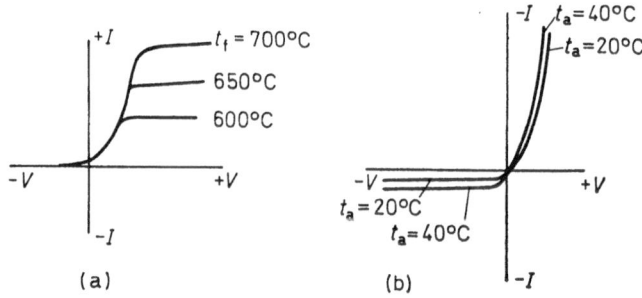

Figure 1.2. Diode characteristics. (*a*) Thermionic. (*b*) *p–n* junction

With three-electrode devices such as triode valves and transistors the cathode or ambient temperatures are assumed to be constant and the parameters of the characteristics are taken as one of the applied voltages or currents. For example the anode characteristics of a triode valve are shown with constant cathode temperature (filament voltage), while the anode current (I_A), is measured for different values of anode–cathode voltage (V_{AK}) taking the grid-cathode voltage (V_{GK}) as the parameter. A 'family' of characteristics is thus obtained for different fixed values of V_{GK} (*Figure 1.3a*). For the transistor collector characteristics, the ambient temperature is constant thus collector current (I_C) is plotted against the collector emitter voltage (V_{CE}) using the base current (I_B) as the parameter (*Figure 1.3b*).

With multi-electrode devices such as pentode valves, other potentials or currents will have to be kept constant for a particular set of characteristics. The anode characteristics of a pentode are quoted for constant filament voltage (V_F) constant screen voltage (V_{G2K}) and constant suppressor cathode voltage (V_{G3K}). The parameter again is V_{G1K}.

2

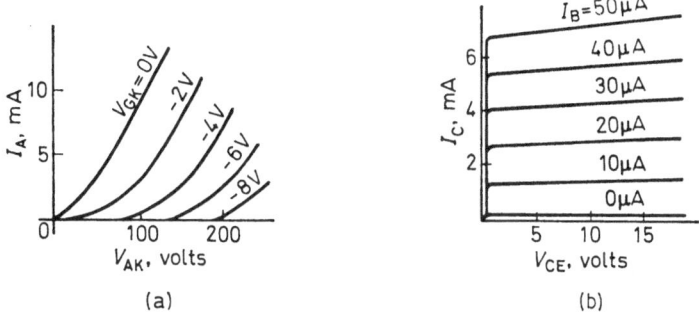

Figure 1.3. Active device characteristics. (*a*) Triode valve. (*b*) Common emitter transistor

D.C. Operating Point

The operating point is given by the values of the two variables and the parameter corresponding to a particular point on the characteristics. Since these three variables are not independent, the operating point is completely defined by any two of the three. For example a particular operating point for a triode valve could be given as V_{AK} 150 V, I_A 3 mA, or I_A 3 mA, V_{GK} -2 V, or as V_{AK} 150 V, V_{GK} -2 V (see *Figure 1.4a*). For a transistor it might be

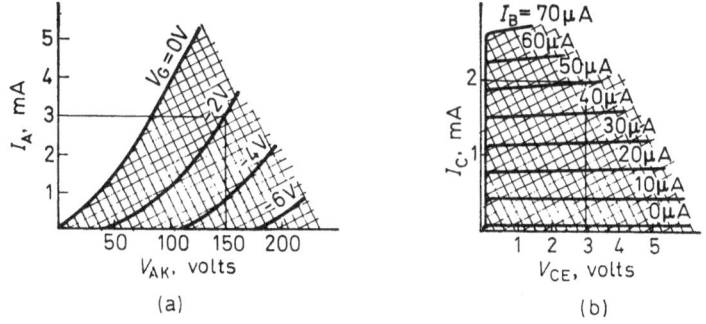

Figure 1.4. Permissible range of d.c. operating point for (*a*) triode valve and (*b*) transistor

given by I_C 2 mA, I_B 50 μA or I_C 2 mA, V_{CE} 3 V etc. (*Figure 1.4b*). The operating point can lie anywhere within the range given by the characteristics. In practice it should not lie outside the shaded area shown, since no information is given in the particular set of characteristics. Thus we can say the characteristics provide one restriction

3

on the possible range of values of the operating point. Further restrictions will be provided by the electrical circuit into which the device is connected.

D.C. Load Lines

The basic problems concerned with graphical solutions may be most easily understood by a consideration of triode valve circuits. These, therefore, will be considered in detail before proceeding to basic transistor circuits. First, consider the case of the triode valve connected in series with a resistive load R_L and a d.c. supply voltage

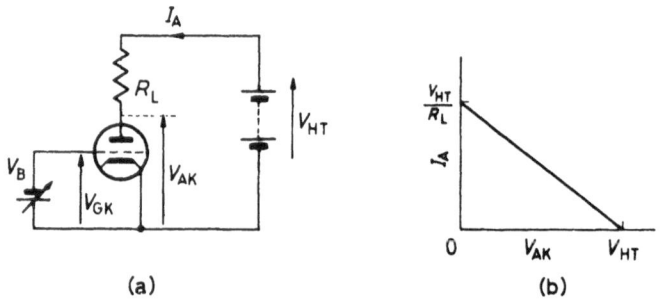

| (a) | (b) |

Figure 1.5. (*a*) D.C. circuit and load for valve. (*b*) Load line resulting from the circuit shown in (*a*)

V_{HT}. The grid will be maintained at a potential negative with respect to cathode by a second battery of V_B volts as shown in *Figure 1.5a*. An equation can now be written relating V_{AK} and I_A.

$$V_{AK} = V_{HT} - I_A R_L$$

This is a straight line law as shown in *Figure 1.5b* having intercepts

if $\qquad\qquad I_A = 0 \qquad V_{AK} = V_{HT}$

if $\qquad\qquad V_{AK} = 0 \qquad I_A = \dfrac{V_{HT}}{R_L}$

Thus in this circuit the operating point may lie only at any point on the straight line. This line is known as a load line since the slope $-1/R_L$ is governed by the load resistor R_L. This load line is a function of the circuit only and if the valve were replaced by another device such as a lamp or transistor it would be unchanged. In the circuit of *Figure 1.5a* we can say that the operating point is restricted by (*a*) the valve characteristics and (*b*) the load line. Under these conditions, if one of the three variables, V_{GK}, V_{AK} or I_A is fixed then the other two can have only one possible pair of values. For example,

4

if V_{GK} is fixed at 0 V, the operating point must lie on the zero volt characteristic. It must also lie on the load line, and the only possible operating point now lies at the intersection between the zero volt characteristic and the load line.

Consider the valve having the characteristics shown in *Figure 1.6* connected in the circuit of *Figure 1.5a* with V_{HT} 300 V, and R_L 20 kΩ.

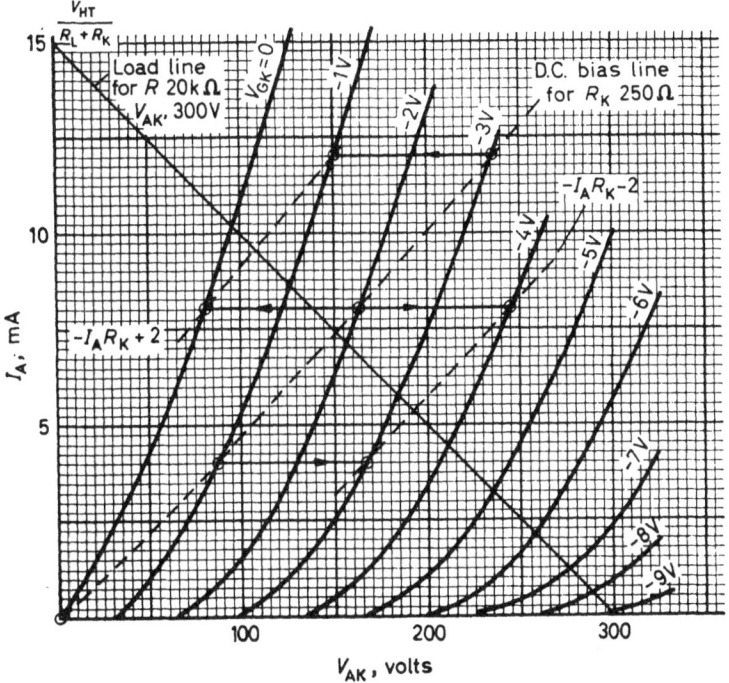

Figure 1.6. Triode characteristics with load and bias lines

The intercepts of the load lines are found:

If $\qquad I_A = 0 \qquad V_{AK} = V_{HT} = 300$ V

If $\qquad V_{AK} = 0 \qquad I_A = \dfrac{V_{HT}}{R_L} = 15$ mA

and the load line is plotted as shown.

Voltage Amplification

The operating point will now be fixed by setting the V_B supply, and thus V_{GK}, to say -2 V. Reading from the graph we find the

5

operating point is V_{AK} 156 V, I_A 7·2 mA. Now suppose V_{GK} is changed from -2 V to -1 V, a change of $+1$ V, the operating point moves to V_{AK} 126 V, I_A 8·7 mA.

Thus a change of $+1$ V in V_{GK} results in a change of $-(156 - 126) = -30$ V on V_{AK}. This effect is known as voltage amplification A_V, and in general

$$A_V = \frac{\Delta V_{out}}{\Delta V_{in}} = \frac{\Delta V_{AK}}{\Delta V_{GK}} = \frac{-30}{+1} = -30$$

Non-linearity

So, for this particular case we can say that the voltage amplification, or the voltage gain, of the circuit is -30. Since a valve is a non-linear device, we cannot expect the voltage gain to be the same for all changes of V_{GK}. To investigate this we shall now change V_{GK} from -2 V to -8 V. The value of V_{AK} changes to 289 V.

Thus $\qquad A_V = \dfrac{+289 - 156}{-8 - (-2)} = \dfrac{-133}{6} = -22·17$

This result is of the same order as the previous one and is only reduced as the spacing between the characteristics becomes less for higher values of V_{AK}.

A.C. AMPLIFICATION

For many electronic applications we are not concerned with changes of direct voltage, but with alternating voltages. If an alternating voltage generator of e_s volts is now connected in series with the V_B battery, the instantaneous value of V_{gk} will be given by $V_{gk} = -2 + E_s \sin \omega t$ where E_s is the peak value of the alternating voltage and ω the angular frequency. It is assumed that the value of

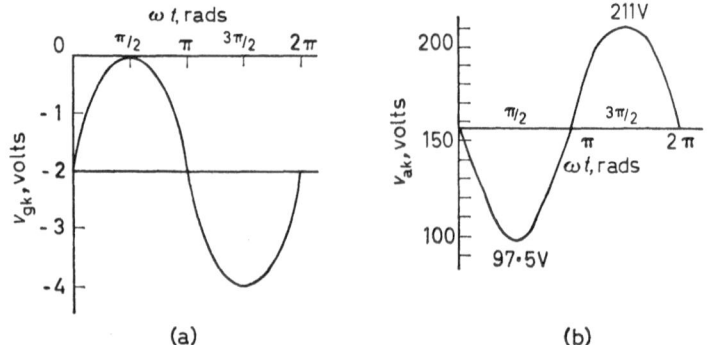

Figure 1.7. Input and output waveforms for simple triode amplifier

6

ω is such that the effect of any stray capacitance or inductance may be neglected. If E_s is 2 V the waveform of V_{gk} is shown in *Figure 1.7a*.

The corresponding values of V_{ak} can be obtained from the graph in *Figure 1.6*, and are shown in *Figure 1.7b*. The resultant V_{ak} is nearly sinusoidal, 180° out of phase with V_{gk} and amplified by a factor of about 28. This output waveform is not quite identical in form to the input waveform. This is the result of the non-linearity of the valve characteristics and is therefore known as non-linear distortion. The larger the signal amplitude the greater the distortion and vice versa. The reader can investigate this for himself by making $V_B = -4$ V and $E_s = 4 \sin \omega t$. The peaks and zeros of the waveform should be sufficient to indicate the degree of distortion obtained in this case.

Biasing

In the circuit discussed above, the mean value of V_{GK} or the grid bias was obtained by the use of a separate d.c. supply. This bias is essential since if it were not present, positive half cycles of the signal

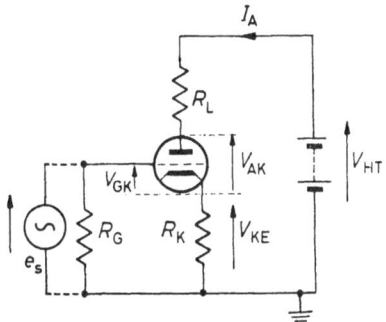

Figure 1.8. Triode amplifier with cathode bias

would cause the grid to become positive with respect to the cathode. This would cause grid current to flow with possible valve damage. In addition the given characteristics show no information for the positive grid region and the variation of operating point could not be predicted. In practice the use of batteries is inconvenient and alternative methods are provided. For most amplifiers, the cathode bias circuit shown in *Figure 1.8* is used. Initially neglecting the a.c. source e_s, the equation for the circuit is

$$V_{AK} = V_{HT} - I_A R_L - I_A R_K$$
$$= V_{HT} - I_A(R_L + R_K) \qquad (1.1)$$

7

Also the voltage at the grid with respect to cathode,

$$V_{GK} = V_{EK} = -V_{KE}$$

where V_{KE} is the voltage at the cathode with respect to earth. R_G has no effect on V_{GK} since grid current is assumed to be zero. Thus R_G maintains the grid at d.c. earth potential. The second equation is therefore

$$V_{GK} = -I_A R_K \qquad (1.2)$$

Equation 1.1 relates V_{AK} to I_A and can be plotted on the anode characteristics to give a load line as before. The intercepts are now V_{HT} and $V_{HT}/(R_L + R_K)$, and once again the operating point V_{AK}, I_A must lie on this line. Since the operating point is not known the grid bias voltage $-I_A R_K$ cannot be found directly. Any given value of I_A would result in a known grid bias. Such values of grid bias can be plotted against I_A on the anode characteristics for any given value of I_A. The point at which this d.c. bias line and the load line intersect is the only operating point satisfying both equations 1.1 and 1.2. A numerical example will illustrate this method.

Example 1.1. Consider the circuit in *Figure 1.8* with values V_{HT} 300 V, R_L 19·75 kΩ, R_K 250 Ω. The valve has characteristics shown in *Figure 1.6*. Determine the d.c. operating point and voltage amplification.

From equation 1.1,

$$V_{AK} = 300 - I_A(19·75 + 0·25) \text{ V}$$
$$= 300 - 20I_A \text{ V}.$$

This gives the same load line as before.

Taking equation 1.2, if $I_A = 0$, $V_{GK} = 0$. If $I_A = 1$ mA, $V_{GK} = -0·25$ V. The second point cannot be plotted since we have no characteristic for $V_{GK} = -0·25$ V. But we have a characteristic for $V_{GK} = -1$ V and the corresponding I_A may be found

$$I_A = \frac{-V_{GK}}{R_K} = \frac{+1}{0·25} = 4 \text{ mA}$$

Similarly for -2 V, $I_A = 8$ mA, for -3 V $I_A = 12$ mA. These four points are shown plotted with the characteristics and the resultant d.c. bias line has been drawn. (Note the bias line is not quite straight and must in general be plotted from more than two points.)

The operating point of the valve in this circuit may now be read off the graph V_{AK}152 V, I_A 7·4 mA.

To evaluate the amplification of this circuit we must find how the operating point moves when the a.c. source e_s is connected between grid and earth (*Figure 1.8*). The voltage between grid and cathode is now given by

$$V_{gk} = -I_a R_k + E_s \sin \omega t$$

If the same value of E_s, 2 V is used as in the previous example then e_s moves between $+2$ V and -2 V. The resulting grid cathode voltage must vary from

$$V_{gk} = -I_A R_K + 2 \tag{1.3}$$

to

$$V_{gk} = -I_A R_K - 2 \tag{1.4}$$

Equations 1.3 and 1.4 represent two further bias lines which may be obtained by shifting the original d.c. bias line by $+2$ V and -2 V respectively. These are shown in *Figure 1.6* and the intersections with the load line give the limits of the variation of the operating point.

From the graph these are found to be

$$e_s = +2, \qquad V_{AK} = 109 \text{ V}, \qquad I_A = 9.6 \text{ mA}$$

$$e_s = -2, \qquad V_{AK} = 194 \text{ V}, \qquad I_A = 5.3 \text{ mA}$$

Voltage gain $A_V = \dfrac{\Delta V_{AK}}{\Delta E_s} = \dfrac{85}{-4} = -21.25.$

Strictly the output voltage should be given by the voltage at the anode with respect to earth. This is given by

$$V_{AE} = V_{AK} + V_{KE} = V_{AK} + I_A R_K$$

The extremes of V_{AE} are $109 + 0.25 \times 9.6 = 111.4$ V, and

$$194 + 0.25 \times 5.3 = 195.3 \text{ V}$$

and

$$A_V = \frac{83.9}{-4} = -20.975$$

Thus the addition of R_K to the circuit can provide the required grid bias but it also results in reduction of the overall amplification V_{ae}/e_s. This is due to an effect known as negative feedback which will be discussed in a later chapter. The explanation in this case, however, is simple. The input signal to the valve is V_{gk} and V_{gk} is the difference between e_s and $i_a R_K$. Taking peak to peak values

$$V_{gk} = 4 - 0.25(9.6 - 5.3)$$
$$= 4 - 1.075$$
$$= 2.925 \text{ V}$$

Valve amplification $= V_{ak}/V_{gk} = 85/2.925 = 29$ which compares very closely with the result previously obtained (28).

Bias Decoupling

At this stage we shall assume that a reduction in amplification is a disadvantage and we shall see how the circuit gain can be made equal to the valve gain. The cause of the reduction was the a.c. voltage $i_a R_K$. If the a.c. voltage can be eliminated without change in the d.c. $I_A R_K$, the bias point V_{GK} will be unchanged, while the a.c. V_{gk} will be $e_s - i_a R_K = e_s - 0 = e_s$.

Provided the signal frequency is not too low a capacitor may be selected having reactance very much less than R_K. If this decoupling capacitor C_K is connected in parallel with R_K, V_{gk} becomes $e_s - i_a Z_k$, and $i_a Z_k$ can be made negligibly small compared with e_s. The exact analysis will be left for a later chapter, but a simple example will illustrate the operation. Considering Example 1.1 above, R_k was 250 Ω. If X_c was, say, 25 Ω, then the $i_a Z_k$ voltage would be only a tenth of its previous value 1·075 V. The resultant ΔV_{GK} would be $4 - 0·1075 = 3·8925$.

V_{ak} now becomes

$$3·8925 \times \text{valve amplification}$$
$$= 3·8925 \times 29$$
$$= 115 \text{ V}$$

This is in fact greater than the value obtained with battery bias (113·5), but in this case we have neglected the effect of non-linearity of characteristics. Finally the value of a suitable capacitor depends on the frequency of e_s, so using a figure of 1 000 Hz,

$$X_c = \frac{1}{2\pi f C}, \quad C = \frac{1}{2\pi f X_c}$$
$$= \frac{10^6}{2\pi 10^3 25} \mu F$$
$$= \frac{20}{\pi} \simeq 6 \mu F$$

The operating voltage of the capacitor is only 2 V and miniature capacitors, 6 V working at 100 μF or more, are readily available should lower frequency operation be required.

A.C. Load Lines

Up to this point we have found that the operating point can move only up and down the d.c. load line. This is true for d.c. or very low frequency changes of V_{gk}. For the cathode bias circuit of *Figure 1.8*, the d.c. relationships are

$$V_{AK} = V_{HT} - I_A(R_L + R_K) \tag{1.5}$$

$$V_{GK} = -I_A R_K \tag{1.6}$$

In the absence of C_K, a.c. relationships may also be written

$$V_{ak} = V_{HT} - i_a(R_L + R_K) \tag{1.7}$$

$$V_{gk} = I_a R_K + e_s \tag{1.8}$$

If C_K is included and the frequency such that X_{CK} is approximately zero, equation 1.8 becomes $V_{gk} = e_s$, and equation 1.7 must become

$$V_{ak} = V_{HT} - i_a R_L \tag{1.9}$$

since to a.c. R_K has been short circuited by C_K.

Equation 1.9 gives the a.c. variations in V_{AK} together with the d.c. level with respect to V_{HT}. The instantaneous amplitude of the a.c. quantities, i_a and V_{ak}, are related only by the load resistance R_L. The a.c. operating point must then move along an a.c. load line of slope $-1/R_L$ (compared with slope $-1/(R_L + R_K)$ for the d.c. load line). If however the signal amplitude is reduced to zero the operating point must return to its d.c. value. The slope of the a.c. load line is not changed by the signal amplitude so it must pass through the d.c. operating point.

In the last example the d.c. load was 20 kΩ and the a.c. load was 19·75 kΩ. It would be difficult to differentiate between these two load lines on the graph so this effect is negligible.

For the circuits shown in *Figure 1.9* the a.c. load lines are very different to the d.c. load lines. The form of the graphical solution

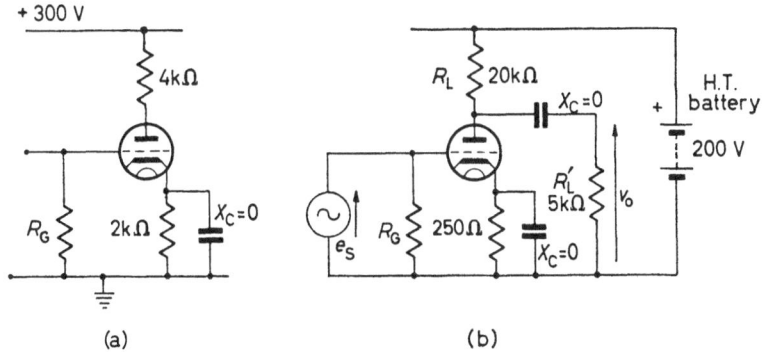

Figure 1.9. Circuits having different a.c. and d.c. loads.
(a) Decoupled cathode bias. (b) RC coupling for next stage

11

is shown in *Figure 1.10*. The first circuit shows the same situation as in the last example but the component values are such that the slopes of the a.c. and d.c. load lines are now very different. The operating point moves up and down the a.c. load line according to the instantaneous value of the applied V_{gk}. In the circuit of *Figure 1.9b* an *RC* coupling network is used to apply the amplified output to, perhaps, a second valve. The calculation of d.c. operating point is unchanged, but the a.c. load is now the parallel combination of R_L and R_L'. These are in parallel to a.c. since the a.c. resistance of the

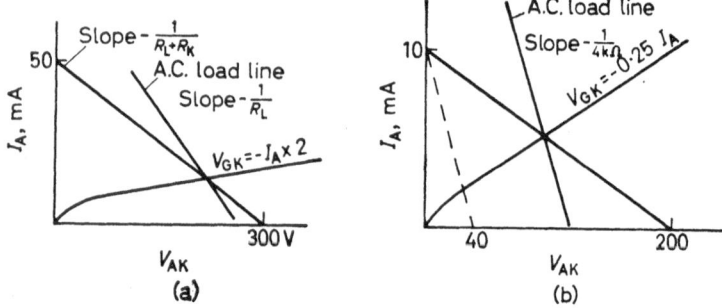

Figure 1.10. D.C. and a.c. load lines for the circuits shown in *Figure 1.9*

HT battery is negligible, and thus to a.c., both R_L and R_L' are connected between anode and earth. Thus the reciprocal of the slope of the a.c. load line is

$$\frac{-R_L R_L'}{R_L + R_L'} = \frac{-20 \times 5}{20 \times 5} = -4\,\text{k}\Omega$$

The dotted line in *Figure 1.10b* shows a convenient method of constructing the a.c. load line, proceeding as follows.

Assuming any convenient value for V_{HT}, in this case 40 V, draw a d.c. load line for the a.c. load value of 4 kΩ. Construct the a.c. load line parallel to this and passing through the operating point. Once again the a.c. operating point will move up and down the a.c. load line according to the applied instantaneous value of v_{gk}, and the peak to peak value of the alternating output voltage can be read off the graph.

A.C. Load Lines with Reactive Loads

A further complication of the a.c. load line occurs when the load is reactive. To simplify this problem, we shall imagine a load

12

consisting of a coil having inductance but negligible resistance. We shall further assume that an alternating v_{gk} will cause an alternating i_a. The circuit and characteristics are shown in *Figure 1.11a* and *b*.

In this circuit the d.c. load line is vertical ($R_L = 0$) and the d.c. operating point is found by the battery bias V_B. The a.c. v_{gk} will cause i_a to vary with time as shown. Remembering that in an inductor the current lags the voltage by 90°, the instantaneous value of the

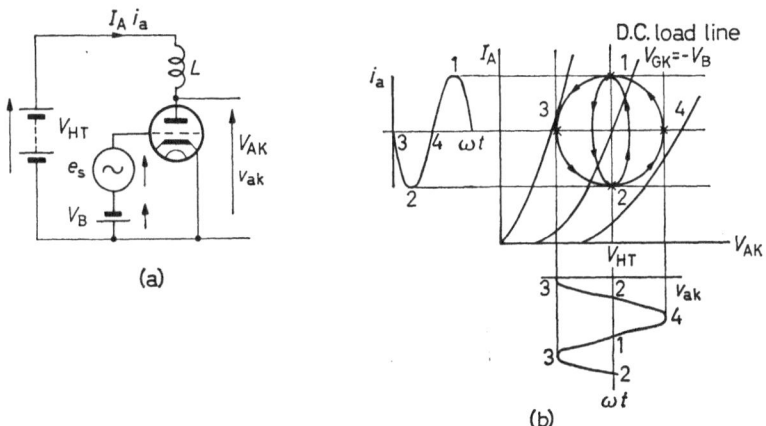

Figure 1.11. (a) Triode valve with reactive load. *(b)* Construction of the resulting elliptical load line

a.c. anode voltage must be zero when the instantaneous i_a is a maximum or minimum. Thus for these values of i_a, the corresponding v_{ak} is V_{HT} giving points (1) and (2) on the graph. When however the a.c. component of i_a passes through zero, the a.c. component of v_{ak} must be maximum or minimum. The values of v_{ak} at these points is given by

$$V_{HT} \pm i_a \text{ peak} \times \omega L$$

where ωL is the coil reactance at the signal frequency. This gives us the two further points (3) and (4). Points could be calculated for intermediate values, but it can be seen that the only single continuous line joining the four points is an ellipse or circle depending on ω. In *Figure 1.11b* the narrow ellipse represents a lower frequency. The a.c. operating point thus moves around the circle in the direction shown. (A capacitive load would result in rotation in the opposite direction.) The peak values of v_{gk} required to produce this load line may now be read from the graph and the voltage gain calculated. In practice the load would not be purely inductive and the procedure for constructing the load line is so complex as to make it rarely useful.

Summarizing: we have found that simple valve circuits not involving reactive components may readily be analysed using graphical methods. The results obtained are as accurate as these methods permit and demonstrate clearly such effects as distortion due to non-linearity. These methods also enable the d.c. operating point for a particular circuit, or the circuit for a required operating point, to be found.

TRANSISTOR CIRCUITS

We shall now see how far these methods can be applied to simple transistor circuits. The commonest form of transistor amplifier is very similar to the valve amplifiers discussed above. The circuit is shown in *Figure 1.12a* and the forms of the output and input

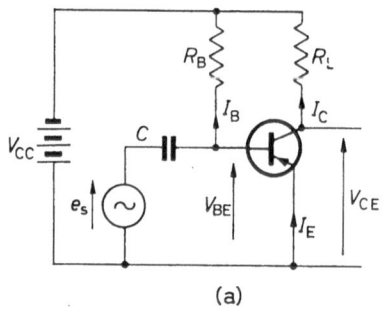

(a)

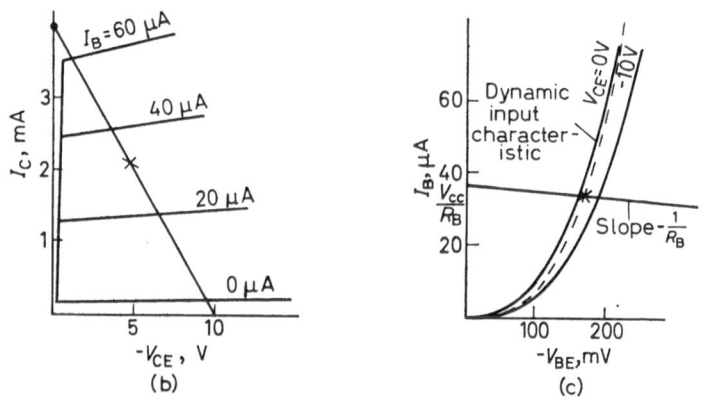

Figure 1.12. The common emitter transistor amplifier. (*a*) D.C. bias circuit and load. (*b*) The output and (*c*) the input characteristics

14

characteristics shown in *Figure 1.12b* and *c* respectively. With a transistor, four variables must be investigated, input and output voltages, and input and output currents. The circuit in *Figure 1.12a* is said to be in the common emitter configuration since the input and output voltages V_{BE} and V_{CE} are measured with respect to the emitter. The input and output currents I_B and I_C flow out of the base and collector respectively, in the conventional direction for p-n-p transistors. The input characteristics show variations of I_B with applied V_{BE} using V_{CE} as the parameter. The output characteristics show variations of I_C with changes of V_{CE} and with I_B as the parameter. As before we may write an equation for the circuit

$$V_{CE} = -V_{CC} + I_C R_L$$

giving a load line with intercepts at $-V_{CC}$ and V_{CC}/R_L. If I_B were known the operating point would now be defined. But the parameter of the input characteristics depends upon V_{CE}, i.e. the operating point. A dynamic input characteristic may be plotted for a particular circuit by reading the V_{CE} values corresponding to I_B values along the load line. These pairs of values may then be plotted on the input characteristics. Now looking at the input circuit we obtain

$$V_{BE} = -V_{CC} + I_B R_B$$

This leads to a further straight line with intercepts at $V_{BE} = -V_{CC}$ and $I_B = V_{CC}/R_B$. The intersection of the line and this dynamic input characteristic then gives the values of V_{BE} and I_B which determine the operating point, on both input and output characteristics. An a.c. signal e_s now applied to the base as shown will result in an input voltage

$$V_{be} = -V_{BE} + \hat{E} \sin \omega t$$

with maximum and minimum values of $-V_{BE} \pm \hat{E}$. The corresponding range of I_B can be determined from the input characteristics using the dynamic curve for the circuit. These in turn lead to the variation of V_{CE} from the output characteristics. This procedure is somewhat involved and is only necessary when calculating maximum power output and distortion for power amplifiers. It is interesting to note that for a sinusoidal input voltage, the base current waveform is distorted by the non-linear input characteristic. The waveform is then further modified by the non-linearity of the output characteristic.

A numerical example of this type of calculation will now be given for the circuit in *Figure 1.12a*.

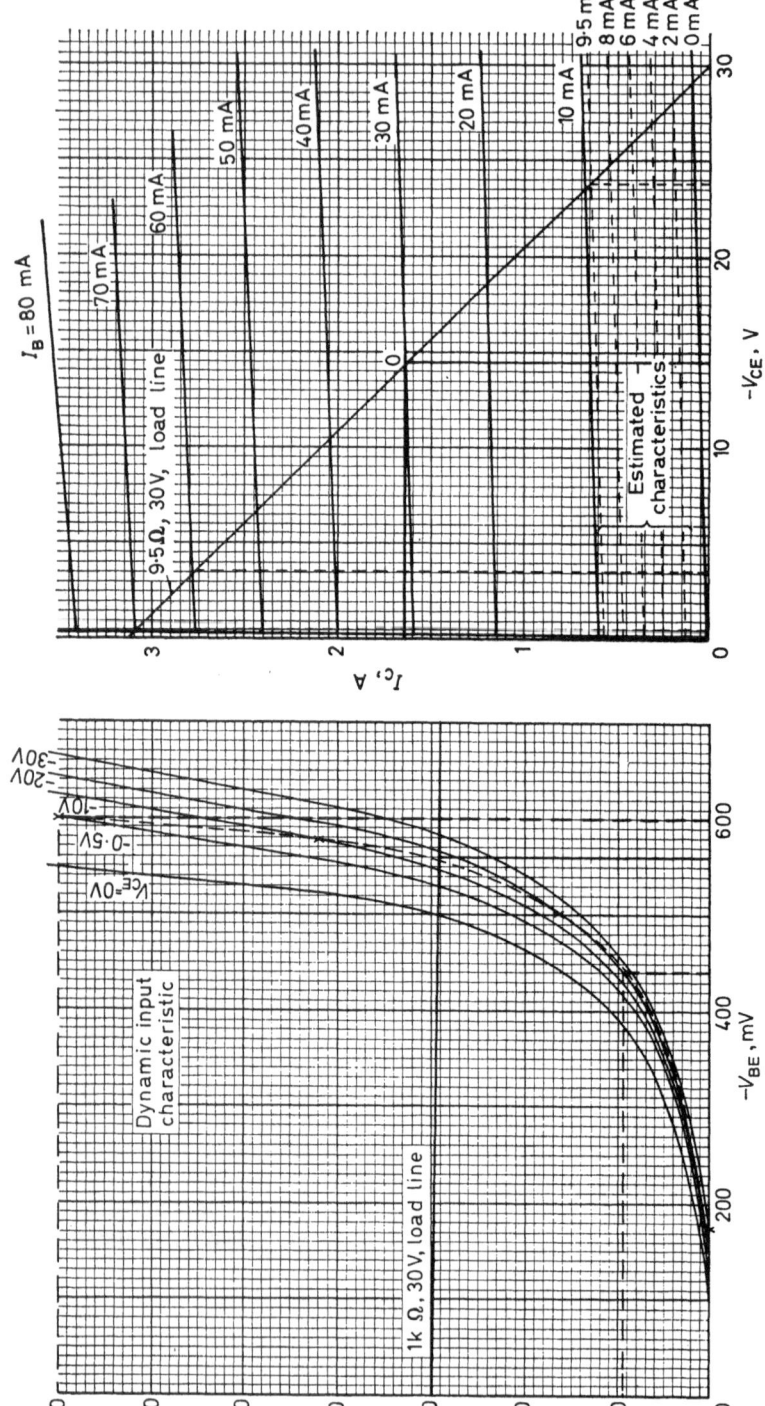

Figure 1.13. The characteristics for Example 1.2

Example 1.2. A simple transistor amplifier is operated from a d.c. supply of 30 V. The load resistor R_L is 9·5 Ω, and the base bias resistor R_B is 1 kΩ. If the transistor characteristics are those shown in *Figure 1.13a* and *b*, determine the d.c. operating point. Find also the a.c. output voltage if (*a*) the a.c. source is 40 mV peak with zero internal resistance, and (*b*) the a.c. source is 3 V peak with 100 Ω internal resistance. Comment on the effect of the characteristic non-linearity in each case.

Following the procedure outlined above we must first insert the load line on the output characteristic. The two intercepts are given by:

If $\qquad I_C = 0 \qquad V_{CE} = -V_{CC} = -30$ V

If $\qquad V_{CE} = 0 \qquad I_C = \dfrac{V_{CC}}{R_L} = \dfrac{30}{9\cdot5} = 3\cdot15$ A

This load line is shown on *Figure 1.13b*.

Using this we can now plot the dynamic input characteristic as follows.

If I_B is 70 mA, $V_{CE} \simeq 0\cdot5$ and this point may be placed on the input curve for $V_{CE} = 0\cdot5$. Similarly when V_{CE} is -10 V, $I_B \simeq 41$ mA which can be plotted on the input curve for $V_{CE} = -10$ V. Two further points for $V_{CE} - 20$ V and -30 V are I_B 17 mA and I_B 0 mA respectively and these may be inserted in the same way. The resulting dynamic curve is shown in *Figure 1.13a*.

The intercepts for the input 'load' line are

$$V_{BE} = 0 \qquad I_B = \frac{V_{CC}}{R_B} = \frac{30 \text{ V}}{1 \text{ k}\Omega} = 30 \text{ mA}$$

$$I_B = 0 \qquad V_{BE} = -V_{CC} = -30 \text{ V}$$

The second point cannot be plotted, so the slope of the line must be used. If ΔV_{BE} is 1 volt, $\Delta I_B = \dfrac{1 \text{ V}}{1 \text{ k}\Omega} = 1$ mA. Therefore at V_{BE} 1 volt, $I_B = 30$ mA $- \Delta I_B = 29$ mA. Thus the input 'load' line may be plotted as shown in *Figure 1.13a*. The intersection between this line and the dynamic input characteristic gives the d.c. operating point.

From *Figure 1.13a*

$$V_{BE} = -560 \text{ mV} \qquad I_B = 29\cdot5 \text{ mA} \simeq 30 \text{ mA}$$

Now since I_B is known, the operating values of V_{CE} and I_C may be found from *Figure 1.13b*

$$V_{CE} = -14\cdot5 \text{ V} \qquad I_C = 1\cdot65 \text{ A}$$

17

When the a.c. source of 40 mV peak and zero internal resistance is connected, V_{BE} varies between -520 mV and -600 mV in a sinusoidal manner. The resulting I_b values obtained from the dynamic input characteristic are

V_{be}	-520 mV	-560 mV	-600 mV	-560 mV
I_b	20 mA	30 mA	70 mA	30 mA

The peak values of the I_b waveform are

$$30 - 20 = 10 \text{ mA} \quad \text{and} \quad 70 - 30 = 40 \text{ mA}$$

This represents a considerable degree of distortion resulting from the curvature of the input characteristic. For comparison purposes, a measure of the distortion is given by the ratio of these two results

$$\therefore \text{ for } I_b, \qquad \text{distortion } D_1 = \frac{40}{10} = 4$$

The corresponding limits of the variation of V_{ce} can now be found from *Figure 1.13b.*

V_{be}	-520 mV	-560 mV	-600 mV	-560 mV
$V_{ce} =$	18·5 V	$-14·5$ V	$-0·5$ V	$-14·5$ V

The peak values of the V_{ce} waveform are

$$18·5 - 14·5 = 4 \text{ V} \quad \text{and} \quad 14·5 - 0·5 = 14 \text{ V}$$

The corresponding distortion factor

$$D_2 = \frac{14}{4} = 3·5$$

This is less than that obtained for the I_b waveform since the output characteristic non-linearity acts in the opposite sense to that of the input characteristic. The a.c. output voltage required is 18 V peak to peak.

In the second case the source has an internal resistance of 100 Ω, and the a.c. base current will be determined by this and the transistor input resistance, R_{in}. At the operating point, the d.c. R_{in} is given by

$$R_{in} = \frac{V_{BE}}{I_B} = \frac{560 \text{ mV}}{30 \text{ mA}} = 18 \ \Omega$$

This is very much less than the source resistance and may therefore be neglected. Thus the peak a.c. base current is given by

18

$e_s/100 = 30$ mA. The resulting variation in I_b values and the corresponding V_{ce} values are

I_b	0 mA	30 mA	60 mA	30 mA
V_{ce}	$-29\cdot3$ V	$-14\cdot5$ V	-3.3 V	$-14\cdot5$ V

The peak values of the V_{ce} waveform are now $29\cdot3 - 14\cdot5 = 14\cdot8$ V and $14\cdot5 - 3\cdot3 = 11\cdot2$ V.

The corresponding distortion factor $D_3 = 14\cdot8/11\cdot2 = 1\cdot33$.

This result is less than before and in this case the negative half cycle is the larger, the phase of the distortion has therefore been reversed. From this result it seems likely that a suitable value of source resistance, say 20 Ω, could result in the input and output non-linear effects cancelling each other and leaving a sinusoidal output.

The a.c. output voltage in case (b) is 26 V peak to peak.

For small signal amplification, calculations are usually simplified by making certain practically valid assumptions. Firstly, V_{BE} is very much less than the d.c. supply voltage V_{CC}, therefore $I_B \simeq V_{CC}/R_B$. Secondly, the a.c. input resistance of the transistor R_{in} is taken as the slope of the input characteristic at the operating point.

$$\therefore \qquad V_{be} = i_b R_{in}$$

These approximations enable the selection of a suitable operating point and an estimate of the required input voltage to be made without difficulty.

Example 1.3. A transistor having the output characteristics shown in *Figure 1.14a* operates with V_{BE} at 0·2 volts, and the slope of the input characteristic at this point is ΔV_{BE} 10 mV, ΔI_B 5 μA. If it is connected in the circuit shown in *Figure 1.14b* determine (a) the d.c. operating point, (b) the output voltage and voltage gain for an input signal of 28·28 mV r.m.s., (c) the current gain, and (d) the maximum output signal for negligible distortion.

First the d.c. (and a.c.) load line can be constructed with intercepts at

$$V_{CE} = -V_{CC} = -6 \text{ V} \quad \text{and} \quad I_C = \frac{V_{CC}}{R_L} = \frac{6}{1} = 6 \text{ mA}$$

(a) The d.c. operating point is now determined by I_B which may be found from

$$\frac{V_{CC} - V_{BE}}{145 \text{ k}\Omega} = \frac{5\cdot8 \times 10^3}{145} \mu\text{A} = 40 \ \mu\text{A}$$

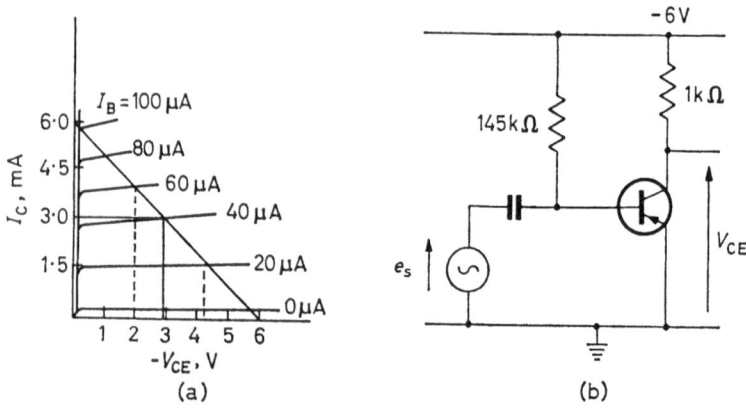

Figure 1.14. The characteristics and circuit for **Example 1.3**

From the characteristics the operating point is now V_{CE} $-2 \cdot 9$ V, I_C 3 mA.

(*b*) The slope of the input characteristic is the a.c. input resistance and is given by

$$\frac{\Delta V_{BE}}{\Delta I_B} \qquad \therefore \text{ a.c. input resistance} = \frac{10 \text{ mV}}{5 \text{ } \mu\text{A}} = 2 \text{ k}\Omega$$

An r.m.s. input voltage of $28 \cdot 28$ mV will result in an r.m.s. base current

$$i_b = \frac{28 \cdot 28 \text{ mV}}{2 \text{ k}\Omega} = 14 \cdot 14 \text{ } \mu\text{A}$$

The peak value of $\qquad i_b = \sqrt{2} \times 14 \cdot 14 \text{ } \mu\text{A}$
$$= 20 \text{ } \mu\text{A}$$

and the peak to peak value $= 20 \times 2 = 40 \text{ } \mu\text{A}$.
Thus from the characteristics, the a.c. operating point moves between $i_b = 60 \text{ } \mu\text{A}$ and $i_b = 20 \text{ } \mu\text{A}$.

The corresponding values of v_{ce} are -2 V and $-4 \cdot 2$ V respectively. The r.m.s. output voltage $= (4 \cdot 2 - 2 \cdot 0)/2\sqrt{2} = 0 \cdot 85$ V, the voltage gain $A_V = 850 \text{ mV}/28 \cdot 28 \text{ mV} = 30$.

As V_{BE} goes positive I_B is reduced and V_{CE} goes negative. Thus as with simple valve amplification we get phase reversal and $A_V = -30$.

(*c*) The value of i_c varies from $1 \cdot 5$ to 4 mA giving an r.m.s. i_c of $(2 - 1 \cdot 5)/2\sqrt{2} = 0 \cdot 885$ mA. The current gain A_I is therefore $0 \cdot 885 \text{ mA}/14 \cdot 14 \text{ } \mu\text{A} = 62 \cdot 5$.

20

(d) The peak output voltage must not cause V_{CE} to exceed -6 or the transistor will cut off. V_{CE} cannot fall below -0.2 V from the characteristics. A peak a.c. v_{ce} of $2.9 - 0.2$ will not cause distortion due to bottoming or cut off. The maximum output signal is therefore $2.7/\sqrt{2} = 1.9$ $V_{r.m.s.}$.

Figure 1.15 shows the distortion that would arise with this circuit if the peak i_b was increased to 60 μA.

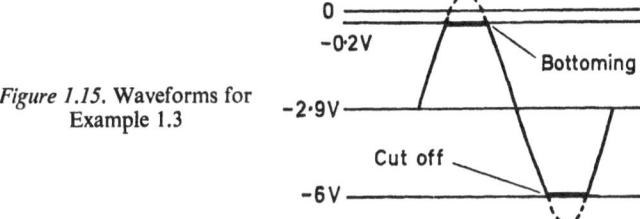

Figure 1.15. Waveforms for Example 1.3

In practice the simple biasing arrangement used in the last two examples has a number of disadvantages. Before we can discuss these we must examine the relationships between the collector, base and emitter currents in a transistor. If a current I_E flows into the emitter lead, transistor action results in αI_E flowing out at the collector, and $(1 - \alpha)I_E$ flowing out of the base. In addition the reverse

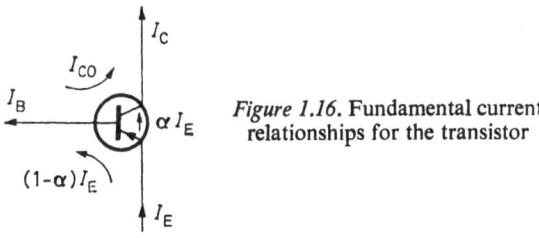

Figure 1.16. Fundamental current relationships for the transistor

biased collector base junction will have a reverse leakage current flowing conventionally from base to collector. This current is known as I_{CO} or I_{CBO} since it is the value of I_C when I_E is zero. These relationships can be simply shown and remembered by the diagram shown in *Figure 1.16*.

By inspection of this diagram the three basic relationships may be found:

$$I_E = I_C + I_B \tag{1.10}$$

$$I_C = \alpha I_E + I_{CO} \tag{1.11}$$

$$I_B = (1 - \alpha)I_E - I_{CO} \tag{1.12}$$

21

In common emitter circuits, we are not concerned with I_E, but we do wish to know how I_C varies with I_B. First let us find the value of I_C when I_B is zero.

From equation 1.12,

$$(1 - \alpha)I_E = I_{CO}$$

$$\therefore \qquad I_E = \frac{I_{CO}}{1 - \alpha}$$

Now applying equation 1.11,

$$I_C = \frac{\alpha I_{CO}}{1 - \alpha} + I_{CO}$$

$$= \frac{\alpha I_{CO} + I_{CO} - \alpha I_{CO}}{1 - \alpha} = \frac{I_{CO}}{1 - \alpha}.$$

This current, the value of I_C when I_B is zero, is known as I_{CO}' (or sometimes I_{CEO} as opposed to I_{CBO} for the common base circuit). Thus

$$I_{CO}' = \frac{I_{CO}}{1 - \alpha} \qquad (1.13)$$

Now since α is nearly equal to unity having typical values from 0·96 to 0·995, I_{CO}' is very much larger than I_{CO}. The values of I_{CO} vary from 1 mA for a high power germanium transistor to 1 μA for a small signal germanium transistor or to 10 nA for small signal silicon transistor. Typically the corresponding value of I_{CO}' would range from 0·1 A to 1 μA respectively.

To find the value of I_C when I_B is not zero we must rearrange equation 1.11 to obtain I_E and equate to equation 1.10.

From equation 1.11,

$$I_E = \frac{I_C - I_{CO}}{\alpha} \qquad (1.14)$$

Equate to 1.10

$$I_C + I_B = \frac{I_C - I_{CO}}{\alpha}$$

$$\alpha I_C + \alpha I_B = I_C - I_{CO}$$

$$I_C(1 - \alpha) = \alpha I_B + I_{CO}$$

and

$$I_C = \frac{\alpha}{1 - \alpha} I_B + \frac{I_{CO}}{1 - \alpha} \qquad (1.16)$$

The second term in the result is I_{CO}' and for convenience we shall put

$$\alpha' = \frac{\alpha}{1 - \alpha} \tag{1.17}$$

and equation 1.16 becomes

$$I_C = \alpha' I_B + I_{CO}' \tag{1.18}$$

Other useful relationships can be obtained by rearranging equations 1.13 and 1.17. These lead to:

$$\alpha = \frac{\alpha'}{1 + \alpha'} \tag{1.19}$$

$$I_{CO} = \frac{I_{CO}'}{1 + \alpha'} \tag{1.20}$$

We are now in a position to discuss the disadvantages of our simple transistor biasing arrangement.

A numerical example will adequately illustrate these points.

Example 1.4. Typical values for an OC 75 transistor are α' 90, I_{CO} 1 μA and V_{BE} 0·2 V; now suppose such an OC 75 was connected into the circuit shown in *Figure 1.17*.

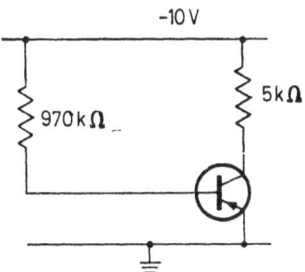

Figure 1.17. Simple bias circuit for Example 1.4

$$I_B = \frac{(10 - 0·2)V}{970 \ K\Omega} = 10·1 \ \mu A$$

$$I_{CO}' = \frac{1}{1 - \alpha} \mu A = \frac{1}{1 - \dfrac{\alpha'}{1 + \alpha'}} \mu A$$

$$= \frac{1}{1 - \dfrac{90}{91}} \mu A = 91 \ \mu A$$

23

$$\therefore \quad I_C = \alpha' I_B + I_{CO}'$$
$$= 90 \times 10\cdot1 + 91 = 1\ 000\ \mu A$$
$$= 1\ \text{mA}.$$

Now $V_{CE} = -10 + (5 \times 1) = -5$ V and since V_{CE} may vary between -10 (cut off) and $-0\cdot2$ (bottoming) this allows maximum variation of a.c. operating point without distortion, with a peak output voltage of $4\cdot8$ V. Unfortunately all OC 75 transistors are not typical and the specified range of α' is from 70 to 130. Taking these extreme values we get

α'	I_{CO}'	I_C	V_{CE}
70	71 μA	0·778 mA	$-6\cdot1$ V
130	131 μA	1·444 mA	$-2\cdot78$ V

Thus to prevent distortion the output signal must be reduced to $3\cdot9$ V peak with α' 70 or $2\cdot08$ V peak with α' 130. Alternatively if the input signal level had been maintained, the output V_{CE} would have been severely distorted by cut off in the first case and bottoming in the second.

Summarizing, the simple bias circuit cannot be designed to suit all transistors with the normal commercial range of α' unless the signal level is severely restricted.

Temperature Effects

A second disadvantage is that transistors are temperature sensitive. The reverse biased leakage currents I_{CO} (collector base) and I_{EO} (emitter base) are the result of the intrinsic properties of the semi-

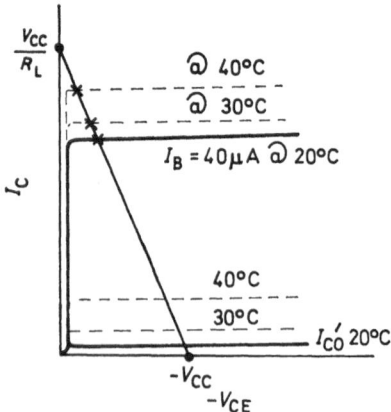

Figure 1.18. The effect of temperature on the output characteristics

conductor material. Both currents approximately double for each 10°C increase in temperature. In the simple circuit discussed above, $I_C = \alpha' I_B + I_{CO}'$. I_B is constant and I_{CO}' doubles as I_{CO} doubles. As can be seen from *Figure 1.18*, the characteristic corresponding to $I_B = 40 \mu A$ moves up with I_{CO}'. The operating point must lie on the load line and it therefore moves towards the bottomed or saturated condition with increase in temperature. In this case, at 20°C the permissible peak a.c. v_{ce} is approximately $V_{CC}/3$ while at 40°C it has been reduced to zero.

In Example 1.4, I_B was maintained at a constant value. An alternative approach might be to maintain V_{BE} constant. It can be shown that the emitter current can be expressed in terms of the emitter base voltage by

$$I_E = I_{EO}[\exp (KV_{BE}/T) - 1] \qquad (1.21)$$

where K is a constant, and T is the absolute temperature in °Kelvin.

Now $\qquad I_B = I_E(1 - \alpha) - I_{CO}$

$\therefore \qquad I_B = (1 - \alpha)I_{EO}[\exp (KV_{BE}/T) - 1] - I_{CO} \qquad (1.22)$

In this expression a change of 10°C will have little effect on T which will be of the order of 300°K. When V_{BE} is very small I_{CO} will have a significant effect but when V_{BE} is larger we can neglect I_{CO} as

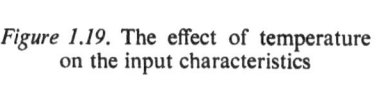

Figure 1.19. The effect of temperature on the input characteristics

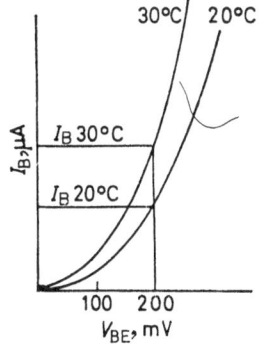

we can one, compared with the exponential term. Thus for any particular value of V_{BE}, if temperature rises by 10°C, I_{EO} and thus I_B doubles. The effect on the output characteristics would be very similar to that shown in *Figure 1.18*. Again the operating point would move towards bottoming. The effect on the input characteristics is shown in *Figure 1.19*.

The commonest form of bias circuit employs a compromise aimed at stabilizing the emitter current and therefore the collector current regardless of changes of transistor (hence α' and required V_{BE}) or temperature. The base is supplied from a source of medium resistance instead of one having high resistance (tending to constant current) or zero resistance giving constant voltage.

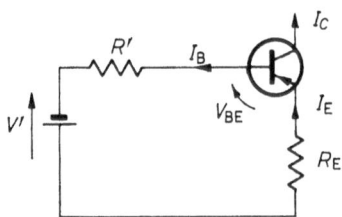

Figure 1.20. Stabilization of operating point

The simplest form of this is shown in *Figure 1.20*. The equation for this loop is given by

$$V_{BE} = V' - I_B R' - I_E R_E \qquad (1.23)$$

but $\qquad I_B = (1 - \alpha)I_E - I_{CO}$

$\therefore \qquad V_{BE} = V' - I_E[R_E + (1 - \alpha)R'] + I_{CO}R' \qquad (1.24)$

This equation represents a 'load line' on the V_{EB}/I_E characteristics, with intercepts at $V' + I_{CO}R'$, and at $\dfrac{V + I_{CO}R'}{R_E + R'(1 - \alpha)}$.

Figure 1.21 shows the V_{EB}/I_E characteristics for 20°C and 30°C. The load lines are drawn on this for the same temperatures. The

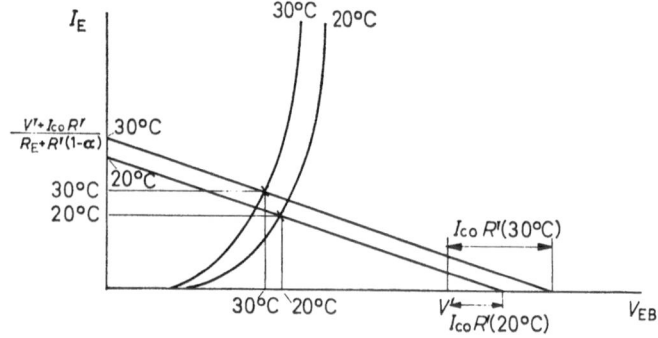

Figure 1.21. Graphical analysis of stabilization

26

resulting change of I_E is shown by the intersections marked. The movement of the load line is due entirely to $I_{CO}R'$, thus if R' is made small this movement will be limited.

Also if the line could be made nearly horizontal the change in I_E could be reduced. Thus the conditions for a stable emitter current are, R_E large, V' much greater than V_{EB} and R' small.

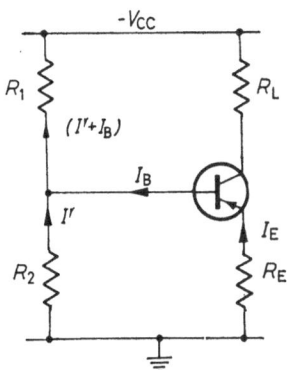

Figure 1.22. Potential divider, emitter resistor stabilization

In practice V' is obtained from a potential divider circuit across the main V_{CC} supply, as shown in *Figure 1.22.* By Thèvenin's theorem (see Chapter 2) it can be shown that

$$R' = \frac{R_1 R_2}{R_1 + R_2} \tag{1.25}$$

and that

$$V' = \frac{V_{CC} R_2}{R_1 + R_2} \tag{1.26}$$

The operating point V_{CE}, I_C can be found by calculating V' and R' from equations 1.25 and 1.26 and substituting in equation 1.23. This may then be rewritten

$$V' - V_{BE} = I_B R' + (I_B + I_C) R_E$$

then collecting terms and substituting for I_B from equation 1.18

$$V' - V_{BE} = I_C R_E + \frac{1}{\alpha'} (I_C - I_{CO}')(R' + R_E)$$

$$= \frac{I_C}{\alpha'} [R' + R_E(1 + \alpha')] - \frac{I_{CO}'}{\alpha'} (R' + R_E)$$

27

From which

$$I_{\rm C} = \frac{\alpha'(V' - V_{\rm BE})}{R' + R_{\rm E}(1 + \alpha')} + \frac{I_{\rm CO}'(R' + R_{\rm E})}{R' + R_{\rm E}(1 + \alpha')} \quad (1.27)$$

$V_{\rm CE}$ may then be found from

$$V_{\rm CE} = -V_{\rm CC} + I_{\rm C}R_{\rm L} + I_{\rm E}R_{\rm E}$$
$$\simeq -V_{\rm CC} + I_{\rm C}(R_{\rm L} + R_{\rm E}) \quad (1.28)$$

An alternative approach to the stabilizing action of the circuit can be seen by consideration of the circuit (*Figure 1.22*). For R' to be small, R_1 and R_2 must be small making I' much greater than $I_{\rm B}$. Thus the voltage between base and earth is approximately V' and nearly constant. Any change in $I_{\rm C}$ due to temperature or α' changes, also cause a change in $I_{\rm E}$. Suppose $I_{\rm C}$ tends to rise, the voltage across $R_{\rm E}$, $I_{\rm E}R_{\rm E}$, will also rise, reducing $V_{\rm BE}$. This in turn reduces $I_{\rm B}$ tending to maintain $I_{\rm C}$ and $I_{\rm E}$ at their original values.

Stability Factors

Three measures of circuit stability are often encountered. These are

$$K = \frac{{\rm d}I_{\rm C}}{{\rm d}I_{\rm CO}'} \quad (1.29)$$

$$S = \frac{{\rm d}I_{\rm C}}{{\rm d}I_{\rm CO}} \quad (1.30)$$

$$K_{\alpha'} = \frac{{\rm d}I_{\rm C}}{{\rm d}\alpha'} \quad (1.31)$$

K may be found directly from equation 1.27 giving

$$K = \frac{R' + R_{\rm E}}{R' + R_{\rm E}(1 + \alpha')} \quad (1.32)$$

by assuming $V_{\rm BE} \ll V'$ and that $V' - V_{\rm BE} =$ constant.

For good stability K is small and in the limit when $R' \ll R_{\rm E}$,

$$K = \frac{1}{1 + \alpha'}$$

For the worst case $R' \gg R_{\rm E}$ and $K = 1$.

K is also sometimes written

$$\frac{1}{1 + M\alpha'} \qquad (1.33)$$

where

$$M = \frac{R_{\mathrm{E}}}{R_{\mathrm{E}} + R'} \qquad (1.34)$$

S may be found in a similar way from equation 1.24 and substituting for I_{E} in terms of I_{C} and I_{CO}. On rearranging and differentiating S can be found

$$S = \frac{R' + R_{\mathrm{E}}}{R_{\mathrm{E}} + R'(1 - \alpha)} \qquad (1\cdot35)$$

In this case if $R_{\mathrm{E}} \gg R'$, S tends to one and $dI_{\mathrm{C}} = dI_{\mathrm{CO}}$.

If $R' \gg R_{\mathrm{E}}$, S tend to $\dfrac{1}{1 - \alpha}$

$$dI_{\mathrm{C}} = \frac{dI_{\mathrm{CO}}}{1 - \alpha} = dI_{\mathrm{CO}}{}'$$

Design Considerations

The limitations on R_{E} being large is that the volt drop $I_{\mathrm{E}}R_{\mathrm{E}}$ must be supplied by the supply battery V_{CC} (equation 1.28). Thus for a 6 V battery and operating point V_{CE} 3 V, I_{C} 1 mA, $R_{\mathrm{L}} + R_{\mathrm{E}}$ is given by $\dfrac{3 \text{ V}}{1 \text{ mA}} = 3 \text{ k}\Omega$. If the load is 2 k$\Omega$, R_{E} can only be 1 kΩ. For higher current transistors the values will of course be much lower.

The minimum size of R' is determined by the minimum permissible input impedance to the amplifier. This will be discussed in later chapters, and typical values of R_1 and R_2 will be used in the following examples.

Finally to find $K_{\alpha'}$ we must take equation 1.27 and substitute from equations 1.33 and 1.34

$$I_{\mathrm{C}} = \frac{\alpha'(V' - V_{\mathrm{BE}})}{(R' + R_{\mathrm{E}})(1 + M\alpha')} + \frac{\alpha'I_{\mathrm{CO}}}{1 + M\alpha'} \qquad (1.36)$$

Note $I_{\mathrm{CO}}{}' \backsimeq \alpha'I_{\mathrm{CO}}$

$$\therefore \qquad I_{\mathrm{C}} = \frac{\alpha'}{1 + M\alpha'}\left[\frac{V' - V_{\mathrm{BE}}}{R' + R_{\mathrm{E}}} + I_{\mathrm{CO}}\right]$$

$$K_{\alpha'} = \frac{dI_{\mathrm{C}}}{d\alpha'} = \frac{(1 + M\alpha') - \alpha'M}{(1 + M\alpha')^2}\left[\frac{V' - V_{\mathrm{BE}}}{R' + R_{\mathrm{E}}} + I_{\mathrm{CO}}\right]$$

This may be rewritten

$$K_{\alpha'} = \frac{1}{1 + M\alpha'} \times \frac{1}{\alpha'} \left[\frac{\alpha'(V' - V_{BE})}{(R' + R_E)(1 + M\alpha')} + \frac{\alpha' I_{CO}}{1 + M\alpha'} \right]$$

From equations 1.33 and 1.36

$$K_{\alpha'} = \frac{KI_C}{\alpha'} \qquad (1.37)$$

K and S are the slopes of approximately straight line relationships. The formulae (equations 1.32, 1.33, 1.35) may be used therefore for large changes in I_C as well as small ones.

$K_{\alpha'}$ however is the slope of a curved relationship and equation 1.37 should only be used for small changes in α'. For large changes, values should be inserted in equations 1.27 or 1.36 whichever is most convenient.

Another stabilizing circuit is occasionally encountered where the base resistor is connected to the collector. This is shown in *Figure 1.23*.

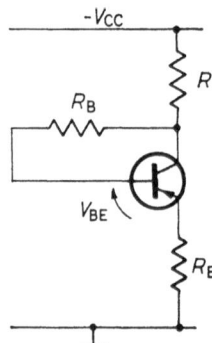

Figure 1.23. Collector feedback, emitter resistor stabilization

The equation may be written in the same way

$$V_{CC} - V_{BE} = (I_C + I_B)R_L + I_B R_B + I_E R_E$$
$$= I_C(R_L + R_E) + I_B(R_L + R_E + R_B)$$

Now by making the same substitutions as before, and differentiating we find

$$K = \frac{dI_C}{dI_{CO}} = \frac{R_B + R_L + R_E}{R_B + (R_L + R_E)(1 + \alpha')} \qquad (1.38)$$

In this circuit, any increase in I_C increases the voltage across R_L. This reduces the voltage across R_B, reducing I_B and thus I_C.

Bias Decoupling

In both stabilizing circuits (*Figures 1.22* and *1.23*) the stabilization is performed by d.c. negative feedback. As with the triode valve amplifier, feedback will reduce the gain of the stage. Decoupling must be used to overcome this effect. *Figure 1.24* shows how this is done.

In *Figure 1.24a* if $X_{CE} \ll R_E$ there can be no a.c. voltage across R_E. This behaves in the same way as the R_K in the valve amplifier.

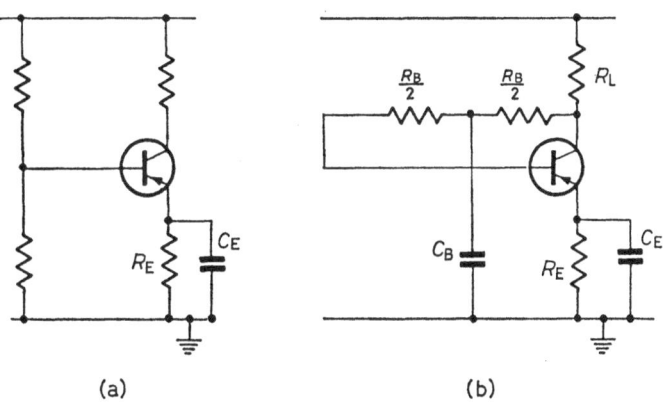

(a)　　　　　　　　　　　　　(b)

Figure 1.24. A.C. decoupling circuits

In *Figure 1.24b* the same applies, but in addition a.c. currents could be fed back through R_B. So if $X_{CB} \ll R_{B/2}$, such a.c. currents can flow to earth.

The exact mechanism of these a.c. circuits will be discussed in later chapters.

We shall now consider some numerical examples on the operating conditions for a given circuit and on the design of a suitable circuit to meet required operating conditions.

Example 1.5. The single stage amplifier shown in *Figure 1.25* employs a transistor having α' in the range 50–95. The reverse collector leakage current I_{CO} is quoted as $2 \, \mu A$ at 20°C and the amplifier may be used in the temperature range 20°C to 50°C. Determine the range of possible d.c. operating points, and hence the maximum permissible r.m.s. output current. Assume V_{BE} is 200 mV.

First we must find the range of d.c. operating point at 20°C due to the spread of α'.

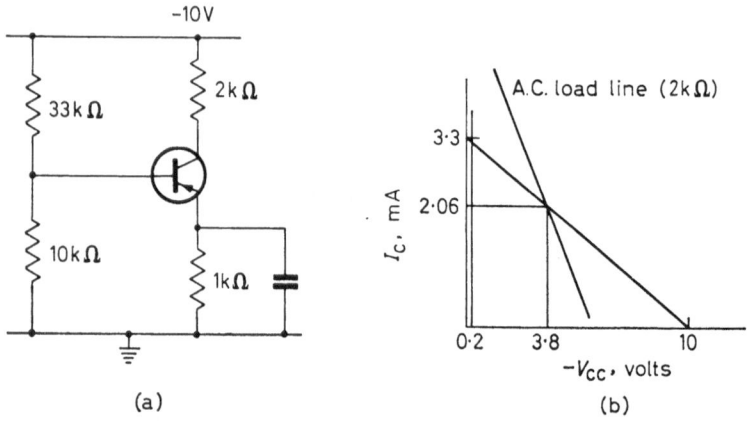

(a)

(b)

Figure 1.25. Circuit and graph for Example 1.5

For T_1, α' 50,

$$I_{CO}' = (1 + \alpha')I_{CO} = 0 \cdot 002(51) \text{ mA}$$
$$= 0 \cdot 102 \text{ mA}$$

For T_2, α' 95,

$$I_{CO}' = 0 \cdot 002(96) = 0 \cdot 192 \text{ mA}$$

Next, to find I_C we need V' and R' for this circuit.

$$V' = \frac{10 \times 10}{33 + 10} = 2 \cdot 323 \text{ V} \qquad R' = \frac{10 \times 33}{10 + 33} = 7 \cdot 67 \text{ k}\Omega$$

Using equation 1.27

$$I_C = \frac{\alpha'(V' - V_{BE}) + I_{CO}'(R' + R_E)}{R' + R_E(1 + \alpha')}$$

For T_1,

$$I_C = \frac{50 \times 2 \cdot 125 + 0 \cdot 102 \times 8 \cdot 67}{7 \cdot 67 + 51} \text{ mA}$$

$$I_C = 1 \cdot 825 \text{ mA}$$

For T_2,

$$I_C = \frac{95 \times 2 \cdot 125 + 0 \cdot 192 \times 8 \cdot 67}{7 \cdot 67 + 96} \text{ mA}$$

$$I_C = 1 \cdot 95 \text{ mA}$$

Increase in ambient temperature can only *increase* I_C, so maximum I_C will occur if T_2 is used at 50°C.

For T_2 at 50°C,

$$I_{CO} = 0.002 \times 2 \times 2 \times 2 \text{ mA}$$
$$= 0.016 \text{ mA}$$

and
$$\Delta I_{CO} = 0.014 \text{ mA}$$
$$\Delta I_{CO}' = (1 + \alpha')\Delta I_{CO}$$
$$= 96 \times 0.014 \text{ mA}$$
$$= 1.34 \text{ mA}$$

From equations 1.29 and 1.32

$$K = \frac{\Delta I_C}{\Delta I_{CO}'} = \frac{R' + R_E}{R' + R_E(1 + \alpha')}$$

$$K = \frac{8.67}{7.67 + 96} = 0.084$$

$$\Delta I_{C\Phi} = K\Delta I_{CO}' = 0.084 \times 1.34 \text{ mA}$$
$$= 0.113 \text{ mA}$$

$\therefore$ Maximum $I_C = 1.95 + 0.113 \text{ mA}$
$$= 2.063 \text{ mA}$$

From equation 1.28,

$$V_{CE} \simeq -V_{CC} + I_C(R_E + R_L)$$

For T_1 at 20°C,

Maximum $V_{CE} = -10 + 1.825(1 + 2) \text{ V}$
$$= -5.525 \text{ V}$$

For T_2 at 50°C,

Minimum $V_{CE} = -10 + 2.063(1 + 2) \text{ V}$
$$= -3.811 \text{ V}$$

To avoid distortion due to bottoming V_{CE} must not become less than V_{BE}. To avoid distortion due to cut off V_{CE} cannot be greater than V_{CC}.

To avoid bottoming with T_2, the peak a.c. v_{ce} must not exceed $3.811 - 0.2 = 3.6 \text{ V}$.

With T_1, cut off will occur if peak a.c. v_{ce} exceeds $10 - 5.525 = 4.475 \text{ V}$.

Thus to allow for all possibilities, peak a.c. v_{ce} has a maximum value of 3.6 V.

We must now consider the effect of the decoupling capacitor. If $X_C \ll R_E$ at signal frequencies, the a.c. load is only 2 kΩ. We can therefore draw an a.c. load line of 2 kΩ passing through the operating point as shown in *Figure 1.25b*. Cut off on the a.c. load line will occur at a lower value of V_{CE}.

To find the value that would cause cut-off distortion we can say

$$\frac{\Delta V_{CE}}{\Delta I_C} = 2 \text{ k}\Omega$$

and

$$\Delta V_{CE} = 2\text{k}\Omega \times 2.06 \text{ mA}$$

$$= 4.12 \text{ V}$$

Our peak a.c. v_{ce} is only 3·6 V so there is still no risk of cut-off distortion.

Finally, the maximum r.m.s. output current is given by

$$I_{r.m.s.} = \frac{V_{peak}}{R_L\sqrt{2}} = \frac{3.6}{2\sqrt{2}} \text{ mA} = 1.27 \text{ mA}$$

Example 1.6. A transistor whose operating point at normal room temperature is to be $V_{CE} - 3$ V, I_C 2 mA, has I_{CO} 2 μA and α0·98. The available d.c. supply is 6 V. The peak a.c. signal current of 2 mA is to produce an R.M.S. output voltage of 1·414 V. If the minimum V_{CE} to avoid bottoming distortion is 300 mV, and the transistor is subjected to a possible temperature rise of 40°C, design a suitable bias circuit. Assume that I_{CO} doubles for each 10°C rise in temperature and that any emitter resistor will be suitably decoupled at the signal frequency of 1 kHz.

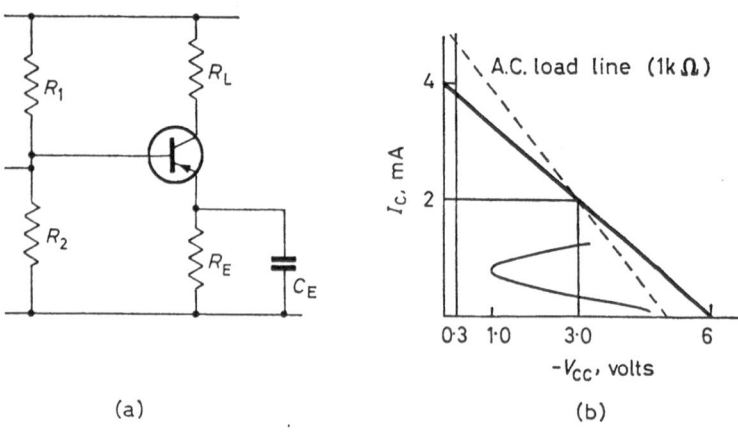

(a)   (b)

Figure 1.26. Circuit and graph for Example 1.6

34

We shall use the common form of circuit shown in *Figure 1.26a*. The problem is to determine the values of R_L, R_E, R_1, R_2 and C_E. R_L is determined by the output conditions from

$$R_L = \frac{v_{ac}}{i_{ac}} = \frac{1\cdot414}{2 \times 0\cdot707} = 1 \text{ k}\Omega.$$

R_E may now be found since V_{CE} is given.

$$6 = 3 + I_C R_L + I_E R_E \text{ V}$$

So $$I_E R_E = 3 - 2 = 1 \text{ V}$$

But $$I_E = \frac{I_C - I_{CO}}{\alpha} = \frac{1\cdot998}{0\cdot98} \simeq 2 \text{ mA}$$

Therefore $$R_E = \frac{1 \text{ V}}{2 \text{ mA}} = 500 \text{ }\Omega$$

Now since $I_E \simeq I_C$ we can draw the d.c. load line as shown in *Figure 1.26b*. The a.c. load line, the a.c. signal voltage, and the 300 mV bottoming line are also shown.

To determine K, we must find ΔI_C. First we find the permissible ΔV_{CE}: at normal room temperature, the operating V_{CE} is -3 V and the peak signal v_{ce} is 2 V, so the minimum instantaneous v_{ce} is $-3 + 2 = -1$ V.

Now V_{CE} must not fall below $-0\cdot3$ V (bottoming) so permissible shift of V_{CE} is given by $\Delta V_{CE} = $ d.c. $V_{CE} + $ peak a.c. $v_{ce} + V$ bottoming, $-3 + 2 + 0\cdot3 = -0\cdot7$ V.

$$\therefore \quad \Delta I_C = \frac{\Delta V_{CE}}{R_L + R_E} \quad \text{(Note this is a d.c. change with temperature)}$$

$$= \frac{0\cdot7}{1\cdot5} \text{ mA} = 466 \text{ }\mu\text{A}$$

$$\Delta I_{CO} = 2^4 I_{CO} - I_{CO} = 30 \text{ }\mu\text{A}$$

$$\therefore \quad \Delta I_{CO}' = \frac{30}{1 - \alpha} = 1\,500 \text{ }\mu\text{A}$$

$$\therefore \quad K = \frac{\Delta I_C}{\Delta I_{CO}'} = \frac{466}{1\,500} = 0\cdot311$$

Using equations 1.33 and 1.34

$$K = \frac{1}{1 + M\alpha'} \quad \text{where} \quad M = \frac{R_E}{R_E + R'} \quad \text{and} \quad \alpha' = \frac{\alpha}{1 - \alpha}$$

We have $$\alpha' = \frac{0 \cdot 98}{1 - 0 \cdot 98} = 49$$

$$\therefore \quad M = \frac{1}{\alpha'} \left(\frac{1}{K} - 1 \right) = \frac{2 \cdot 22}{49} = 0 \cdot 0453$$

and $$R' = R_E \left(\frac{1}{M} - 1 \right) = 21 \cdot 1 R_E$$

$$= 10 \cdot 5 \text{ k}\Omega$$

To find the values of R_1 and R_2, we use the Thèvenin equivalent circuit and equation 1.23

$$V_{BE} = V' - I_B R' - I_E R_E$$

Now $$I_B = \frac{I_C - I_{CO}'}{\alpha'} = \frac{2 - 0 \cdot 1}{49} = 39 \ \mu\text{A}$$

Taking $$I_E = I_C$$

$$V' = V_{BE} + 0 \cdot 5 \times 2 + 10 \cdot 5 \times 0 \cdot 039$$
$$= 0 \cdot 3 + 1 + 0 \cdot 41 = 1 \cdot 71 \text{ V}$$

But $$V' = \frac{6 R_2}{R_1 + R_2} \quad \text{and} \quad R' = \frac{R_1 R_2}{R_1 + R_2}$$

So $$1 \cdot 71 = \frac{6 R_2}{R_1 + R_2}$$

and $$R_1 + R_2 = \frac{6}{1 \cdot 71} R_2$$

and $$10 \cdot 5 = \frac{R_1 R_2}{6 R_2 / 1 \cdot 71} = \frac{R_1}{3 \cdot 5}$$

$$\therefore \quad R_1 = 37 \text{ k}\Omega$$

and $$R_2 = \frac{R_1}{\dfrac{6}{1 \cdot 71} - 1} = 14 \cdot 7 \text{ k}\Omega$$

Finally, for adequate decoupling $X_{CE} \ll R_E$ at 1 kHz,

$$X_{CE} = \frac{1}{2\pi 10^3 C} \quad \text{so} \quad C = \frac{1}{2\pi 10^3 X_{CE}}$$

Let $$X_{CE} = 10 \ \Omega$$

Thus $$C = \frac{10^6}{2\pi 10^4} \text{ F} = \frac{50}{\pi} \ \mu\text{F}$$

36

In practice a 100 μF capacitor would be used allowing the amplifier to be used at lower frequencies.

We have not considered the effect of coupling circuits or reactive loads on the graphical solution of transistor amplifiers, but as with valves, these will make the solution exceedingly difficult if not impossible.

Summarizing, in this chapter we have seen how simple valve and transistor circuits may be analysed by graphical methods. We have found the limitations of these methods, and we have seen how the d.c. operating conditions may be found. We have investigated suitable circuits to give the correct d.c. conditions, and in the case of transistors we have seen how this operating point may be stabilized against changes of temperature and transistor.

EXAMPLES

Example 1.7. A triode valve having the characteristics given below is connected in series with (a) an H.T. supply of 100 V and a resistive load of 8 kΩ, or (b) an H.T. supply of 300 V and load of 60 kΩ. Find the d.c. operating point in each case, if with (a) V_{GK} is -0.5 V and with (b) I_A is 1·5 mA.

V_{AK} (volts)	0	25	50	75	100	125	150	175	200	225	250	275	300
I_A (mA) 0 for V_{GK} (volts) 0·5	0	3·3 1·0	6·6 3·3	10·3 6·1	14·0 9·5	 13·5	 17·5						
1·0 1·5		0·25 0	1·3 0·5	3·3 1·7	6·1 3·5	9·5 6·5	13·5 9·75	17·5 13·5					
2·0 2·5			0	0·7 0·2	2·0 1·0	4·1 2·5	6·3 4·4	10·0 6·9	 10·2	 13·6			
3·0 3·5				0	0·4 0·15	1·4 0·6	2·8 1·5	4·85 2·9	7·4 5·0	10·4 7·6	 10·6		
4·0 4·5					0	0·2 0·1	0·75 0·4	1·6 0·9	3·2 2·0	5·3 3·5	8·0 5·75	8·5	
5·0 5·5							0·15	0·5 0·2	1·25 0·7	2·3 1·5	4·0 2·8	6·1 4·5	 6·5
6·0 6·5							·	0	0·3 0·1	0·75 0·4	1·7 1·0	3·0 2·0	4·6 3·3
7·0										0·1	0·4	1·2	2·3

Ans. (a) 63 V, 4·7 mA. (b) 208 V, -5 V.

Example 1.8. Using the circuits described in Example 1.7, determine the voltage gain. Assume an a.c. signal 0·5 V peak in each case. How is the gain for case (b) modified if the bias V_{GK} is changed to -0.5 V?

Ans. (a) 25·5. (b) 24 or 45·5.

Example 1.9. A triode valve using the characteristics given for Example 1.7 is connected in series with an anode load resistor R_L and a cathode resistor R_K and an H.T. of 250 V. Find the values of R_L and R_K if the required operating point is (*a*) V_{AK} 100 V, I_A 6 mA. (*b*) V_{AK} 150 V, I_A 9·75 mA.

Ans. (*a*) 24·8 kΩ, 166 Ω. (*b*) 10 kΩ, 154 Ω.

Example 1.10. The circuit described in Example 1.9 employs R_L 17·15 kΩ, R_K 250 Ω and an H.T. supply of 300 V. Find the output voltage and voltage gain if (*a*) $e_s = 1 \sin \omega t$ and (*b*) $e_s = 1·5 \sin \omega t$. Find the new value of v_0 and A_v if a capacitor C_K is connected in parallel with R_K such that its reactance at the signal frequency is negligible.

Ans. (*a*) 43, 21·5. (*b*) 64·5, 21·5. (*a*) 61, 30·5. (*b*) 96, 32.

Example 1.11. A triode valve having the characteristics given for Example 1.7 is connected in the cathode follower circuit shown in *Figure 1.27*. Determine the voltage gain. *Ans.* 0·65.

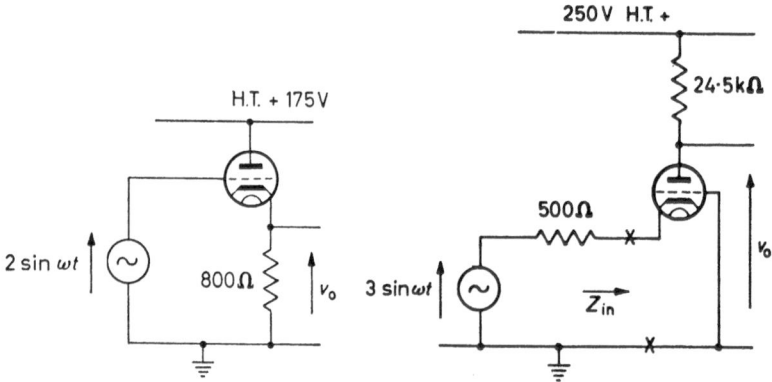

Figure 1.27. Circuit for Example 1.11   *Figure 1.28.* Circuit for Example 1.12

Example 1.12. If the triode valve shown in *Figure 1.27* is now connected in the circuit shown in *Figure 1.28*, calculate the voltage gain and the terminal input impedance. *Ans.* 19·2, 790 Ω.

Example 1.13. A single stage valve amplifier employs a resistive load of 4 kΩ. The d.c. grid bias is provided by an R_K of 1kΩ. The characteristics are such that the 200 V H.T. results in an operating anode current of 20 mA. Assuming that anode current excursions to less than 5 mA d.c. level will result in excessive distortion, calculate (*a*) the maximum peak a.c. anode current and hence the peak anode voltage; (*b*) the maximum peak a.c. anode voltage when R_K is adequately decoupled.

Ans. (*a*) 15 mA, 60 V. (*b*) 60 V.

Example 1.14. A single stage amplifier is operated with battery bias and an H.T. of 200 V resulting an anode current of 2·5 mA through the 40 kΩ load. The anode characteristic for V_{GK} 0 V is approximately linear having an r_a of 15 kΩ. Either positive V_{GK} or I_A less than 0·4 mA results in excessive distortion. Find the maximum peak alternating anode voltage and state the limiting factor, (*a*) with the circuit as described and (*b*) if an external load of 5·7 kΩ is coupled to the anode through a capacitor of negligible reactance.

Ans. (*a*) 45 V positive V_{GK}. (*b*) 11 V low I_A.

Example 1.15. A triode having the characteristics given for Example 1.7 is operated with a bias voltage of −1·5 V and an H.T. supply of 150 V. The anode load, a coil of 0·2 H, is assumed to be purely inductive. If the frequency is 5 kHz, draw the operating load line for a peak alternating anode current of 7·75 mA. Hence estimate the voltage gain and phase shift. *Ans.* 30 $\angle$ 315°.

Example 1.16. A transistor having the characteristics given below is connected in the circuit shown in *Figure 1.12a*. Determine the operating point on the input and output characteristics taking R_L and R_B as 500 Ω and 47 kΩ respectively and V_{CC} as −6 V. Calculate new values for R_L and R_B to change the operating point to I_C 8 mA, V_{CE} −2·7 V.

−V_{BE} (mV)		0	50	100	150	200	250	300
I_B (μA) for V_{CE} (volts)	0	0	22	90	200	345		
	−1·5		3	18	60	135	235	360
	−3·0		1	14	50	117	213	332
	−4·5			10	42	105	198	313
	−6·0			8	39	98	190	300

$-V_{CE}$ (Volts)		0	0·2	0·4	1·0	6·0
I_C (mA) for I_B (μA)	0	0	0·2	0·2	0·2	0·2
	40		1·2	1·4	1·5	1·9
	80		2·4	2·8	2·9	3·9
	120		3·6	4·1	4·3	5·7
	160		4·8	5·6	5·8	7·6
	200		6·0	6·9	7·3	9·5
	240		7·4	8·4	8·8	11·5

Ans. 123 μA, −204 mV, 5·1 mA, −3·43 V, 413 Ω, 28·8 kΩ.

Example 1.17. An alternating signal of 50 mV peak is applied to the base of the simple amplifier circuit in Example 1.16. Calculate the input impedance, the voltage gain, and the current gain.

Ans. 488 Ω, 36, 35.

Example 1.18. A certain transistor is found to have α 0·97 and I_{CO} 4 μA. If I_E is 2 mA, calculate α′, $I_{CO}′$, I_C, and I_B.

Ans. 32·3, 133 μA, 1·944 mA, 56 μA.

Example 1.19. A transistor operating at 5 mA is known to have α′ 160, and $I_{CO}′$ 0·2 mA.

Calculate α, I_{CO}, I_B and I_E.

Ans. 0·994, 1·24 μA, 30 μA, 5·03 mA.

Example 1.20. A transistor having α 0·99 and I_{CO} 3 μA is connected in the potential divider emitter resistor bias circuit shown in *Figure 1.22*. The d.c. supply is 12 V and the circuit components are R_1 33 kΩ, R_2 22 kΩ, R_L 1·8 kΩ, and R_E 1 kΩ. If V_{BE} is taken as 0·3 V, calculate the operating point and the stability factor K.

Ans. 3·93 mA, 1 V, 0·125.

Example 1.21. A transistor has a stable bias condition fixed by the collector feedback emitter resistor circuit shown in *Figure 1.23*. R_L is 3·3 kΩ, R_B 140 kΩ, R_E 500 Ω, and the supply battery 10 V. If the transistor has α′ 45, $I_{CO}′$ 0·1 mA and V_{BE} −0·2 V, calculate the operating point and the stability factor K.

Ans. 1·45 mA, 4·5 V, 0·46.

Example 1.22. The design requirement for a single stage transistor amplifier include a collector current of 1 mA, a 4 kΩ load, and a stability K of 0·05. It is required to amplify signals widely differing in amplitude in the frequency range 100 Hz to 10 kHz. The available power supply is 12 V and the silicon transistor has α' 120 and negligible I_{CO}. Assuming V_{BE} to be 0·7 V, calculate the values of the remaining components if a potential divider emitter resistor circuit is to be used.

Ans. 2 kΩ, 45·6 kΩ, 13·9 kΩ, 10 μF.

Example 1.23. A transistor amplifier uses potential divider emitter resistor bias with adequate decoupling. The components are R_L 1 kΩ, R_E 1 kΩ, R_1 40 kΩ, R_2 12 kΩ and V_{CC} is −6 V. If the transistor has negligible I_{CO} and V_{BE} of −0·2 V, calculate the maximum peak alternating output voltage (*a*) if α' is 50, and (*b*) if α' is 150.

Ans. 1 V, 1·11 V.

Example 1.24. The power transistor shown in *Figure 1.29* has a maximum collector dissipation 16·4 W, α' 150, and at 20°C, an

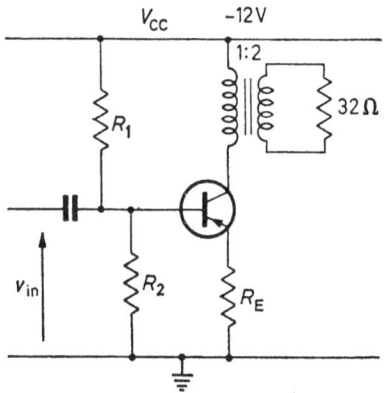

Figure 1.29. Circuit for Example 1.24

I_{CO} of 500 μA. The available d.c. supply is 12 V and the selected operating point is V_{CE} −8 V, I_C 2 A. The transformer has 2 : 1 turns ratio and 0·5 Ω primary resistance. At the signal frequency, the shunt primary reactance is very much greater than the reflected load impedance. Assuming the transistor to have V_{BE} −0·2 V, and further, that it will bottom at V_{CE} −0·4 V, determine: (*a*) the remaining bias components if the maximum temperature is 50°C (*b*) If the effect of distortion is ignored, the maximum output power. (*c*) the approximate input voltage to obtain this output.

Ans. 1½ Ω, 195 Ω, 70 Ω, 3·6 W, 1 V r.m.s.

2

FUNDAMENTALS OF NETWORK ANALYSIS

Equivalent circuits for electronic devices will be shown to consist of simple current or voltage generators, together with impedances or admittances. In Chapter 1, we found that such devices could only operate correctly if they were connected to suitable load and bias circuits consisting of impedances or admittances. In addition, any signal to be amplified was supplied from an a.c. current or voltage source together with its associated impedance or admittance.

Thus if we are to analyse these equivalent circuits, we must be able to analyse complex networks of impedances, admittances and generators. The basic rules for such analysis are fortunately simple and will probably be familiar to the reader. These are Ohm's law and Kirchhoff's laws. Correct application of Ohm's law demonstrates a number of important relationships which will be shown in the first section of this chapter. Kirchhoff's laws lead to the two most important tools for network solutions, mesh analysis and nodal analysis. These methods and the solution of the resultant equations will be considered in Section 2 of this Chapter. Section 3 will state some additional theorems which frequently simplify analysis and Section 4 will explain the analysis of a common form of network, the four terminal or two port network.

Section 1

FUNDAMENTALS

Ohm's law states that the current I flowing in an electrical circuit is directly proportional to the electrical pressure or voltage V applied to the circuit. The constant of proportionality is known as the circuit admittance Y, having dimension mhos,

i.e. $$I = VY \qquad (2.1)$$

An alternative way of expressing this is: the potential difference V across an electrical circuit is directly proportional to the current I

flowing through the circuit. In this case, the constant of proportionality is known as the circuit impedance Z, having dimension ohms, i.e.

$$V = IZ \tag{2.2}$$

By inspection we can see that

$$Z = \frac{V}{I} = \frac{1}{Y} \quad \text{or} \quad Y = \frac{I}{V} = \frac{1}{Z} \tag{2.3}$$

For direct currents and voltages the impedance of a circuit is the resistance R and the admittance is the conductance G. For steady state alternating or sinusoidal currents and voltages, the impedance and admittance are complex and become

$$Z = R + jX \tag{2.4}$$

$$Y = G + jB \tag{2.5}$$

where X and B are known as the circuit reactance and susceptance respectively. Note, in general

$$B \neq \frac{1}{X} \quad \text{and} \quad G \neq \frac{1}{R}$$

If a circuit has $Z = R + jX$,

$$Y = \frac{1}{Z} = \frac{1}{R + jX} = \frac{R - jX}{R^2 + X^2}$$

$$\therefore \qquad G = \frac{R}{R^2 + X^2} \qquad B = \frac{-X}{R^2 + X^2} \tag{2.6}$$

and only if R is zero, $B = -\frac{1}{X}$.

Notation

Before we can proceed to further relationships, we must consider the sense of measurement of voltage and current.

Consider first the simple d.c. circuit shown in *Figure 2.1*. If the voltmeter and ammeter are centre zero and connected with the

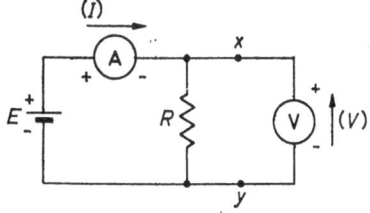

Figure 2.1. Circuit notation for direct currents and voltages

positive and negative terminals as shown, the voltmeter will read $+E$ volts and the ammeter $+I$ amps where $I = (E/R)$ amps. Conventional current flow is from positive to negative so the sense of measurement of $+I$ is as shown by the arrow (I).

The voltmeter measures the voltage at x with respect to y and will measure

$$+V = +IR = +E \text{ volts}$$

in the sense shown by the second arrow (V).

If however, the ammeter connections and thus the (I) arrow were reversed, the meter would read $-I$ amps and we could say that $I = -(E/R)$.

If the sense of voltage measurement was also reversed we should find $V_{yx} = -V_{xy}$ and $V_{yx} = IR = -(E/R) \times R = -E$ as would be expected from the circuit.

This example leads to the deduction of a useful rule:

The potential difference that is produced by a current flowing in an impedance will be positive $(+IZ)$ if the sense of the potential difference is taken in the opposite direction to the chosen sense of current flow.

This rule applies to alternating quantities in exactly the same way. Consider the part of a circuit shown in *Figure 2.2*, given that current

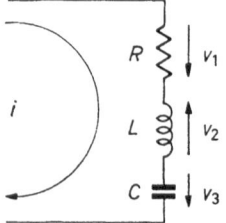

Figure 2.2. Circuit notation for alternating currents and voltages

is flowing. The instantaneous value of the current will be

$$i = \hat{I} \sin (\omega t \pm \phi)$$

where ϕ gives the phase with respect to some unknown reference.

In terms of this current i we can say that

$$v_1 = -iZ_1 \qquad v_2 = iZ_2 \qquad v_3 = -iZ_3$$

$$= -iR \qquad\quad = +j\omega Li \qquad = -i\left(\frac{-j}{\omega C}\right)$$

$$= \frac{ji}{\omega C}$$

So for greater clarity, rewording the rule:

In terms of a specified current, a volt drop due to an impedance-current product will be positive only if taken in the opposite sense or direction to that of the current.

Use of Ohm's Law

Now applying Ohm's law to a number of simple series and parallel circuits.

From *Figure 2.3a*

$$e = v_1 + v_2$$
$$= iZ_1 + iZ_2$$
$$= i(Z_1 + Z_2)$$

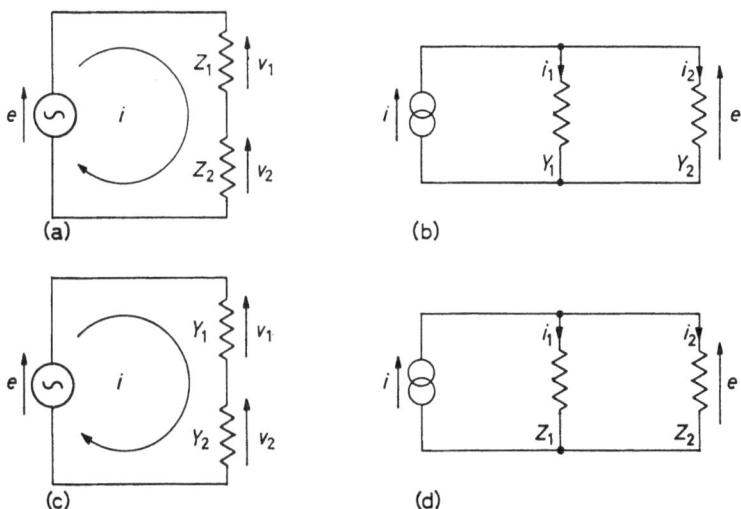

Figure 2.3. Series and parallel circuit arrangements

Total impedance $Z_T = \dfrac{e}{i} = Z_1 + Z_2$.

$\therefore$ Impedance in series may be simply added. (2.7)

Also $\qquad v_1 = iZ_1 = \dfrac{e}{Z_1 + Z_2} \times Z_1$

and $\qquad v_2 = iZ_2 = \dfrac{eZ_2}{Z_1 + Z_2}$

$\therefore$ Potential division between series impedances is in the direct ratio of the impedance values. (2.8)

Now referring to *Figure 2.3b*

$$i = i_1 + i_2$$
$$= eY_1 + eY_2$$
$$= e(Y_1 + Y_2)$$

Total admittance $\qquad Y_T = \dfrac{i}{e} = Y_1 + Y_2$

∴ $\qquad$ Admittances in parallel may be simply added. $\qquad$ (2.9)

Also $\qquad i_1 = eY_1 = \dfrac{i}{Y_1 + Y_2} \times Y_1$

and $\qquad i_2 = eY_2 = \dfrac{iY_2}{Y_1 + Y_2}$

∴ Current division between parallel admittances is in the direct ratio of the admittance values. $\qquad$ (2.10)

The reader should now compare 2.9 and 2.10 with 2.7 and 2.8. The similarity between the results in a series impedance system and a parallel admittance system is known as duality. The duality of a *VZI* system with an *IYV* system will become more obvious with further examples.

Consider *Figure 2.3c*

$$e = v_1 + v_2 = \frac{i}{Y_1} + \frac{i}{Y_2} = i\left(\frac{1}{Y_1} + \frac{1}{Y_2}\right)$$

Total admittance $\qquad Y_T = \dfrac{i}{e} = \dfrac{1}{\dfrac{1}{Y_1} + \dfrac{1}{Y_2}}$

∴ $\qquad \dfrac{1}{Y_T} = \dfrac{1}{Y_1} + \dfrac{1}{Y_2}$

∴ Total admittance of a number of admittances in series is given by the reciprocal of the sum of the reciprocals of the individual admittances. $\qquad$ (2.11)

This rule is frequently simplified to

$$Y_T = \frac{Y_1 Y_2}{Y_1 + Y_2} \text{ for two admittances in series}$$

or $\qquad Y_T = \dfrac{Y_1 Y_2 Y_3}{Y_1 Y_2 + Y_2 Y_3 + Y_3 Y_1}$ for three in series etc.

Also $\quad v_1 = \dfrac{i}{Y_1} = \dfrac{eY_T}{Y_1} = \dfrac{e}{Y_1} \times \dfrac{Y_1Y_2}{Y_1 + Y_2} = \dfrac{eY_2}{Y_1 + Y_2}$

Similarly $\qquad v_2 = \dfrac{i}{Y_2} = \dfrac{eY_T}{Y_2} = \dfrac{eY_1}{Y_1 + Y_2}$

$\therefore$ Potential division between series admittance is in the inverse ratio of the admittance values. $\qquad$ (2.12)

Now from *Figure 2.3d*

$$i = i_1 + i_2 = \frac{e}{Z_1} + \frac{e}{Z_2} = e\left(\frac{1}{Z_1} + \frac{1}{Z_2}\right)$$

Total impedance $\qquad Z_T = \dfrac{e}{i} = \dfrac{1}{\dfrac{1}{Z_1} + \dfrac{1}{Z_2}}$

or $\qquad \dfrac{1}{Z_T} = \dfrac{1}{Z_1} + \dfrac{1}{Z_2}$ and $Z_T = \dfrac{Z_1Z_2}{Z_1 + Z_2}$ etc.

$\therefore$ Total impedance of a number of impedances in parallel is given by the reciprocal of the sum of the reciprocals of the individual impedances. $\qquad$ (2.13)

Finally, $\qquad i_1 = \dfrac{e}{Z_1} = \dfrac{iZ_T}{Z_1} = \dfrac{i}{Z_1}\dfrac{Z_1Z_2}{(Z_1 + Z_2)} = \dfrac{iZ_2}{Z_1 + Z_2}$

and $\qquad\qquad\qquad i_2 = \dfrac{iZ_1}{Z_1 + Z_2}$

$\therefore$ Current division between parallel impedances is in the inverse ratio of the impedance values $\qquad$ (2.14)

An example will illustrate the use of the above rules.

Example 2.1. Figures 2.4(a) and *(b)* show the same circuit. In *(a)* the branches are given in their admittance values while in *(b)* impedances

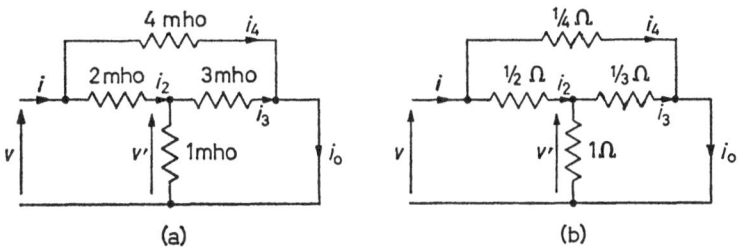

(a) $\qquad\qquad\qquad\qquad\qquad$ (b)

Figure 2.4. Circuit for Example 2.1

47

are given. Using both (a) and (b) determine the input admittance i/v, the value of i_0 in terms of i, and the potential difference v' across the $1\,\Omega$ branch in terms of v.

First consider circuit (a); the 4 mho branch is in parallel with (the 2 mho branch in series with the 3 mho and 1 mho branches in parallel).

Applying rules 2.9 and 2.11,

$$Y_T = 4 + \frac{2(3 + 1)}{2 + (3 + 1)} = 4 + \frac{8}{6} = 5\tfrac{1}{3} \text{ mhos}$$

For circuit (b) we must use rules 2.7 and 2.13 since all branches are quoted as impedances.

Now

$$Z_T = \frac{\dfrac{1}{4}\left(\dfrac{1}{2} + \dfrac{1 \times \tfrac{1}{3}}{1 + \tfrac{1}{3}}\right)}{\dfrac{1}{4} + \dfrac{1}{2} + \dfrac{1 \times \tfrac{1}{3}}{1 + \tfrac{1}{3}}}\,\Omega$$

multiplying numerator and denominator by $\tfrac{4}{3}$;

$$Z_T = \frac{\tfrac{1}{6} + \tfrac{1}{12}}{\tfrac{1}{3} + \tfrac{2}{3} + \tfrac{1}{3}} = \tfrac{3}{16}\Omega$$

$$\therefore \qquad Y_T = \frac{1}{Z_T} = \frac{16}{3} = 5\tfrac{1}{3} \text{ mhos}$$

the same result as that found when working in admittances.

Now, to find i_0 from circuit (a); i_0 is the sum of the currents flowing in the 3 mho and 4 mho branches.

$$\therefore \qquad i_0 = i_4 + i_3$$

The 4 mho branch is connected directly across v, therefore

$$i_4 = 4v \text{ amps}$$

i_3 is found by calculating i_2 and dividing i_2 between the 3 mho and 1 mho branches (rule 2.10).

$$i_2 = v\left[\frac{2 \times (3 + 1)}{2 + 3 + 1}\right] = \frac{8}{6}v \text{ amp}$$

$$\therefore \qquad i_3 = \frac{8v}{6} \times \frac{3}{3 + 1} = v \text{ amp}$$

$$\therefore \qquad i_0 = v + 4v = 5v \text{ amp}$$

48

To find i_0 from the impedance circuit (b) we follow the same procedure using rules 2.7, 2.13 and 2.14.

As before,

$$i_0 = i_4 + i_3$$

$$= \frac{v}{\frac{1}{4}} + \frac{v}{\frac{1}{2} + \frac{\frac{1}{3} \times 1}{\frac{1}{3} + 1}} \times \frac{1}{\frac{1}{3} + 1}$$

$$= 4v + \frac{v}{\frac{1}{2} \times \frac{4}{3} + \frac{1}{3}}$$

$$= 4v + \frac{6v}{4 + 2} = 5v \text{ amp}$$

Finally, to find v' from each circuit, we can ignore the 4 mho branch and note that v' is the potential across the 3 mho and 1 mho branches in parallel.

For circuit (a) using rules 2.9, 2.11 and 2.12

$$v' = v \times \frac{2}{2 + 3 + 1} = \frac{v}{3} \text{ volts}$$

and for circuit (b) using rules 2.7, 2.8 and 2.13,

$$v' = v \times \frac{\frac{1 \times \frac{1}{3}}{1 + \frac{1}{3}}}{\frac{1}{2} + \frac{1 \times \frac{1}{3}}{1 + \frac{1}{3}}}$$

$$= \frac{v \times \frac{1}{3}}{\frac{2}{3} + \frac{1}{3}} = \frac{v}{3} \text{ volts}$$

Section 2

From Example 2.1, it can be seen that simple circuit problems may be solved by the rules based on Ohm's law. There are three disadvantages to this approach. Firstly, with more involved circuits, the resulting expressions become exceedingly unwieldly and the possibility of a mistake increases. Secondly, if several different unknowns are required, as in Example 2.1, a separate solution is required for each. Finally, if the circuit involves more than one generator, the combined effect cannot be determined. Two most important methods of circuit analysis are based on Kirchhoff's law. These are really only common sense and will be explained in these terms.

KIRCHHOFF'S LAWS

1. The Current Law

In everyday language this would be defined as 'That which goes in must come out'. Consider the part circuit shown in *Figure 2.5a*.

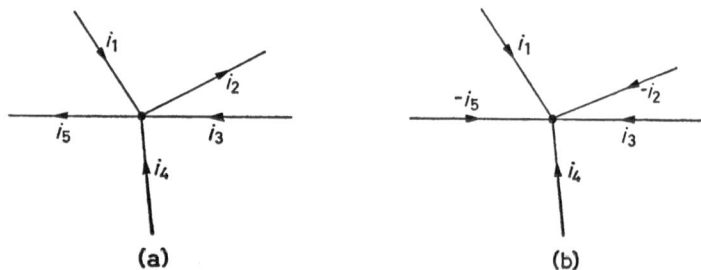

(a)   **(b)**

Figure 2.5. Summation of currents at a node

This shows a circuit *node* or junction between a number of branches. It is not shown as a capacitor and can therefore store no charge. Thus as a quantity of charge enters the node, an equal quantity must leave it. But moving charge is current, so we can say: the sum of the currents entering a node must be equal to the sum of the currents leaving that node. In *Figure 2.5a* this relationship is given by

$$i_1 + i_3 + i_4 = i_2 + i_5$$

An alternative definition is: the algebraic sum of the currents entering a node is equal to zero.

Again referring to *Figure 2.5a*, currents i_2 and i_5 leaving the node are equivalent to currents $-i_2$ and $-i_5$ entering the node as in *Figure 2.5b*. From this alternative definition,

$$i_1 + i_3 + i_4 - i_2 - i_5 = 0$$

which leads to the same result as that obtained by the first definition.

A useful analogy is traffic flow at a roundabout, where unless there is an accident or breakdown, the total number of vehicles entering the roundabout is equal to the total number of vehicles leaving it.

2. The Voltage Law

For this law, in everyday language, we could say that however far one falls in altitude, one must climb by the same amount to reach the original starting point.

A particular point in an electrical circuit can possess only a single value of potential or voltage. It might be at earth or zero potential or it might be at say, 5 000 V but it cannot be both at the same time. If we start at such a point (at say $+10$ V) and move around the circuit, we might climb to a high positive voltage or fall to negative voltage, but when we return to the starting point, it must be at $+10$ V.

Change in potential may either be due to currents flowing in impedances (IZ volt drops) or due to generators or sources of e.m.f.

Thus we can say that for any closed loop in an electrical circuit, the sum of the rises in potential due to generators, must be equal to the sum of the falls in potential due to IZ volt drops. Consider *Figure 2.6a*.

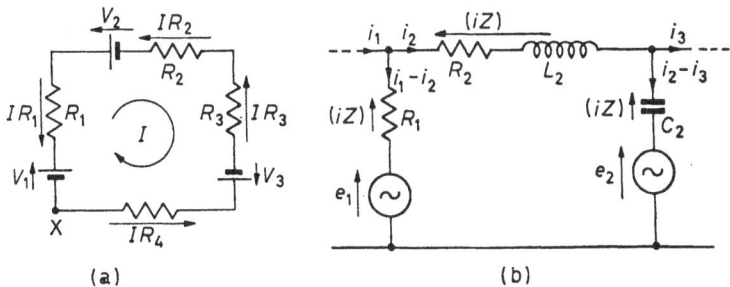

(a)   (b)

Figure 2.6. Summation of voltages around a loop or mesh

Starting from point X and travelling in a clockwise direction, we shall first sum the changes in potential due to the generators. We climb through V_1, fall through V_2 and climb again through V_3.

Thus the sum of the rises in potential is $V_1 - V_2 + V_3$. To find the sense of the IZ volt drops, we need to know the sense or direction of I. This could only be found if values for V_1, V_2 and V_3 were given. We therefore guess the direction and then if on inserting values, we obtain a negative answer, we can change the direction for the solution. In this case we guess that I flows in a clockwise direction as shown. Now to obtain positive IZ products, the potential must be measured in the opposite direction. The positive sense of these IZ products have also been inserted in *Figure 2.6a*. Now starting from point X we can add the changes in potential until we return to the starting point where we must return to our starting potential, i.e. the total change in potential is zero.

$$+V_1 - IR_1 - V_2 - IR_2 - IR_3 + V_3 - IR_4 = 0$$

or the algebraic sum of the potential difference taken around a closed loop is zero.

Alternatively we can say that the sum of the potential rises due to generators is equal to the IZ volt drops when taken around a closed loop in the same direction.

For the circuit of *Figure 2.6a*, this would be given by

$$V_1 - V_2 + V_3 = I(R_1 + R_2 + R_3 + R_4)$$

from which I could be obtained if numerical values were given.

In practice the currents in the different branches may have different values as shown in the a.c. circuit in *Figure 2.6b*. The method of writing the equations is exactly the same for a.c., and the solution will give the value of the unknown currents in the sense shown and their phase with respect to the given a.c. generators. The positive sense of IZ drops has been shown and the equation may be written

$$e_1 + R_1(i_1 - i_2) - (R_2 + j\omega L_2)i_2 - \left(\frac{-j}{\omega C_2}\right)(i_2 - i_3) - e_2 = 0$$

Further examples of writing loop or mesh equations and junction or nodal equations will be found when we see how Kirchhoff's laws are applied to mesh and nodal analysis.

MESH ANALYSIS

For mesh analysis, all branches are shown as impedances and all sources are voltage generators. Unknown currents are selected (i.e. named i_1, i_2, etc.) and their sense indicated. The loop or mesh equations are written and solved giving all the branch currents in terms of the source voltages and branch impedances. This technique will be best understood by considering some examples.

Example 2.2. Using mesh analysis determine the current flowing in the 2 Ω branch in the circuit shown in *Figure 2.7*.

The first problem is to insert the unknown currents. Any notation can be used provided Kirchhoff's current law is obeyed at the junctions. One possible set of unknowns can be inserted as follows.

Let the two batteries have currents I_1 and I_2 as shown and let the required current in the 2 Ω branch be I_x. Now, following the current law, the 4 Ω branch must carry $(I_2 - I_x)$ and the 5 Ω branch $(I_1 + I_2 - I_x)$.

This is not the simplest method but we shall first obtain the required solution using these unknown currents. There are six

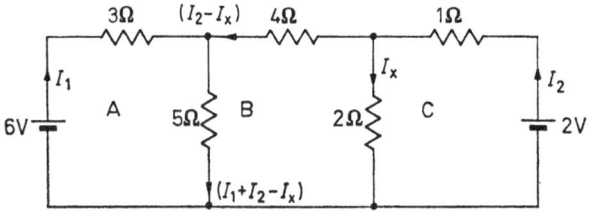

Figure 2.7. Circuit for Example 2.2

possible loops for which equations can be written, but since there are only three unknown currents only three equations are required.

The equations for meshes A, B and C are respectively

A $\qquad 6 = 3I_1 + 5(I_1 + I_2 - I_x)$

B $\qquad 0 = -5(I_1 + I_2 - I_x) - 4(I_2 - I_x) + 2I_x$

C $\qquad -2 = -2I_x - I_2$

Collecting terms

$$6 = 8I_1 + 5I_2 - 5I_x \qquad (2.15)$$

$$0 = -5I_1 - 9I_2 + 11I_x \qquad (2.16)$$

$$2 = 2I_x + I_2 \qquad (2.17)$$

One method of solving these equations is by substitution as follows

From 2.17 $\qquad I_2 = 2 - 2I_x \qquad (2.18)$

substitute in 2.16 $\qquad 5I_2 = -18 + 18I_x + 11I_x$

or $\qquad I_1 = \dfrac{-18}{5} + \dfrac{29}{5} I_x \qquad (2.19)$

substituting for I_1 and I_1 from 2.19 and 2.18 in equation 2.15

$$6 = -\frac{144}{5} + \frac{232}{5} I_x + 10 - 10I_x - 5I_x$$

53

Collecting terms, $6 + \dfrac{144}{5} - 10 = I_x \left(\dfrac{232}{5} - 15 \right)$

simplifying, $24 \cdot 8 = 31 \cdot 4 I_x$

and $I_x = \dfrac{24 \cdot 8}{31 \cdot 5} = 0 \cdot 79 \text{ A}$

Maxwell's Circulating Currents

Now since any notation for the unknown currents may be used, we should investigate the quickest and most convenient method. This is known as the Maxwell's circulating current rule. The circuit for Example 2.2 is redrawn as *Figure 2.8*.

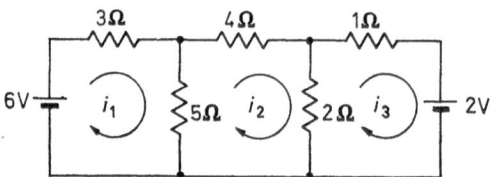

Figure 2.8. Use of Maxwell's circulating currents

Maxwell's circulating currents (i_1, i_2 and i_3) are shown. It is assumed that each closed loop has a current associated only with that loop. Where a branch is peculiar to a loop, the branch current is equal to the loop current. If, however, a branch is common to two loops (the $5 \, \Omega$ branch for example), the branch current is equal to the difference between the two loop currents.

Thus the $5 \, \Omega$ branch current is ($i_1 - i_2$) flowing down or ($i_2 - i_1$) flowing up. Similarly the $2 \, \Omega$ branch current is ($i_2 - i_3$) flowing down or ($i_3 - i_2$) flowing up.

So writing the equation for the i_1 loop we find

$$6 = 3i_1 + 5(i_1 - i_2)$$

and for the i_2 and i_3 loops respectively,

$$0 = 5(i_2 - i_1) + 4i_2 + 2(i_2 - i_3)$$

$$-2 = 2(i_3 - i_2) + i_3$$

Collecting terms and rearranging

$$6 = 8i_1 - 5i_2 \tag{2.20}$$

$$0 = -5i_1 + 11i_2 - 2i_3 \tag{2.21}$$

$$-2 = \qquad\quad -2i_2 + 3i_3 \tag{2.22}$$

These equations could have been obtained more simply by applying the following rule.

For any loop, sum the e.m.f. in the direction of the loop current and equate to the loop current times all the loop branches, minus any adjacent loop current, times the common branch sharing that current.

Following this for the first loop:

The e.m.f. in the direction of the loop current is $+6$ V.
Loop current $\times$ branch impedances $= i_1(3 + 5)$.
Adjacent current $\times$ shared branch $= 5i_2$.
$\therefore$ Equation becomes $6 = 8i_1 - 5i_2$ which is the same as equation 2.20 above.

Further application of this rule will appear in the next example.

A better method for the solution of a number of simultaneous equations is the applications of determinants. For the theory behind this method the reader is referred to any good mathematics textbook. Here, we shall apply the method in full detail so that the reader can regard this method as a useful aid to the solution of circuit problems.

The determinant Δ of the network is made up from the coefficients of the unknowns. In this case

$$\Delta = \begin{vmatrix} 8 & -5 & 0 \\ -5 & 11 & -2 \\ 0 & -2 & 3 \end{vmatrix}$$

To obtain the value of one of the unknown currents, i_1, we write a second determinant Δ_1 which is the same as Δ except for the i_1 column. This is replaced by the constants. Thus for Δ_1

$$\Delta_1 = \begin{vmatrix} 6 & -5 & 0 \\ 0 & 11 & -2 \\ -2 & -2 & 3 \end{vmatrix}$$

Similarly for i_2 and i_3 we should require Δ_2 and Δ_3

$$\Delta_2 = \begin{vmatrix} 8 & 6 & 0 \\ -5 & 0 & -2 \\ 0 & -2 & 3 \end{vmatrix} \quad \text{and} \quad \Delta_3 = \begin{vmatrix} 8 & -5 & 6 \\ -5 & 11 & 0 \\ 0 & -2 & -2 \end{vmatrix}$$

Now $$i_1 = \frac{\Delta_1}{\Delta}, \quad i_2 = \frac{\Delta_2}{\Delta}, \quad i_3 = \frac{\Delta_3}{\Delta}$$

Now all we have to do is to find the numerical values of the determinants. This process is known as expanding the determinants. The expansion of a second order determinant is as follows:

$$\begin{vmatrix} a_1 & a_2 \\ b_1 & b_2 \end{vmatrix} = a_1b_2 - b_1a_2$$

For a third order determinant the expression is as follows:

$$\begin{vmatrix} a_1 & a_2 & a_3 \\ b_1 & b_2 & b_3 \\ c_1 & c_2 & c_3 \end{vmatrix} = a_2\begin{vmatrix} b_2 & b_3 \\ c_2 & c_3 \end{vmatrix} - a_2\begin{vmatrix} b_1 & b_3 \\ c_1 & c_3 \end{vmatrix} + a_3\begin{vmatrix} b_1 & b_2 \\ c_1 & c_2 \end{vmatrix}$$

This is known as expansion on the first row. Alternatively, we may expand on the first column.

$$\begin{vmatrix} a_1 & a_2 & a_3 \\ b_1 & b_2 & b_3 \\ c_1 & c_2 & c_3 \end{vmatrix} = a_1\begin{vmatrix} b_2 & b_3 \\ c_2 & c_3 \end{vmatrix} - b_1\begin{vmatrix} a_2 & a_3 \\ c_2 & c_3 \end{vmatrix} + c_1\begin{vmatrix} a_2 & a_3 \\ b_2 & b_3 \end{vmatrix}$$

Note in each case if a horizontal and a vertical line are drawn through the principal coefficient, the remaining second order determinant or cofactor consists of the remaining terms.

Now applying this to equations 2.20, 2.21 and 2.22 for Example 2.2.

$$\Delta = \begin{vmatrix} 8 & -5 & 0 \\ -5 & 11 & -2 \\ 0 & -2 & 3 \end{vmatrix} = 8(33 - 4) + 5(-15 - 0) + 0$$
$$= 232 - 75 = 157$$

$$\Delta_2 = \begin{vmatrix} 8 & 6 & 0 \\ -5 & 0 & -2 \\ 0 & -2 & 3 \end{vmatrix} = 8(0 - 4) - 6(-15 - 0) + 0$$
$$= 58$$

$$\Delta_3 = \begin{vmatrix} 8 & -5 & 6 \\ -5 & 11 & 0 \\ 0 & -2 & -2 \end{vmatrix} = 8(-22 - 0) + 5(10 - 0) + 6(10 - 0)$$
$$= -66$$

$$\therefore \qquad i_2 = \frac{58}{157}\,\text{A}, \quad \text{and} \quad i_3 = \frac{-66}{157}\,\text{A}$$

Unknown $\qquad I_2 = i_2 - i_3 = \dfrac{58 + 66}{157}\,\text{A}$

$$= \frac{124}{157} = 0\cdot79\,\text{A}$$

This method may appear longer, but with practice it is possible to write down the values for the determinant directly from the original equations. In addition it is the only convenient method if the co-efficients of the unknowns are complex, as is the case in the next example.

Example 2.3. Determine the current supplied by the 6 V generator in the circuit shown in *Figure 2.9.*

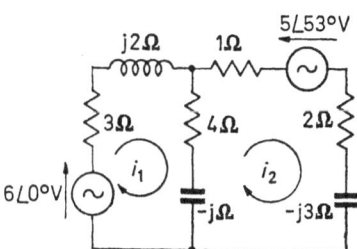

Figure 2.9. Circuit for Example 2.3

First we insert the currents as before and then write the normal mesh equations, expressing all impedances in the complex $(a + jb)$ form.

$$6\underline{/0°} = i_1(3 + j2 + 4 - j) - i_2(4 - j)$$

$$-5\underline{/53°} = -i_1(4 - j) + i_2(4 - j + 1 + 2 - j3)$$

Now collect terms and converting generator e.m.f.s to $(a + jb)$ form

$$6 + j0 = (7 + j)i_1 - (4 - j)i_2 \qquad (2.23)$$

$$-5(\cos 53° + j \sin 53°) = -(4 - j)i_1 + (7 - j4)i_2 \quad (2.24)$$

Evaluating the e.m.f. for equation 2.24,

$$-5(\cos 53° + j \sin 53°) = -5(0·6 + j0·8) \text{ V}$$
$$= -3 - j4 \text{ V}$$

Now solving by determinants

$$i_1 = \frac{\begin{vmatrix} (6 + j0) & -(4 - j) \\ (-3 - j4) & (7 - j4) \end{vmatrix}}{\begin{vmatrix} (7 + j) & -(4 - j) \\ -(4 - j) & (7 - j4) \end{vmatrix}} \text{ A}$$

57

expanding the determinants

$$i_1 = \frac{(6 + j0)(7 - j4) - (3 + j4)(4 - j)}{(7 + j)(7 - j4) - (4 - j)(4 - j)} \text{ A}$$

$$= \frac{42 - j24 - 12 - 4 - j16 + j3}{49 + 4 - j28 + j7 - 16 + 1 + j4 + j4} \text{ A}$$

$$= \frac{26 - j37}{38 - j13} \text{ A}$$

Converting the numerator and denominator into the R/θ form:

$$i_1 = \frac{\sqrt{(26^2 + 37^2)}}{\sqrt{(38^2 + 13^2)}} \frac{\underline{/-\tan^{-1} \frac{37}{26}}}{\underline{/-\tan^{-1} \frac{13}{38}}} \text{ A}$$

$$= \sqrt{\frac{2\,048}{1\,484}} \underline{/-55° - (-19°)} \text{ A}$$

$$= 0·686\underline{/-36°} \text{ A}$$

Thus the current is lagging the 6 V generator voltage by 36°.

Many other examples of mesh analysis will occur in later chapters.

NODAL ANALYSIS

Mesh analysis was developed by the use of Kirchhoff's voltage law. The dual system known as nodal analysis is based on Kirchhoff's current law. Since it is a dual we shall expect to use this method on

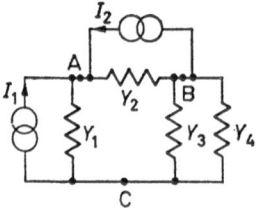

Figure 2.10. Circuit suitable for nodal analysis

circuits consisting of current generators and admittances. We solve for unknown node voltages (as opposed to unknown mesh currents) in terms of current generators and the circuit admittances.

Consider the circuit shown in *Figure 2.10*.

This circuit has three nodes each of which will have a potential or voltage V_A, V_B, V_C. In practice we usually express the voltage at a circuit node with respect to earth or zero. In this case we shall let node C be at earth, i.e. $V_C = 0$.

We shall now apply Kirchhoff's current law in the following manner. Currents entering a node from current generators may be equated to currents leaving a node through admittance branches.

Thus at node A,

$$I_1 + I_2 = (V_A - 0)Y_1 + (V_A - V_B)Y_2$$

and at node B

$$-I_2 = (V_B - 0)(Y_3 + Y_4) + (V_B - V_A)Y_2$$

Collecting up terms,

$$I_1 + I_2 = V_A(Y_1 + Y_2) - V_B Y_2$$
$$-I_2 = -Y_A Y_2 + V_B(Y_2 + Y_3 + Y_4)$$

From this, we can see that these nodal equations could be found in another way. For each node, equate the currents entering the node to that node voltage times the sum of all admittances connected to the node, minus each adjacent node voltage times the connecting branch admittance. This is of course the dual of the rule for formation of the mesh analysis equations on page 55.

Example 2.4. Determine the current flowing in the 5 mho branch of the circuit shown in *Figure 2.11.*

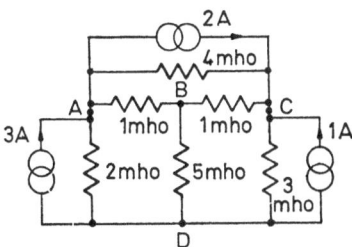

Figure 2.11. Circuit for Example 2.4

This circuit has four nodes, and we shall consider node D to be at zero potential.

For node A,

$$3 - 2 = V_A(1 + 2 + 4) - 1V_B - 4V_C$$

For node B,

$$0 = -1V_A + V_B(1 + 1 + 5) - 1V_C$$

For node C,

$$2 + 1 = -4V_A - 1V_B + V_C(3 + 1 + 4)$$

Rewriting,

$$1 = 7V_A - V_B - 4V_C$$
$$0 = -V_A + 7V_B - V_C$$
$$3 = -4V_A - V_B + 8V_C$$

By determinants,

$$V_B = \cfrac{\begin{vmatrix} 7 & 1 & -4 \\ -1 & 0 & -1 \\ -4 & 3 & 8 \end{vmatrix}}{\begin{vmatrix} 7 & -1 & -4 \\ -1 & 7 & -1 \\ -4 & -1 & 8 \end{vmatrix}} \text{ V}$$

Expanding

$$\therefore \quad V_B = \frac{7(0 + 3) - 1(-8 - 4) - 4(-3 - 0)}{7(56 - 1) + 1(-8 - 4) - 4(1 + 28)} \text{ V}$$

$$V_B = \frac{21 + 12 + 12}{392 - 12 - 116} = \frac{45}{264} = 0{\cdot}17 \text{ V}$$

The current flowing in the 5 mho branch is given by VY.

$$\therefore \qquad I = 5 \times 0{\cdot}17 = 0{\cdot}85 \text{ A}$$

Other examples of nodal analysis will appear in later chapters.

In general if a circuit has mainly parallel branches, nodal analysis will be quicker. If there are more series branches mesh analysis is best. In some instances, use of the better method will result in reducing the number of unknowns. and hence the number of equations. It may be necessary to convert voltage generators to current generators or vice versa. Such conversions may be accomplished by the use of Thèvenin's theorem and Norton's theorem which will be discussed in the next section.

Section 3

In this section a number of useful theorems will be stated and demonstrated without academic proof. Examples using each will also be given.

THE SUPERPOSITION THEOREM

This may be stated in two forms, one in terms of an impedance network, the other in terms of an admittance network.

In any linear network of impedances and generators, the current flowing in one branch is equal to the sum of the currents flowing in that branch due to each generator taken separately with all other generators replaced by their internal impedances.

Example 2.5. By use of the superposition theorem calculate the current flowing in the 5 Ω branch of the circuit shown in *Figure 2.12.* Check this answer by use of mesh analysis.

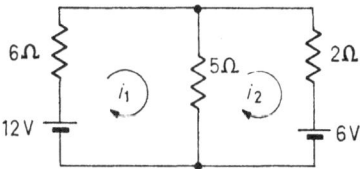

Figure 2.12. Circuit for Example 2.5

Taking the 12 V battery alone, we remove the 6 V battery leaving only its internal impedance of 2 Ω. Now for this circuit

$$I = I_1 = \frac{12}{6 + \dfrac{2 \times 5}{2 + 5}} \times \frac{2}{2 + 5}$$

(Rules 2.7, 2.12 and 2.13 on pages 45 and 47.)

Evaluating,
$$= \frac{24}{42 + 10} = \frac{24}{52} \text{ A}$$

Now taking the 6 V battery alone, we remove the 12 V battery leaving only its internal impedance of 6 Ω. Now applying the same rules,

$$I = I_2 = \frac{6}{2 + \dfrac{5 \times 6}{5 + 6}} \times \frac{6}{5 + 6} \text{ A}$$

$$= \frac{36}{22 + 30} = \frac{36}{52} \text{ A}$$

By the superposition theorem, for the whole circuit,

$$I = I_1 + I_2 = \frac{24 + 36}{52} = 1 \cdot 155 \text{ A}$$

Now checking this result by mesh analysis and using the unknown currents i_1 and i_2 shown in *Figure 2.12,*

$$12 = 11i_1 - 5i_2$$

$$-6 = -5i_1 + 7i_2$$

By determinants,

$$i_1 = \frac{84 - 30}{77 - 25} = \frac{54}{52} \text{ A}$$

$$i_2 = \frac{-66 + 60}{52} = \frac{-6}{52} \text{ A}$$

But the required current $I = i_1 - i_2 = \frac{60}{52}$ A as was found by superposition.

The dual form of this theorem may be stated:

In any network of admittances and current generators the potential across one branch is equal to the sum of the potentials across that branch due to each generator taken separately with all others replaced by their internal admittances.

Example 2.6. By the use of the superposition theorem calculate the potential across the branch Y_L in the circuit in *Figure 2.13.*

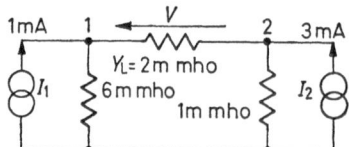

Figure 2.13. Circuit for Example 2.6

Check the answer by use of nodal analysis.

$$V \text{ due to } I_1 \text{ alone } = V_1 = \frac{1}{6 + \dfrac{2 \times 1}{2 + 1}} \times \frac{1}{2 + 1} \text{ V}$$

$$= \frac{1}{20} \text{ V}$$

$$V \text{ due to } I_2 \text{ alone } = V_2 = \frac{-3}{1 + \dfrac{2 \times 6}{2 + 6}} \times \frac{6}{2 + 6} \text{ V}$$

$$V_2 = -\frac{18}{20} \text{ V}$$

Note V_2 is negative following from the direction of I_2 and the required sense of V. Now

$$V = V_1 + V_2 = -\frac{17}{20}\,\text{V}$$

Now checking by nodal analysis:

at node 1, $\qquad\qquad 1 = 8V_1 - 2V_2$

at node 2, $\qquad\qquad 3 = -2V_1 + 3V_2$

$$\therefore \qquad\qquad V_1 = \frac{3+6}{24-4} = \frac{9}{20}$$

and $\qquad\qquad V_2 = \frac{24+2}{20} = \frac{26}{20}$

Now

$$V = V_1 - V_2 = \frac{9-26}{20} = \frac{-17}{20}\,\text{V}$$

which is the same result as that found using the superposition theorem.

Thèvenin's theorem states that any two-terminal network of generators and impedances may be replaced by a single voltage generator in series with a single impedance.

In *Figure 2.14,* the network is shown as a box with two terminals.

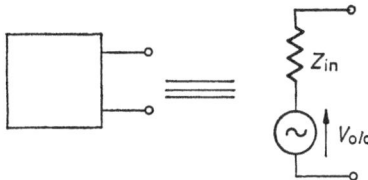

Figure 2.14. Thèvenin's theorem

The two-terminal equivalent is shown on the right. The components of the equivalent are found as follows:

$V_{o/c}$ is the voltage measured across the terminals of the network when no load is connected.

63

Z_{in} is the impedance measured between the terminals when all internal generators are suppressed or replaced by their internal impedances.

The truth of this may be demonstrated by a simple example.

Example 2.7. Determine the value of R_L that will carry a current of $\frac{1}{3}$ A in the circuit shown in *Figure 2.15*.

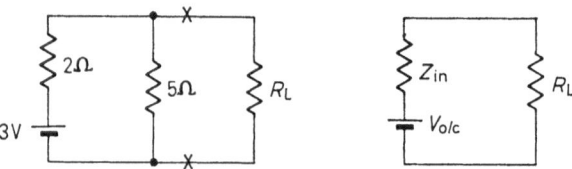

Figure 2.15. Circuit for Example 2.7

First break the circuit at points XX and apply Thèvenin's theorem to the left-hand half of the circuit. Now

$$V_{o/c} = \frac{3 \times 5}{2 + 5} = \frac{15}{7} \text{ V}$$

$$Z_{in} = \frac{2 \times 5}{2 + 5} = \frac{10}{7} \Omega$$

The equivalent circuit is now shown in *Figure 2.15b*. Now,

$$I = \frac{1}{3} = \frac{\frac{15}{7}}{\frac{10}{7} + R_L} = \frac{15}{10 + 7R_L} \text{ A}$$

$$\therefore \qquad 10 + 7R_L = \frac{15}{\frac{1}{3}} = 45 \ \Omega$$

$$\therefore \qquad R_L = \frac{(45 - 10)}{7} = 5 \ \Omega$$

Checking by basic methods

$$I = \frac{1}{3} = \frac{3}{2 + \dfrac{5R_L}{5 + R_L}} \times \frac{5}{5 + R_L} \text{ A}$$

$$\frac{1}{3} = \frac{15}{10 + 2R_L + 5R_L}$$

$$\therefore \qquad 10 + 7R_L = 45 \ \Omega$$

$$R_L = 5 \ \Omega \text{ as before.}$$

Now consider a more difficult a.c. example.

Example 2.8. Determine the equivalent generator for the circuit shown in *Figure 2.16,* and hence find the power that it could supply to a load of $(3 + j2) \ \Omega$.

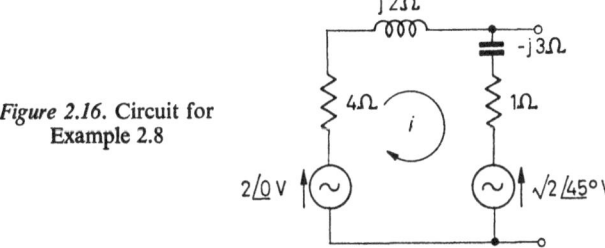

Figure 2.16. Circuit for
Example 2.8

Applying Thèvenin's theorem,

$$Z_{1n} = \frac{(4 + j2)(1 - j3)}{5 - j} = \frac{4 + 6 + j2 - j12}{5 - j} \ \Omega$$

$$= \frac{10}{26}(1 - j)(5 + j) = \frac{10}{26}(5 + 1 - j5 + j) \ \Omega$$

$$= (2\cdot31 - j1\cdot54) \ \Omega$$

To find $V_{o/c}$, write the mesh equation for the closed loop:

$$2 - \sqrt{2}\underline{/45°} = i(5 - j)$$

$$\therefore \qquad 2 - 1 - j = i(5 - j)$$

and

$$i = \frac{1 - j}{5 - j}$$

Now

$$V_{o/c} = \sqrt{2}\underline{/45°} + (1 - j3)i$$

$$= 1 + j + \frac{(1 - j3)(1 - j)}{5 - j} \text{ V}$$

$$= 1 + j + \frac{(-2 - j4)(5 + j)}{26} \text{ V}$$

$$= (26 + j26 - 10 + 4 - j2 - j20)\frac{1}{26} \text{ V}$$

$$= \frac{20}{26} + j\frac{4}{26} \text{ V} = (0\cdot77 + j0\cdot154) \text{ V}$$

When the resulting Thèvenin equivalent is connected to the load of $(3 + j2)\,\Omega$, the load current may be found;

$$i = \frac{0.77 + j0.154}{3 + j2 + 2.31 - j1.54}\,\text{A}$$

$$= \frac{0.77 + j0.154}{5.31 + j0.46}\,\text{A}$$

Since power can be dissipated only in resistance, the load power is given by

$$P = |i|^2\,3\,\text{W}$$

$$= \frac{0.77^2 + 0.154^2}{5.31^2 + 0.46^2} \times 3\,\text{W}$$

$$= \frac{0.612 \times 3}{28.61} = 0.0645\,\text{W}$$

or $\qquad\qquad P = 65\,\text{mW}$

Norton's theorem is similar to Thèvenin's theorem except the equivalent is expressed as a current generator in parallel with an admittance.

In *Figure 2.17* the network is shown as a box and the Norton

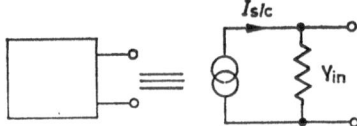

Figure 2.17. Norton's theorem

equivalent circuit is shown on the right. The components of this equivalent are found as follows:

$I_{s/c}$ is the current that would flow in a short-circuit connected across the terminals.

Y_{in} is the admittance measured between the terminals with all generators suppressed. Y_{in} is therefore the reciprocal of the Thèvenin equivalent Z_{in}.

Example 2.9. Repeat example 2.7 using Norton's theorem.
To determine $I_{s/c}$, note that with a short circuit across XX in *Figure 2.15*, no current will flow in the 5 Ω resistor.

$\therefore \qquad\qquad I_{s/c} = \tfrac{3}{2}\,\text{A}$

$\qquad\qquad Y_{in} = \tfrac{1}{2} + \tfrac{1}{5} = \tfrac{7}{10}\,\text{mhos}$

Since current divides in the direct ratio of parallel admittances

$$\frac{1}{3} = \frac{\frac{3}{2} \times Y_L}{Y_L + \frac{7}{10}}$$

$$\tfrac{1}{3} Y_L + \tfrac{7}{30} = \tfrac{3}{2} Y_L$$

$$\therefore \qquad Y_L = \frac{\frac{7}{30}}{\frac{3}{2} - \frac{1}{3}} = \frac{7}{45 - 10} \text{ mhos}$$

and

$$R_L = \frac{1}{Y_L} = \frac{35}{7} = 5\,\Omega$$

Example 2.10. Repeat Example 2.8 using Norton's theorem. For the circuit shown in *Figure 2.16*

$$Y_{in} = \frac{1}{4 + j2} + \frac{1}{1 - j3} = \frac{(1 - j3) + (4 + j2)}{4 + 6 - j12 + j2} \text{ mhos}$$

$$= \frac{5 - j}{10(1 - j)} \text{ mhos}$$

$$I_{s/c} = \frac{2}{4 + j2} + \frac{1 + j}{1 - j3} = \frac{2(1 - j3) + (1 + j)(4 + j2)}{10(1 - j)} \text{ A}$$

$$= \frac{2 - j6 + 4 - 2 + j4 + j2}{10(1 - j)} = \frac{4}{10(1 - j)} \text{ A}$$

When the load is connected, current divides in the direct ratio of parallel admittances.

Now

$$Y_L = \frac{1}{Z_L} = \frac{1}{3 + j2} = \frac{3 - j2}{13} \text{ mhos}$$

$$\therefore \qquad I_L = \frac{\dfrac{4}{10(1 - j)} \times \dfrac{3 - j2}{13}}{\dfrac{3 - j2}{13} + \dfrac{5 - j}{10(1 - j)}} \text{ A}$$

Multiplying numerator and denominator by $130(1 - j)$,

$$I_L = \frac{4(3 - j2)}{10(1 - j)(3 - j2) + 13(5 - j)}$$

$$= \frac{4(3 - j2)}{30 - 20 - j30 - j20 + 65 - j13} \text{ A}$$

$$= \frac{4(3 - j2)}{75 - j63} \text{ A}$$

Load power $= |I_L|^2 R_L$

$$= \frac{16(9 + 4) \times 3}{75^2 + 63^2} \text{ W}$$

$= 65$ mW (i.e. the same result as that obtained using Thèvenin's theorem)

Section 4

FOUR-TERMINAL NETWORKS

A common form of network, occurring frequently in electronic circuits, is known as a four-terminal network. Such networks have two pairs of terminals or two ports to which sources, loads, or other networks may be connected.

The diagrammatic form of this is shown in *Figure 2.18*.

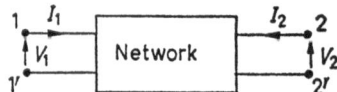

Figure 2.18. A four-terminal network

The currents and voltages at the two ports are conventionally taken in the directions shown. This allows for each port to be taken as either input or output. Terminals 1' and 2' are frequently, but not always, common.

One description of such networks is in terms of network para-meters. This approach is similar to the description of two-terminal networks by means of Thèvenin's and Norton's theorems.

There are four variables V_1, I_1, V_2 and I_2 associated with the network. For any particular set of parameters, two of these variables are considered as being independent while the other two are depen-dent. Since any two may be taken as independent, there are six possible sets of parameters.

Z Parameters

Let I_1 and I_2 be the independent variables. In general we may write two equations

$$V_1 = I_1 P + I_2 Q \qquad (2.25)$$

$$V_2 = I_1 R + I_2 S \qquad (2.26)$$

To show that such equations are possible, consider the simple circuit shown in *Figure 2.19*.

Applying Kirchhoff's laws we can see that

$$V_1 = I_1(Z_1 + Z_2) + I_2Z_2 \qquad (2.27)$$

$$V_2 = I_1Z_2 + I_2Z_2 \qquad (2.28)$$

which have the form of equations 2.25 and 2.26 above.

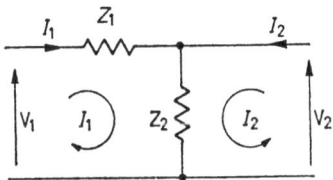

Figure 2.19. Solution of a four-terminal network by mesh analysis

Dimensionally the parameters of the network P, Q, R and S must be impedances since in each case the parameter multiplied by current results in voltage.

The equations may therefore be written

$$V_1 = I_1Z_{11} + I_2Z_{12} \qquad (2.29)$$

$$V_2 = I_1Z_{21} + I_2Z_{22} \qquad (2.30)$$

The double suffix notation indicates which pair of variables is related by the particular parameter, i.e. Z_{11} relates V_1 to I_1, Z_{12} relates V_1 to I_2, etc. In each case the first number is given by the dependent and the second by the independent variable.

The parameters of a particular network may be found by writing the mesh equations and rearranging so that the parameters can be found by inspection.

By comparing equations 2.27 and 2.28 with equations 2.29 and 2.30 above, we can see that the Z parameters of the circuit shown in *Figure 2.19* are given by

$$Z_{11} = Z_1 + Z_2, \quad Z_{12} = Z_2, \quad Z_{21} = Z_2, \quad Z_{22} = Z_2$$

A more general method of determining the parameters for a network is as follows.

Since equations 2.29 and 2.30 must be true for all values of the independent variables, they must be true for either I_1 or I_2 equal to zero. Suppose I_2 is zero. This can only occur if terminals 2,2' are open circuit.

Equations 2.29 and 2.30 now become

$$V_1 = I_1 Z_{11}, \qquad V_2 = I_1 Z_{21}$$

From which Z_{11} and Z_{21} may be defined.

$$Z_{11} = \frac{V_1}{I_1}\bigg|_{I_2=0} \quad \text{and} \quad Z_{21} = \frac{V_2}{I_1}\bigg|_{I_2=0}$$

Similarly, if we let terminals 1,1' be open circuit making I_1 zero, we obtain

$$Z_{12} = \frac{V_1}{I_2}\bigg|_{I_1=0} \quad \text{and} \quad Z_{22} = \frac{V_2}{I_2}\bigg|_{I_1=0}$$

Example 2.11. Find the Z parameters of the circuit shown in *Figure 2.20.*

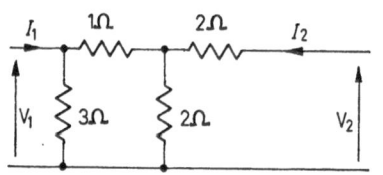

Figure 2.20. Circuit for Example 2.11

Let $I_2 = 0$.
Applying the rules based on Ohm's law

$$V_1 = I_1 \times \frac{3(1+2)}{3+1+2} = 1\tfrac{1}{2}I_1$$

$$\therefore \qquad Z_{11} = \frac{V_1}{I_1}\bigg|_{I_2=0} = 1\tfrac{1}{2}\ \Omega$$

Also

$$V_2 = I_1 \times \frac{3}{3+1+2} \times 2 = I_1$$

$$\therefore \qquad Z_{21} = \frac{V_2}{I_1}\bigg|_{I_2=0} = 1\ \Omega$$

Now let $I_1 = 0$

$$V_2 = I_2 \times \left[2 + \frac{2(1+3)}{2+1+3}\right] = 3\tfrac{1}{3}I_2$$

$$\therefore \qquad Z_{22} = 3\tfrac{1}{3}\ \Omega$$

Also

$$V_1 = I_2 \times \frac{2}{2+3+1} \times 3 = I_2$$

$$\therefore \qquad Z_{12} = 1\ \Omega$$

Note $Z_{12} = Z_{21}$. This is always true for a passive network. A passive network is one containing no elements such as valves or transistors.

It is frequently convenient to show the Z parameter equations in the form of an equivalent circuit.

From equation 2.29

$$V_1 = I_1 Z_{11} + I_2 Z_{12}$$

Since this represents the sum of two voltages, we can see that the input side of our equivalent circuit must contain two components. As I_1 flows into this part of the circuit, Z_{11} can be shown simply as an impedance. I_2 does not flow in this part of the circuit, so Z_{12} must be shown as a voltage generator of $I_2 Z_{12}$ volts. Similarly Z_{22} can be shown as an impedance in the output circuit but Z_{21} must appear as a voltage generator of $Z_{21} I_1$ volts. The resulting equivalent circuit is shown in *Figure 2.21*.

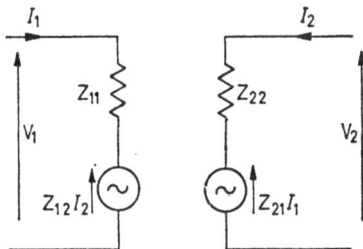

Figure 2.21. The Z parameter equivalent circuit

No additional information is given by this equivalent circuit but its use sometimes makes the formation of complete network equations much easier.

General Solutions in Terms of Z Parameters

A common problem is to find the input impedance and voltage gain of a network when a load Z_L is connected to one port as shown in *Figure 2.22*.

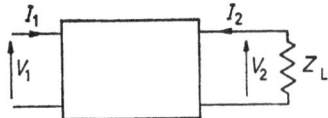

Figure 2.22. A loaded four-terminal network

71

From the sense of V_2 and I_2 we can say,

$$V_2 = -I_2 Z_L$$

Equation 2.30 now becomes

$$-I_2 Z_L = I_1 Z_{21} + I_2 Z_{22} \tag{2.31}$$

Rewriting equation 2.29 and rearranging equation 2.31,

$$V_1 = I_1 Z_{11} + I_2 Z_{12}$$

$$0 = I_1 Z_{21} + I_2 (Z_{22} + Z_L)$$

Solving for I_1 by determinants

$$I_1 = \frac{V_1 (Z_{22} + Z_L) - 0}{Z_{11}(Z_{22} + Z_L) - Z_{21} Z_{12}}$$

The input impedance given by

$$\frac{V_1}{I_1} = \frac{Z_{11}(Z_{22} + Z_L) - Z_{21} Z_{12}}{Z_{22} + Z_L}$$

$$\therefore \qquad Z_{\text{in}} = Z_{11} - \frac{Z_{21} Z_{12}}{Z_{22} + Z_L} \tag{2.32}$$

Also

$$I_2 = \frac{0 - Z_{21} V_1}{Z_{11}(Z_{22} + Z_L) - Z_{21} Z_{12}}$$

From which the transfer admittance

$$\frac{I_2}{V_1} = \frac{-Z_{21}}{Z_{11}(Z_{22} + Z_L) - Z_{21} Z_{12}} \tag{2.33}$$

Now since $V_2 = -I_2 Z_L$, the voltage gain Av is given by

$$\frac{V_2}{V_1} = \frac{Z_L Z_{21}}{Z_{11}(Z_{22} + Z_L) - Z_{21} Z_{12}} \tag{2.34}$$

Equations 2.32 and 2.33 are known as the general solutions for the network in terms of the Z parameters. There are two further general solutions, which are obtained by connecting an impedance Z_s across terminals 1,1′. The equations now become

$$0 = I_1 (Z_{11} + Z_s) + I_2 Z_{12}$$

$$V_2 = I_1 Z_{21} + I_2 Z_{22}$$

The reader should check that these are correct and solve for the output impedance V_2/I_2 and the reverse voltage gain V_1/V_2. The results that should be obtained are

$$Z_{\text{out}} = Z_{22} - \frac{Z_{21}Z_{12}}{Z_{11} + Z_{\text{s}}} \qquad (2.35)$$

and $$\frac{V_1}{V_2} = \frac{Z_{12}Z_{\text{s}}}{Z_{22}(Z_{11} + Z_{\text{s}}) - Z_{12}Z_{21}} \qquad (2.36)$$

Example 2.12. A certain electronic device is represented by the equivalent circuit shown in *Figure 2.23*.

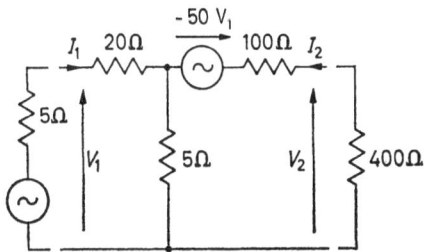

Figure 2.23. Circuit for Example 2.12

Determine the Z parameters for the device and hence find (*a*) the input impedance (V_1/I_1) when the output is loaded with $400\ \Omega$, (*b*) the voltage gain (V_2/V_1), when loaded with $400\ \Omega$ and (*c*) the output impedance (V_2/I_2) if the source impedance is $5\ \Omega$, and the load is disconnected.

To find the Z parameters we apply the standard technique, first letting $I_2 = 0$ and then $I_1 = 0$. Writing equations by inspection then leads to the required parameters.

Let $I_2 = 0$. Since the output terminals are open circuit, I_1 flows through the $20\ \Omega$ and $5\ \Omega$ resistances only.

$\therefore$ $$V_1 = I_1(20 + 5)\ \text{V}$$

$\therefore$ $$Z_{11} = \frac{V_1}{I_1}\bigg|_{I_2=0} = 25\ \Omega$$

There will be no potential difference across the $100\ \Omega$ resistor ($I_2 = 0$), so V_2 is the sum of the generated voltage, $-50V_1$, and the volt drop across the $5\ \Omega$ resistor.

$\therefore$ $$V_2 = -50V_1 + 5I_1$$

73

But $\qquad\qquad V_1 = 25I_1$

$\therefore\qquad\qquad V_2 = -50 \times 25I_1 + 5I_1$

and $\qquad\qquad Z_{21} = \dfrac{V_2}{I_1}\bigg|_{I_2=0} = -1\ 245\ \Omega$

Now let $I_1 = 0$. In this case I_2 flows through the $100\ \Omega$ and $5\ \Omega$ resistors, and there is no potential difference across the $20\ \Omega$ resistor. Thus

$$V_1 = 5I_2$$

and $\qquad\qquad Z_{12} = \dfrac{V_1}{I_2}\bigg|_{I_1=0} = 5\ \Omega$

and $\qquad\qquad V_2 = 105I_2 - 50V_1$

$$= 105I_2 - 50 \times 5I_2$$

$\therefore\qquad\qquad Z_{22} = \dfrac{V_2}{I_2}\bigg|_{I_1=0} = -145\ \Omega$

Summarizing $Z_{11} = 25\ \Omega$, $Z_{12} = 5\ \Omega$, $Z_{21} = -1\ 245\ \Omega$ and $Z_{22} = -145\ \Omega$.

For the remaining solutions we need only apply equations 2.32, 2.34 and 2.35.

(a) $\quad Z_{1\mathrm{n}} = Z_{11} - \dfrac{Z_{12}Z_{21}}{Z_{22} + Z_{\mathrm{L}}} = 25 + \dfrac{5 \times 1\ 245}{400 - 145} = 49{\cdot}4\ \Omega$

(b) Voltage gain

$$\frac{V_2}{V_1} = \frac{Z_{\mathrm{L}}Z_{21}}{Z_{11}(Z_{22} + Z_{\mathrm{L}}) - Z_{21}Z_{12}} = \frac{-400 \times 1\ 245}{25(400 - 145) + 1\ 245 \times 5}$$

This last expression is conveniently simplified by dividing numerator and denominator by $1\ 245$, thus

$$\frac{V_2}{V_1} = \frac{-400}{\dfrac{25 \times 255}{1\ 245} + 5}$$

$$= \underline{-39{\cdot}5}$$

(c) The output impedance of a circuit will be discussed in detail in later chapters, but simply it is the impedance of the Thèvenin equivalent generator determined at the output terminals of a circuit in the absence of a load. In this case

$$\text{Output impedance} = \frac{V_2}{I_2} = Z_{22} - \frac{Z_{21}Z_{12}}{Z_{11} + Z_s} \; \Omega$$

$$= -145 + \frac{5 \times 1\,245}{25 + 5} \; \Omega$$

$$= 63 \; \Omega$$

h Parameters

So far in this section we have considered only the Z parameters of a four-terminal network. These were obtained by selecting I_1 and I_2 as the independent variables. The other possible choices, V_1 and V_2, V_1 and I_2, I_1 and V_2, V_1 and I_1, and V_2 and I_2 each lead to a separate set of parameters. We shall investigate one of these in detail, and noting that the results are very similar to those of the Z parameters, we shall step directly to the final result for the other sets. The use of the different types will then be illustrated in this and later chapters.

Now let I_1 and V_2 be the independent variables. Our two equations must have the form

$$V_1 = I_1 P + V_2 Q$$

$$I_2 = I_1 R + V_2 S$$

Notice in this case our four parameters cannot have the same dimensions. While P relates V_1 to I_1 and is therefore an impedance, Q relates V_1 to V_2 and is simply a number. Similarly R, a current ratio, is a number and S, relating I_2 to V_2, is an admittance. When the parameters are mixed in this fashion, they are known as a hybrid set of parameters and are given the symbol h. Our equations thus become

$$V_1 = I_1 h_{11} + V_2 h_{12} \tag{2.37}$$

$$I_2 = I_1 h_{21} + V_2 h_{22} \tag{2.38}$$

To define our parameters we can now let either I_1 or V_2 be zero, leading to the following relationships

$$h_{11} = \frac{V_1}{I_1}\bigg|_{V_2=0} \; \Omega \qquad h_{12} = \frac{V_1}{V_2}\bigg|_{I_1=0}$$

$$h_{21} = \frac{I_2}{I_1}\bigg|_{V_2=0} \qquad h_{22} = \frac{I_2}{V_2}\bigg|_{I_1=0} \; \text{mho}$$

75

Note that h_{22} is the reciprocal of Z_{22} since both are determined with the input open circuit ($I_1 = 0$). Z_{11} is not the same as h_{11} since Z_{11} is an open circuit parameter ($I_2 = 0$) while h_{11} is a short circuit parameter ($V_2 = 0$).

Example 2.13. Determine the *h* parameter equivalent circuit of the network shown in *Figure 2.24*.

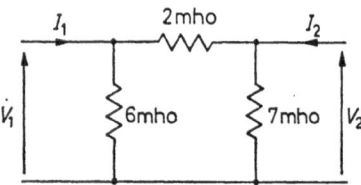

Figure 2.24. Circuit for Example 2.13.

Following the same method as we used for *Z* parameter calculations, first let V_2 equal zero. This implies a short circuit across the output terminals. Note I_2 is the current flowing in the short circuit.

For h_{11} we must determine I_1 in terms of V_1 and since the 7 mho conductance is short circuited

$$I_1 = V_1(6 + 2) \text{ A}$$

$$h_{11} = \left. \frac{V_1}{I_1} \right|_{V_2=0} = \frac{1}{8} \Omega$$

For h_{21} we must find I_2 in terms of I_1. In this case, this is simple current division between parallel conductances. Thus

$$I_2 = -I_1 \times \frac{2}{2+6} = -\frac{1}{4} I_1 \text{ A}$$

The minus is required since the parameter convention requires that I_2 flows into the network (see *Figure 2.24*). Now

$$h_{21} = \left. \frac{I_2}{I_1} \right|_{V_2=0} = -\frac{1}{4}$$

For h_{12} and h_{22} we must open-circuit the input to make $I_1 = 0$.

First determine I_2 in terms of V_2 by finding the total conductance at the output terminals.

$$h_{22} = \left. \frac{I_2}{V_2} \right|_{I_1=0} = 7 + \frac{2 \times 6}{2 + 6} = 8\tfrac{1}{2} \text{ mho}$$

76

Finally to find V_1 in terms of V_2, we have potential division across series conductances so,

$$V_1 = V_2 \times \frac{2}{2+6} \text{ V}$$

$$h_{12} = \left.\frac{V_1}{V_2}\right|_{I_1=0} = \frac{1}{4}$$

Note once again the passive network leads to the same numerical value for h_{21} and h_{12} but the current convention results in

$$h_{12} = -h_{21}$$

As with Z parameters, it is often convenient to use an equivalent circuit. Rewriting the h parameter equations

$$V_1 = I_1 h_{11} + V_2 h_{12}$$

$$I_2 = I_1 h_{21} + V_2 h_{22}$$

We can see that the first equation is similar to the Z parameter equations in that it is the sum of two voltages. Since this part of our equivalent circuit will carry I_1, the voltage $h_{11}I_1$ will appear across an impedance h_{11} Ω. The other voltage $V_2 h_{12}$ must be produced by a voltage generator.

The second equation is the sum of two currents so our equivalent must have two parallel components. The current $V_2 h_{22}$ will flow in an admittance of h_{22} mho when V_2 is applied across it. The second current $h_{21}I_1$ can be provided only by a current generator in parallel with the admittance h_{22}.

The resulting equivalent circuit is shown in *Figure 2.25*.

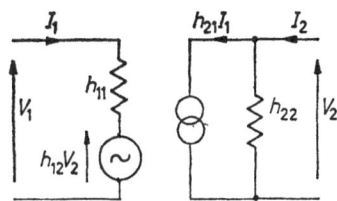

Figure 2.25. The h parameter equivalent circuit

General Solutions in terms of h parameters

If the four terminal network is loaded, general solutions may be derived in terms of the h parameters. In this case the second equation is for I_2, so the required substitution is $I_2 = -V_2 Y_L$. Now rewriting and rearranging the equations, we obtain

$$V_1 = I_1 h_{11} + V_2 h_{12} \qquad (2.39)$$

$$0 = I_1 h_{21} + V_2 (h_{22} + Y_L) \qquad (2.40)$$

Solving by determinants for I_1,

$$I = \frac{V_1 (h_{22} + Y_L)}{h_{11}(h_{22} + Y_L) - h_{21}h_{12}}$$

From which

$$\frac{V_1}{I_1} = Z_{1n} = h_{11} - \frac{h_{12}h_{21}}{h_{22} + Y_L} \qquad (2.41)$$

Solving for V_2

$$V_2 = \frac{-V_1 h_{21}}{h_{11}(h_{22} + Y_L) - h_{12}h_{21}}$$

From which

$$\frac{V_2}{V_1} = \text{voltage gain} = \frac{-h_{21}}{h_{11}(h_{22} + Y_L) - h_{12}h_{21}} \qquad (2.42)$$

Similarly by putting $V_1 = -I_1 Z_s$, we can find,

$$\frac{I_2}{V_2} = \text{output admittance} = h_{22} - \frac{h_{21}h_{12}}{h_{11} + Z_s} \qquad (2.43)$$

and

$$\frac{I_1}{I_2} = \text{reverse current gain} = \frac{-h_{12}}{h_{22}(h_{11} + Z_s) - h_{21}h_{12}} \qquad (2.44)$$

Equations 2.41, 2.42 and 2.43 are important since they will be very useful for the solution of transistor circuits.

The reader should now compare equations 2.41, 2.42, 2.43 and 2.44 with equations 2.32, 2.33, 2.35 and 2.36. Each set of equations has exactly the same form. The only difference is the result for which each equation is true.

We have now derived two sets of parameters, Z and h, by taking I_1, I_2 and I_1, V_2 as our independent variables. Since the steps for developing the y and g parameters are precisely the same we shall summarize only the important results.

Y Parameters

Independent variables V_1, V_2.
Equations:

$$I_1 = V_1 Y_{11} + V_2 Y_{12} \qquad (2.45)$$

$$I_2 = V_1 Y_{21} = V_2 Y_{22} \qquad (2.46)$$

Conditions for defining parameters—short circuit input or output.
Definitions:

$$Y_{11} = \frac{I_1}{V_1}\bigg|_{V_2=0} \text{ mho} \qquad Y_{12} = \frac{I_1}{V_2}\bigg|_{V_1=0} \text{ mho}$$

$$Y_{21} = \frac{I_2}{V_1}\bigg|_{V_2=0} \text{ mho} \qquad Y_{22} = \frac{I_2}{V_2}\bigg|_{V_1=0} \text{ mho}$$

Equivalent circuit:

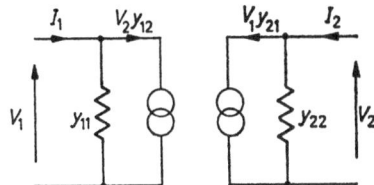

Figure 2.26. The y parameter equivalent circuit

General solutions:

$$\text{Input admittance} = \frac{I_1}{V_1} = Y_{11} - \frac{Y_{12} Y_{21}}{Y_{22} + Y_L} \qquad (2.47)$$

$$\frac{V_2}{I_1} = \frac{-Y_{21}}{Y_{11}(Y_{22} + Y_L) - Y_{21} Y_{12}} \qquad (2.48)$$

From which

$$\text{Current gain} = \frac{I_2}{I_1} = \frac{+Y_{21} Y_L}{Y_{11}(Y_{22} + Y_L) - Y_{21} Y_{12}} \qquad (2.49)$$

$$\text{Output admittance} = \frac{I_2}{V_2} = Y_{22} - \frac{Y_{12} Y_{21}}{Y_{11} + Y_s} \qquad (2.50)$$

$$\text{Reverse current gain} \ \frac{I_1}{I_2} = \frac{Y_{12} Y_s}{Y_{22}(Y_{11} + Y_s) - Y_{21} Y_{12}} \qquad (2.51)$$

g Parameters

Independent variables: V_1, I_2.

Equations:

$$I_1 = V_1 g_{11} + I_2 g_{12} \qquad (2.52)$$

$$V_2 = V_1 g_{21} + I_2 g_{22} \qquad (2.53)$$

Conditions for defining parameters—short circuit input or open circuit output.

Definitions:

$$g_{11} = \left.\frac{V_1}{I_1}\right|_{I_2=0} \text{mho} \qquad g_{12} = \left.\frac{I_1}{I_2}\right|_{V_1=0} \text{ratio}$$

$$g_{21} = \left.\frac{V_2}{V_1}\right|_{I_2=0} \text{ratio} \qquad g_{22} = \left.\frac{V_2}{I_1}\right|_{V_1=0} \Omega$$

Equivalent circuit:

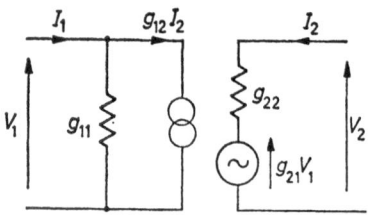

Figure 2.27. The g parameter equivalent circuit

General solutions:

$$\text{Input admittance} = \frac{I_1}{V_1} = g_{11} - \frac{g_{12}g_{21}}{g_{22} + Z_L} \qquad (2.54)$$

$$\text{Current gain} = \frac{I_2}{I_1} = \frac{-g_{21}}{g_{11}(g_{22} + Z_L) - g_{12}g_{21}} \qquad (2.55)$$

$$\text{Output impedance} = \frac{V_2}{I_2} = g_{22} - \frac{g_{21}g_{12}}{g_{11} + Y_s} \qquad (2.56)$$

$$\text{Reverse voltage gain} = \frac{V_1}{V_2} = \frac{-g_{12}}{g_{22}(g_{11} + Y_s) - g_{21}g_{12}} \qquad (2.57)$$

Before applying these results to some examples, one further rule should be discussed. It may be necessary to convert from one set of parameters to another. This is simply achieved by drawing the equivalent circuit for the available parameters, writing the equations for the required parameters and solving by conventional methods.

Example 2.14. For the network shown in *Figure 2.28* determine (*a*) the *y* parameters and (*b*) by conversion the *g* parameters. Using

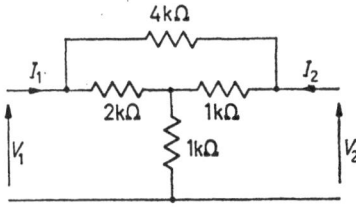

Figure 2.28. Circuit for Example 2.14

each form calculate the current gain when the network is loaded with an impedance of 1 kΩ.

To determine the *y* parameters, we must let either V_1 or V_2 be zero by assuming a short circuit across the required pair of terminals. First let $V_2 = 0$. Converting to admittances we have 0·25 mmho in parallel with [0·5 mmho in series with (1 mmho in parallel with 1 mmho)]

$$y_{11} = \frac{I_1}{V_1}\bigg|_{V_2=0} = 0\cdot25 + \frac{0\cdot5(1+1)}{0\cdot5+1+1}$$

$$y_{11} = 0\cdot65 \text{ mmho}$$

Also the current in the short circuit is $-I_2$, so

$$-I_2 = 0\cdot25V_1 + V_1\left[\frac{0\cdot5\times2}{2\cdot5}\right]\times\frac{1}{2}\text{ A}$$

$$\therefore \quad y_{21} = \frac{I_2}{V_1}\bigg|_{V_2=0} = -(0\cdot25+0\cdot2)\text{ mmho}$$

and $\qquad y_{21} = -0\cdot45 \text{ mmho}$

Now put $V_1 = 0$.
Calculating y_{22} in the same way as y_{11},

$$y_{22} = \frac{I_2}{V_2}\bigg|_{V_1=0} = 0\cdot25 + \frac{1(1+0\cdot5)}{1+1+0\cdot5}\text{ mmho}$$

$$y_{22} = 0\cdot85 \text{ mmho}$$

81

Also $\quad -I_1 = 0{\cdot}25V_1 + V_1 \left[\dfrac{1 \times 1{\cdot}5}{1 + 1{\cdot}5}\right] \times \dfrac{0{\cdot}5}{1{\cdot}5}$ A

$$y_{12} = \left.\frac{I_1}{V_2}\right|_{V_1=0} = -0{\cdot}45 \text{ mmho}$$

The required equivalent circuit is shown in *Figure 2.29*.

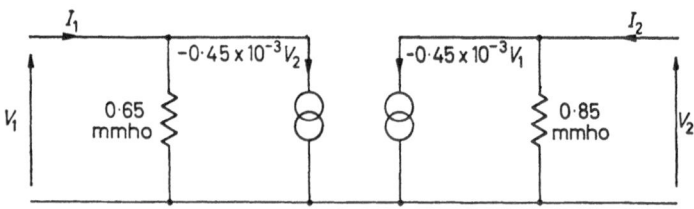

Figure 2.29. Solution for Example 2.14

To find the g parameters we must let either V_1 or I_2 be zero by either short circuiting the input terminals or open circuiting the output terminals.

Let $I_2 = 0$.

Under these conditions the whole of the y_{21} generator current must flow in the admittance y_{22}. Thus

$$V_2 = \frac{-(-0{\cdot}45 \times 10^{-3}V_1}{0{\cdot}85 \times 10^{-3}}$$

and $\qquad \left.\dfrac{V_2}{V_1}\right|_{I_2=0} = g_{21} = 0{\cdot}53$

We can now express the y_{12} generator current in terms of V_1, and write an equation for I_1.

$$I_1 = 0{\cdot}65 \times 10^{-3}V_1 - 0{\cdot}45 \times 10^{-3} \times 0{\cdot}53V_1$$

and $\qquad \left.\dfrac{I_1}{V_1}\right|_{I_2=0} = g_{11} = (0{\cdot}65 - 0{\cdot}258) \times 10^{-3} \text{ mho}$

$$= 0{\cdot}412 \text{ mmho}$$

To obtain g_{12} and g_{22}, let $V_1 = 0$.
This eliminates the y_{21} generator. Now

$$g_{22} = \left.\frac{V_2}{I_2}\right|_{V_1=0} = \frac{1}{0{\cdot}85 \times 10^{-3}} = 1{\cdot}178 \text{ k}\Omega$$

Since we have a short circuit on the input terminals all the y_{12} generator current flows as I_1.

$$g_{12} = \left.\frac{I_1}{I_2}\right|_{V_1=0} = (-0.45 \times 10^{-3}) \times \frac{1}{0.85 \times 10^{-3}}$$
$$= -0.53$$

To find the current gain using each set of parameters we can apply formulae from equations 2.49 and 2.55.

First using y parameters

Current gain, $\qquad A_1 = \dfrac{y_{21}Y_L}{y_{11}(y_{22} + Y_L) - y_{21}y_{12}}$

Working in mmho, $\quad A_1 = \dfrac{-0.45 \times 1}{0.65(1 + 0.85) - 0.45 \times 0.45}$

$$\simeq -0.45$$

Now using equation 2.55

$$A_1 = \frac{-g_{21}}{g_{11}(g_{22} + Z_L) - g_{21}g_{12}}$$
$$= \frac{-0.53}{0.412 \times 10^{-3}(1 + 1.178) \times 10^3 + 10.53 \times 0.53}$$
$$\simeq -0.45$$

Thus properties of four-terminal networks such as voltage and current gain, input and output impedances or admittances may be found using whichever parameters are available.

Interconnection of Four-terminal Networks

We shall now see how these parameters are of use when two or more four-terminal networks are interconnected in various configura-tions.

These configurations are

    Series input and output,
    Parallel input and output,
    Series input, parallel output, and
    Parallel input, series output.

We shall see that if the individual networks are described by the appropriate parameters, the combined network parameters will be the sum of the separate network parameters.

Figure 2.30 shows two networks interconnected series input, series output. Network A has Z' parameters and network B has Z'' parameters.

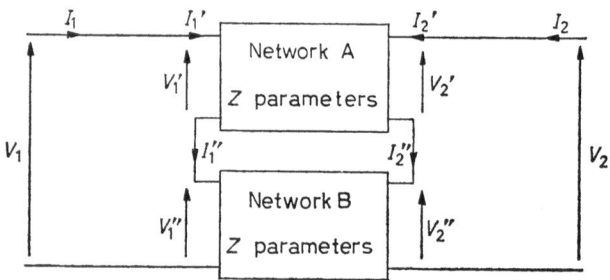

Figure 2.30. Two four-terminal networks connected in series

Inspection of the circuit shows that

$$I_1 = I_1' = I_1''$$

and $\quad I_2 = I_2' = I_2''$

Also $\quad V_1 = V_1' + V_1'' = I_1Z_{11}' + I_2Z_{12}' + I_1Z_{11}'' + I_2Z_{12}''$

and $\quad V_2 = V_2' + V_2'' = I_1Z_{21}' + I_2Z_{22}' + I_1Z_{21}'' + I_2Z_{22}''$

Collecting terms

$$V_1 = I_1(Z_{11}' + Z_{11}'') + I_2(Z_{12}' + Z_{12}'')$$

$$V_2 = I_1(Z_{21}' + Z_{21}'') + I_2(Z_{22}' + Z_{22}'')$$

But these are the equations for the combined network, and the overall Z parameters are given by the sum of the individual Z parameters.

Example 2.15. A network having Z parameters Z_{11} 5 Ω, Z_{22} 3 Ω, and $Z_{21} = Z_{12}$ 2 Ω is connected in series with a 4 Ω resistor as shown in *Figure 2.31a*. Determine the overall Z parameters and hence calculate the output impedance if the network is supplied from a source of internal impedance 3 Ω.

First we must find the Z parameters of the sub-network shown in *Figure 2.31b*.

84

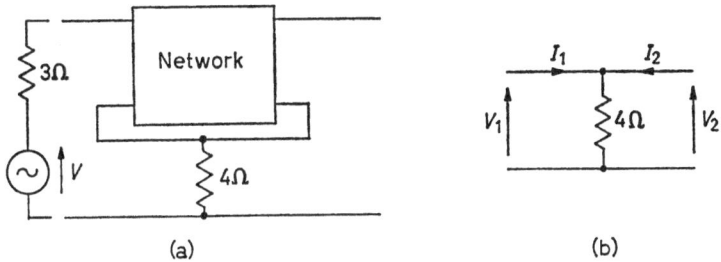

Figure 2.31. Circuit for Example 2.15

Since these are open circuit parameters, by inspection

$$Z_{11} = Z_{22} = 4\,\Omega$$

$$Z_{12} = \frac{V_1}{I_2}\bigg|_{I_1=0} \quad \text{but} \quad V_1 = I_2 \times 4$$

$$\therefore \qquad Z_{12} = 4\,\Omega$$

and since the network is symmetrical $Z_{21} = 4\,\Omega$.

Now the overall Z parameters are:

$$Z_{11} = 5 + 4 = 9\,\Omega \qquad Z_{12} = 2 + 4 = 6\,\Omega$$

$$Z_{21} = 2 + 4 = 6\,\Omega \qquad Z_{22} = 3 + 4 = 7\,\Omega$$

To find the output impedance, apply equation 2.35.

$$Z_{\text{out}} = Z_{22} - \frac{Z_{21}Z_{12}}{Z_{11} + Z_s} = 7 - \frac{36}{9 + 3}\,\Omega$$

$$= 4\,\Omega$$

Parallel Parallel

Figure 2.32 shows two networks connected parallel input, parallel output. Network A has y' parameters, and network B has y'' parameters.

With this connection we can see that:

$$V_1 = V_1' = V_1'' \quad \text{and} \quad V_2 = V_2' = V_2''$$

Also $\quad I_{\mathfrak{t}} = I_1' + I_1'' = V_1 y_{11}' + V_2 y_{12}' + V_1 y_{11}'' + V_2 y_{12}''$

and $\quad I_2 = I_2' + I_2'' = V_1 y_{21}' + V_2 y_{22}' + V_1 y_{21}'' + V_2 y_{22}''$

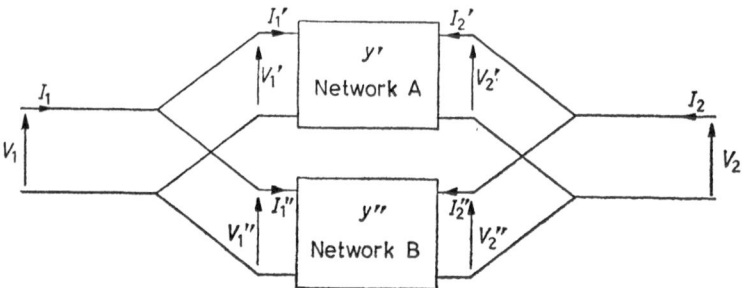

Figure 2.32. Two four-terminal networks connected in parallel

so collecting terms

$$I_1 = V_1(y_{11}' + y_{11}'') + V_2(y_{12}' + y_{12}'')$$

and $\quad I_2 = V_1(y_{21}' + y_{21}'') + V_2(y_{22}' + y_{22}'')$

But these are the y parameter equations for the combined network, and the overall y parameters are given by the sums of the individual y parameters.

Series Parallel

Figure 2.33 shows two networks connected series input and parallel output.

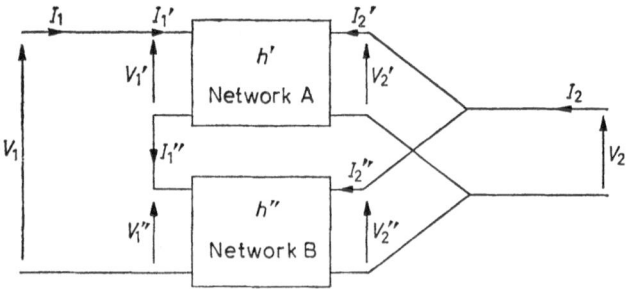

Figure 2.33. Two four-terminal networks connected series input, parallel output

Network A has h' parameters and network B has h'' parameters. In this case

$$I_1 = I_1' = I_1'' \qquad\qquad \text{and} \quad V_2 = V_2' = V_2''$$

86

Also $V_1 = V_1' + V_1'' = I_1h_{11}' + V_2h_{12}' + I_1h_{11}'' + V_2h_{12}''$

and $I_2 = I_2' + I_2'' = I_1h_{21}' + V_2h_{22}' + I_1h_{21}'' + V_2h_{22}''$

Collecting terms

$$V_1 = I_1(h_{11}' + h_{11}'') + V_2(h_{12}' + h_{12}'')$$

$$I_2 = I_1(h_{21}' + h_{21}'') + V_2(h_{22}' + h_{22}'')$$

Thus for the combined network the overall h parameters are given by the sum of the individual h parameters.

Parallel Series

This is exactly the reverse connection for the previous case, and provided the g parameters for the individual networks are used then the overall parameters for the combined network are given by the sums of the individual g parameters. The derivation of this is left to the reader as further practice.

No further examples will be given at this stage since practical applications of this work will not become apparent until later chapters.

SUMMARY

Summarizing the work of this chapter, the reader should ensure that he is proficient in the use of the methods discussed in the first and second sections. These are the fundamentals without which he cannot proceed to the topics to be discussed throughout the book. The theorems in Section 3, particularly those of Thèvenin and Norton, are equally essential.

The general background from Section 4 is very useful and although problems may be solved directly with the derived general solutions we shall also work from first principles in many cases, using the parameters only to construct an equivalent circuit. The work on interconnected networks provides an alternative approach to the solution of feedback problems but is not essential for any of the work in later chapters.

EXAMPLES

Example 2.16. Repeat Example 2.1 with the circuit modified by interchanging the 4 mho conductance with the 1 mho conductance.

Ans. $Y_T \frac{23}{9}$ mho, $i_0 \frac{15}{23}1$, $v' \frac{2}{9} v$.

Example 2.17. For the circuit shown in *Figure 2.34*, determine

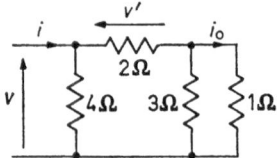

Figure 2.34. Circuit for Example 2.17

(*a*) the input impedance $v/1$, (*b*) the current i_0 in terms of the input current i, and (*c*) the voltage v' in terms of the input voltage v.

Ans. (*a*) 1·63 Ω. (*b*) 0·445i. (*c*) 0·727v.

Example 2.18. The admittance circuit shown in *Figure 2.35* is

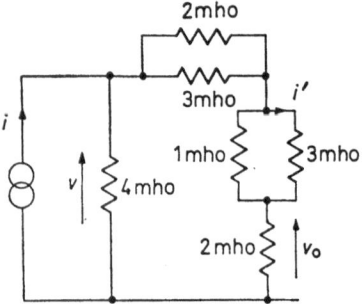

Figure 2.35. Circuit for Example 2.18

supplied from the current source i of 0·2 A. Calculate (*a*) the input voltage v, (*b*) the voltage ratio v_0/v, and (*c*) the current i'.

Ans. (*a*) 39·6 mV. (*b*) 0·526. (*c*) 31·2 mA.

Example 2.19. Using mesh analysis calculate the voltage v' across the 3 ohm resistor shown in *Figure 2.36*.

Ans. 0·756 V.

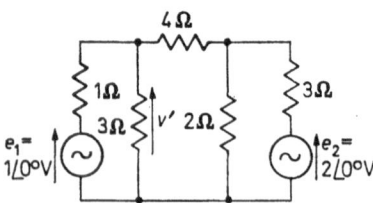

Figure 2.36. Circuit for Example 2.19

88

Example 2.20. For the circuit shown in *Figure 2.37*, determine the potential at the node marked X using mesh analysis.

Ans. 2·2 V.

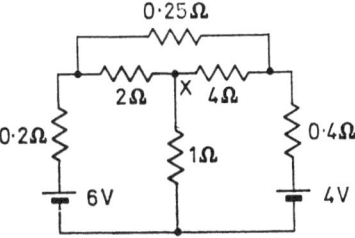

Figure 2.37. Circuit for Example 2.20

Example 2.21. If the circuit shown in *Figure 2.9* is modified by changing the centre branch to $(2 - j2)\ \Omega$, and by reversing the connections to the 5 V generator, calculate the resulting current from the 6 V generator.

Ans. $1\ \angle\ 7°\ 18'$ A.

Example 2.22. Using mesh analysis, calculate the current i_2 in the circuit shown in *Figure 2.38*.

Ans. $\dfrac{-8 + j26}{37}$ mA.

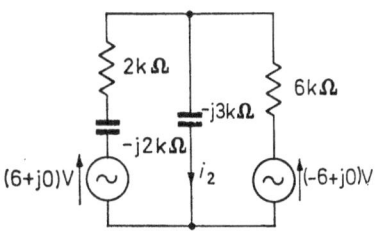

Figure 2.38. Circuit for Example 2.22

Example 2.23. Repeat Example 2.19 by converting impedances to admittances, voltage sources to current sources, and using nodal analysis.

Example 2.24. Repeat Example 2.20 using nodal analysis.

Example 2.25. Repeat Example 2.21 using nodal analysis.

Example 2.26. Convert the circuit shown in *Figure 2.39* to the

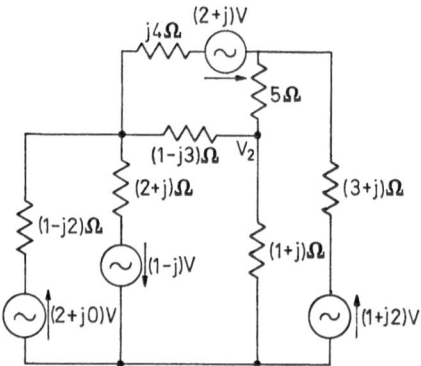

Figure 2.39. Circuit for Example 2.26

admittance-current generator form, and determine the node voltage V_2 using nodal analysis.

Ans. $0.737 \angle 121° 26'$.

Example 2.27. Repeat Example 2.21 using the superposition theorem.

Example 2.28. By repeated applications of Thèvenin's theorem, determine the components of the equivalent Thèvenin generator seen at the terminals T, T' on the circuit shown in *Figure 2.40*.

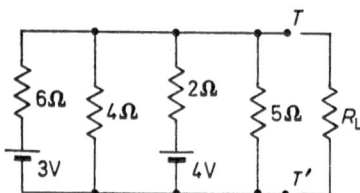

Figure 2.40. Circuit for Example 2.28

Hence calculate the value of R_L to be connected to these terminals such that the current supplied to it is 1 A.

Ans. $1.35\ \Omega$.

Example 2.29. Repeat Example 2.3 using Thèvenin's theorem.

Example 2.30. Repeat Example 2.21 using Thèvenin's theorem.

Example 2.31. Repeat Example 2.26 using Thèvenin's theorem.

Example 2.32. Repeat Example 2.28 using Norton's theorem.

Example 2.33. The circuit shown in *Figure 2.41* is the equivalent circuit for part of an amplifier to be used at an angular frequency

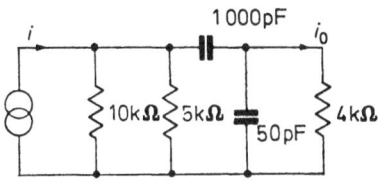

Figure 2.41. Circuit for Example 2.33

of 10^6 rad/sec. Using Norton's theorem, calculate the output current i_0 in terms of the source current i.

Ans. $0 \cdot 44 \angle 2° 30' i$.

Example 2.34. If the circuit shown in *Figure 2.20* is modified by interchanging the 3 Ω and the 1 Ω branches, determine the new z parameters. Hence find the input impedance V_1/I_1 when the output is loaded with 4 Ω.

Ans. z_{11} $\frac{5}{6}$ Ω, z_{22} $3\frac{1}{3}$ Ω, $z_{21} = z_{12}$ $\frac{1}{3}$ Ω. Z_{1n} $\frac{9}{11}$ Ω.

Example 2.35. Determine the h parameters for the circuit in *Figure 2.42*.

Ans. h_{11} $\frac{18}{7}$ Ω, h_{22} $\frac{13}{56}$ mho, $h_{12} = -h_{21} = \frac{11}{14}$.

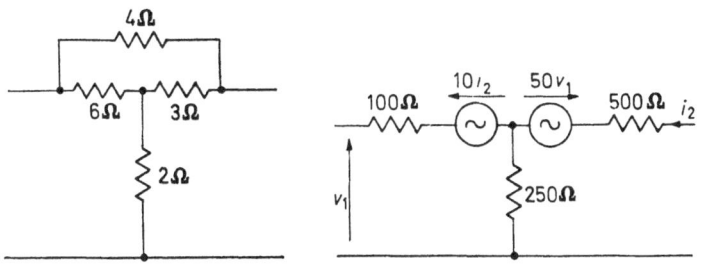

Figure 2.42. Circuit for Example 2.35

Figure 2.43. Circuit for Example 2.36

Example 2.36. Determine the h parameters of the network shown in *Figure 2.43*.

Ans. $14 \cdot 3$ Ω, $1 \cdot 9 \times 10^{-2}$, $-1 \cdot 29$, 73 μmho.

Example 2.37. A network having h_{11} 1 000 Ω, h_{12} 10⁻³, h_{21} 50, h_{22} 200 μmho, is loaded with 4 kΩ and driven by a source e_s of internal impedance 600 Ω. Calculate the value of e_s if the load voltage is to be 3 V. Find also the overall output impedance of the circuit including the load.

Ans. 40 mV, 2·39 kΩ.

Example 2.38. The T network shown in *Figure 2.44* represents an active device. Determine the y parameters for the device and

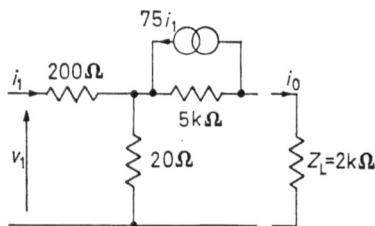

Figure 2.44. Circuit for Example 2.38

hence calculate the input impedance v_1/i_1 and current gain i_2/i_1 when it is loaded with 2 kΩ.

Ans. 600, −2·3, 45 × 10³, 23 μmho. Z_{1n} 1·25 kΩ, A_1 54.

Example 2.39. Repeat Example 2.38 using g parameters.

Example 2.40. A certain active device has the following h parameters: h_{11} 500 Ω, h_{12} 10⁻³, h_{21} 75, h_{22} 200 μmho. Determine (*a*) the y parameters, (*b*) from the results of (*a*), the g parameters, (*c*) from the results of (*b*) the z parameters. Check these results by finding the input impedance in each case when loaded with 2 kΩ.

Ans. y, 2 × 10⁻³, −2 × 10⁻⁶, 0·15, 50 × 10⁻⁶.

g, 8 × 10⁻³, −0·04, −3 000, 2 × 10⁴.

z, 125, 5, −375 × 10³, 5 000.

Z_{1n}, 393 Ω.

Example 2.41. A four-terminal network having z_{11} 100 Ω, z_{12} 10 Ω, z_{21} −5 000 Ω, z_{22} 500 Ω, is connected in series with a 100 Ω resistor as in *Figure 2.31.* If the combination is loaded with 1 000 Ω and supplied from a source of impedance 200 Ω, determine the terminal input and output impedances and the voltage gain V_2/V_1.

Ans. 37 Ω, 1 944 Ω, −5·7.

Example 2.42. The network shown in *Figure 2.45* is known to have

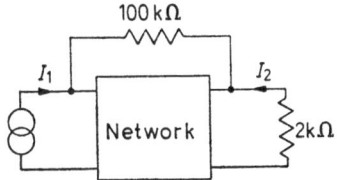

Figure 2.45. Circuit for Example 2.42

h_{11} 2 000 Ω, h_{12} 10^{-4}, h_{21} 250, h_{22} 300 μmho. Compare the current ratio I_2/I_1 with and without the 100 kΩ resistor connected as shown. (Networks in parallel.)

Ans. 157, 38.

Example 2.43. Figure 2.46 shows two interconnected four-terminal

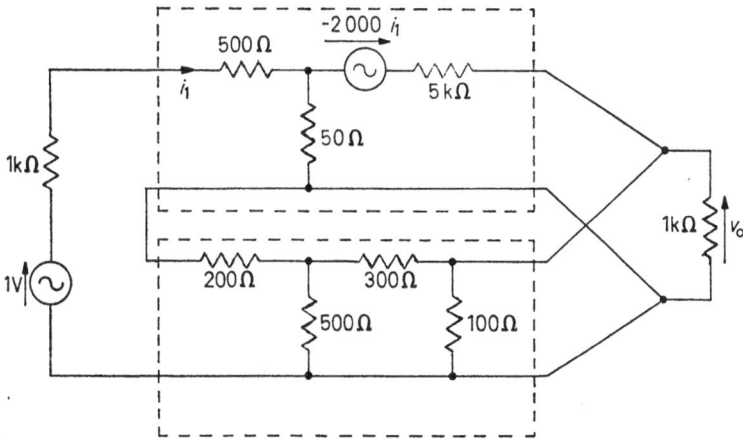

Figure 2.46. Circuit for Example 2.43

networks. Calculate the appropriate parameters for each network and hence find the voltage v_0.

Ans. 9·8 mV.

3

LOW FREQUENCY, SMALL SIGNAL EQUIVALENT CIRCUITS FOR VALVES AND TRANSISTORS

In Chapter 1 we found that the a.c. operation of valve and transistor circuits could be investigated by graphical methods. For more complex circuits these methods become exceedingly difficult and time consuming, and an alternative approach is desirable. This approach is to replace the active device by an equivalent circuit and then to analyse the resulting arrangement using the network methods discussed in Chapter 2.

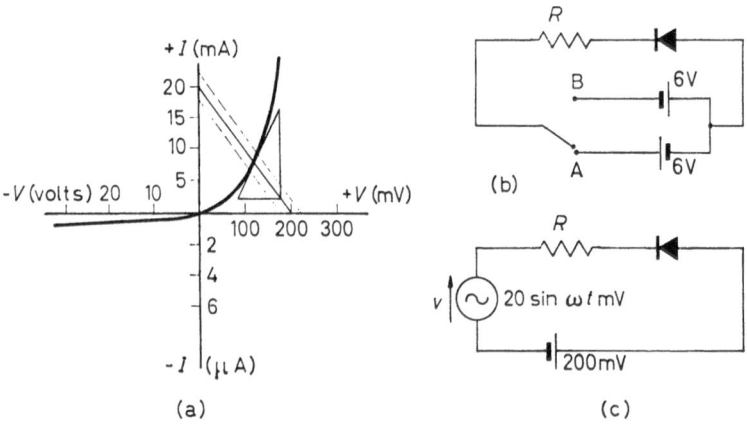

Figure 3.1. (*a*) Junction diode characteristics. (*b*) Diode circuit for forward and reverse bias. (*c*) A.C. circuit with forward biased diode

In order that we may understand the idea of an equivalent circuit, let us first consider a diode having the characteristics shown in *Figure 3.1a*. First consider the circuit shown in *Figure 3.1b*. If the switch is in position B, the diode is forward biased, i.e. the first

94

quadrant on the characteristics. The approximate resistance of the diode is given by

$$\frac{V}{I} = \frac{200 \times 10^{-3}}{20 \times 10^{-3}} = 10 \ \Omega$$

If the resistance of R is much greater than this, say 300 Ω, the variation in total series resistance due to the non-linearity of the characteristics will be negligible. Also the diode resistance is negligible compared with R which therefore determines the circuit current.

If, on the other hand, the switch is moved to A, the diode is now reverse biased and operates in the third quadrant of the characteristics. Here the diode resistance is given by

$$\frac{V}{R} = \frac{20}{10^{-6}} = 20 \ \text{M}\Omega$$

Now this value is very much greater than R and will determine the circuit current. By comparison with the first case when the circuit current was 20 mA, this current, 0·3 μA, is approximately zero. Thus in this circuit our diode equivalent could be a switch, as in *Figure 3.2a*, open for an applied voltage of one polarity and closed

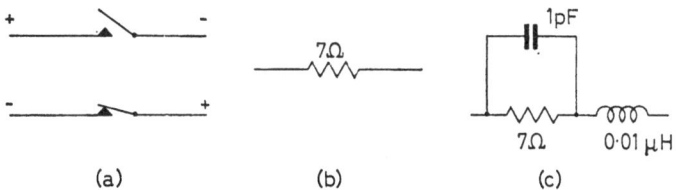

(a) (b) (c)

Figure 3.2. Diode equivalent circuits. (*a*) D.C. (*b*) Small signal a.c. for low frequencies and (*c*) for high frequencies

for the opposite case. Note the required condition for the use of such an equivalent circuit is that: $R_r \gg R \gg R_f$ where R_r and R_f are the reverse and forward bias diode resistance respectively.

An alternative situation is shown in *Figure 3.1c*. Here suppose R is 10 Ω and the d.c. supply voltage is 200 mV. This circuit will impose a load line on the characteristics as shown, giving an operating point of 110 mV, 7·5 mA. If the a.c. generator has a peak value of 20 mV, the load line will move between the two dotted lines shown. Over the range of resulting operating points the characteristic is approximately a straight line. Thus the a.c. resistance which is given by V_{ac}/i_{ac} can be obtained from the slope of the characteristic at the d.c. operating point. At this point the value of the resistance is

$$\frac{105 \times 10^{-3}}{15 \times 10^{-3}} = 7 \, \Omega$$

Thus for this particular case the a.c. equivalent circuit is a 7 Ω resistor, provided the d.c. bias current is 7 mA, and provided the a.c. signal is sufficiently small so that the characteristic may be assumed linear. We can call the 7 Ω resistor a small signal equivalent circuit for the diode. The term small signal implies that the a.c. voltage and current variations are sufficiently small so that over the operating region the characteristics can be assumed linear. Small signal for one condition may mean a few millivolts while another will lead to a reasonable approximation with signals of hundreds of volts amplitude. A further term in the chapter heading is low frequency. This is necessary since a diode by its construction will also have shunt capacitance and to a very small degree, series inductance. At very high frequency the equivalent circuit becomes that shown in *Figure 3.2c.*

Note that these additional components are present at all frequencies but their effect may be neglected at frequencies where $X_C \gg 7 \, \Omega$ and $X_L \ll 7 \, \Omega$.

In considering valve and transistor equivalents in this chapter we shall ignore the effects of shunt capacitance and series inductance which will be considered in a later chapter.

VALVE EQUIVALENT CIRCUITS

To find a small signal equivalent circuit for a valve we must consider the characteristics to be linear over the operating region. First let us imagine we have a valve with linear characteristics as shown in *Figure 3.3.* The relationships between the various electrode voltages and currents can now be specified in terms of valve 'constants'. These are known as r_a, the anode slope resistance; g_m the mutual conductance or transconductance, and μ the amplification factor.

Figure 3.3 shows a set of linear I_A/V_{AK} or anode characteristics and a corresponding I_A/V_{GK} or mutual characteristic.

A load line of slope $-1/R_L$ is shown passing through the d.c. operating point Q.

The inverse slope of the anode characteristics AC/AB is the first constant r_a. The slope of the mutual characteristic EO/DO is the

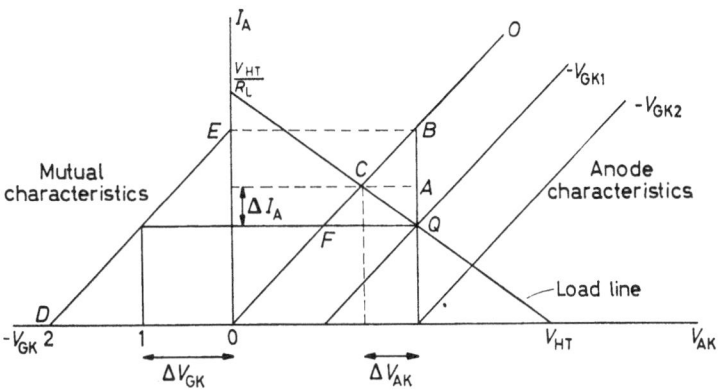

Figure 3.3. 'Ideal' triode characteristics for development of a.c. equivalent circuits

mutual conductance g_m. The ratio of change in anode voltage to change in grid voltage for constant anode current FQ/V_{gk} is μ the amplification factor. (Note this is not a geometrical relationship on *Figure 3.3.*)

These three constants are related since

$$r_a = \frac{\Delta V_{ak}}{\Delta I_a}\bigg|_{\Delta V_{ak}=0} \qquad (3.1)$$

$$g_m = \frac{\Delta I_a}{\Delta V_{gk}}\bigg|_{\Delta V_{ak}=0} \qquad (3.2)$$

and $\quad r_a \times g_m = \frac{\Delta V_{ak}}{\Delta I_a} \times \frac{\Delta I_a}{\Delta V_{gk}} = \frac{\Delta V_{ak}}{\Delta V_{gk}}\bigg|_{\Delta I_a=0} = \mu \qquad (3.3)$

To find an equivalent circuit in terms of these constants and R_L only we must find an expression for $\Delta V_{ak}/\Delta V_{gk}$ on load, i.e. when I_a is not constant. From the graph, a change of ΔV_{gk} on load results in $\Delta I_a = QA$. But

$$\Delta I_a = QB - AB \qquad (3.4)$$

and $\qquad \dfrac{QB}{\Delta V_{gk}} = g_m$ (mutual characteristic)

and $\qquad \dfrac{AC}{AB} = r_a = \dfrac{\Delta V_{gk}}{AB}$

Rewriting equation 3.4,

$$\Delta I_a = g_m \Delta V_{gk} - \frac{\Delta V_{ak}}{r_a}$$

97

Now
$$\frac{\Delta V_{ak}}{\Delta I_a} \text{ on load} = R_L$$

$\therefore$
$$\Delta I_a = g_m \Delta V_{gk} - \frac{R_L}{r_a} \Delta I_a$$

So
$$\Delta I_a \left(1 + \frac{R_L}{r_a}\right) = g_m \Delta V_{gk}$$

and
$$\Delta I_a = \frac{g_m \Delta V_{gk}}{1 + \frac{R_L}{r_a|}}$$

But
$$\Delta V_{ak} = \Delta I_a R_L = \frac{g_m \Delta V_{gk} R_L}{1 + \frac{R_L}{r_a}}$$

$$= \frac{g_m r_a R_L \Delta V_{gk}}{r_a + R_L}$$

Putting
$$g_m r_a = \mu$$

Voltage gain
$$= \frac{\Delta V_{ak}}{\Delta V_{gk}} = \frac{\mu R_L}{r_a + R_L}$$

Note this equation does not show the expected phase reversal. This is because magnitude of changes have been considered. If direction is taken into account:

Voltage gain
$$A_v = \frac{-\mu R_L}{r_a + R_L} \qquad (3.5)$$

or
$$A_v = \frac{-g_m r_a R_L}{r_a + R_L} \qquad (3.6)$$

Any equivalent circuit for a valve must, when loaded with a resistor R_L, result in the voltage gain given by equation 3·5 or 3·6.

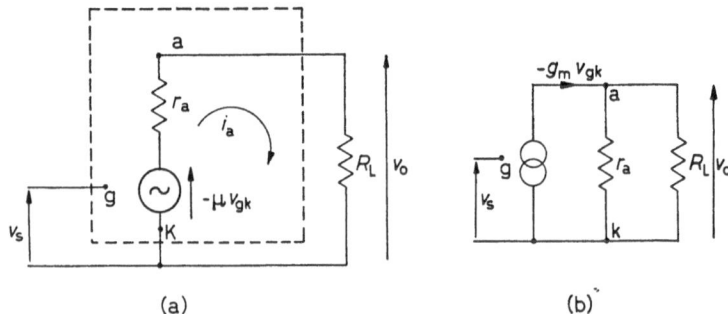

Figure 3.4. Small signal equivalent circuits for loaded triode valve. (*a*) Constant voltage equivalent. (*b*) Constant current equivalent

98

Consider the circuit shown in *Figure 3.4a*. The section enclosed by the dashed line represents the valve. v_{gk}, the voltage at the grid with respect to cathode is the same as the input signal voltage v_s. Applying mesh analysis to this circuit we obtain:

$$-\mu v_s = i_a(r_a + R_L)$$

$$\therefore \qquad i_a = \frac{-\mu v_s}{r_a + R_L}$$

and

$$v_0 = i_a R_L = \frac{-\mu v_s R_L}{r_a + R_L}$$

Thus the voltage gain

$$A_v = \frac{v_0}{v_s} = \frac{-\mu R_L}{r_a + R_L}$$

which is identical to the expression in equation 3.5.

The alternative circuit in *Figure 3.4b* can be solved by nodal analysis. As before $v_{gk} = v_s$, so

$$-g_m v_s = v_0\left(\frac{1}{r_a} + \frac{1}{R_L}\right)$$

and

$$v_0 = \frac{-g_m v_s r_a R_L}{r_a + R_L}$$

and

$$A_v = \frac{-g_m r_a R_L}{r_a + R_L}$$

Note that putting $g_m r_a = \mu$ makes these two results the same.

These two equivalent circuits satisfy the requirements and we shall consider an alternative approach for obtaining them. By

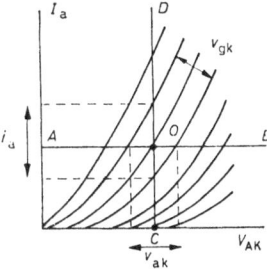

Figure 3.5. Triode characteristics for application of Thèvenin's and Norton's theorems

examining the characteristics with reference to a.c. changes at a d.c. operating point we can obtain an a.c. Thèvenin or Norton equivalent circuit for the valve.

The characteristics shown in *Figure 3.5* are those of a valve having the d.c. operating condition indicated by point *O*.

According to Thèvenin's theorem, any linear circuit may be represented by a voltage generator in series with an impedance.

The voltage generator has a value equal to the open circuit output voltage. If a circuit is open then the current is zero. In this case we are concerned with a.c. condition, so if the alternating current i_a is zero, the current is constant and the operating point can only move along the line AB.

The open circuit output voltage is v_{ak} and given by $v_{ak} = -\mu v_{gk}$. See equation 3.3.

The minus sign arises since a positive change in V_{GK} causes a negative change in V_{AK}.

The Thèvenin series resistance is that measured with all generators suppressed. Thus to a.c., v_{gk} must be zero, hence the resistance to be measured is that of the anode characteristic line passing through point O.

Now from equation 3.1

$$\frac{v_{gk}}{i_a} = r_a$$

So our Thèvenin equivalent is that shown in *Figure 3.6a* and by comparison with *Figure 3.4a* we can see that the valve equivalent is

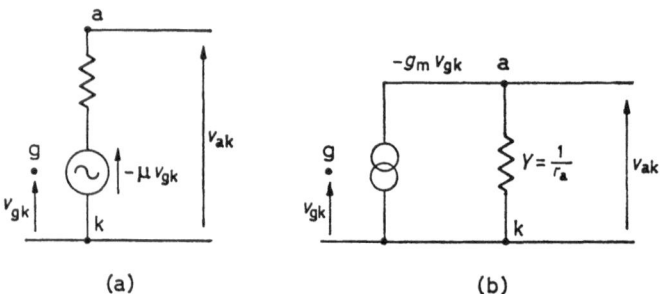

Figure 3.6. Triode small signal equivalent circuits. (*a*) Thèvenin equivalent and (*b*) Norton equivalent

shown in *Figure 3.6a* while the loaded circuit equivalent is shown in *Figure 3.4a*.

Applying Norton's theorem in the same way, our current generator is the current that would flow in a short circuit. This means that the a.c. voltage must be zero, i.e. the operating point can only move along the line CD on our characteristics.

Now the current variation i_a resulting from an a.c. input v_{gk} is $g_m v_{gk}$. (See equation 3.2.)

The parallel admittance with generators suppressed is given by i_a/v_a when v_{gk} is zero, i.e. it is $1/r_a$. The resulting Norton equivalent is shown in *Figure 3.6b*.

Again, by comparison with *Figure 3.4b* we can see that *Figure 3.6b* is the equivalent for the unloaded valve.

This approach is more useful than the graphical approach used first, since it can be easily extended to the case of the transistor. The application of these valve equivalents to more complex circuits will be discussed in Chapter 4.

TRANSISTOR EQUIVALENT CIRCUITS

The transistor is a three-terminal device, but since one terminal is usually common to both input and output connections, it may be treated as a four-terminal network. In practice the emitter is most frequently used as this common point and the transistor is then said to be in the common emitter configuration. Under these circumstances the applied voltages are measured at the base and collector with respect to the emitter, and the input and output currents are the base and collector currents. Alternative connections also used are common base and common collector. For common base the voltages are measured at emitter and collector with respect to base and the currents are emitter and collector currents. With the common collector configuration the voltages are measured at the base and emitter with respect to collector and the currents are the base and emitter currents.

Thus we can expect to find at least three different equivalent circuits representing the transistor. There are in fact many more possibilities; for each configuration we can obtain z, g, h, or y parameter equivalent circuits and other possibilities are known as T and π equivalents.

h Parameter Equivalent Circuits

For low frequencies the most popular equivalents are those based on the h parameter equations. This is because these are the parameters that can most easily be measured. Separate measurements may be made for each configuration or one set can be measured and the other two found by network manipulation.

Since the common emitter circuit is the most important we shall investigate this in full.

LOW FREQUENCY, SMALL SIGNAL EQUIVALENT CIRCUITS

Figure 3.7 shows the transistor connected in the common emitter configuration as a four-terminal network. First writing the general h parameter equations:

$$V_1 = I_1 h_{11} + V_2 h_{12}$$
$$I_2 = I_1 h_{21} + V_2 h_{22}$$

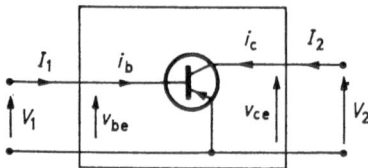

Figure 3.7. 'Black Box' representation of transistor in the common emitter configuration

Now replacing the variables by those specifically applicable to this transistor configuration

$$v_{be} = i_b h_{11} + v_{ce} h_{12}$$

$$i_c = i_b h_{21} + v_{ce} h_{22}$$

Finally since these parameters refer only to a common emitter connected transistor we can introduce a new notation for the parameters

$$v_{be} = i_b h_{ie} + v_{ce} h_{re} \tag{3.7}$$

$$i_c = i_b h_{fe} + v_{ce} h_{oe} \tag{3.8}$$

In each case the e stands for emitter and the i, r, f and o for input, reverse, forward and output respectively. For common base and collector these would become h_{ib} and h_{ic} etc.

Now remembering the definitions for the parameters obtained in Chapter 2 we can write

$$h_{ie} = \frac{v_{be}}{i_b}\bigg|_{v_{ce}=0} \qquad h_{re} = \frac{v_{be}}{v_{ce}}\bigg|_{i_b=0}$$

$$h_{fe} = \frac{i_c}{i_b}\bigg|_{v_{ce}=0} \qquad h_{oe} = \frac{i_c}{v_{ce}}\bigg|_{i_b=0}$$

Remember that all these voltages and currents are a.c. quantities measured about a particular d.c. operating point. Thus the statement $v_{ce} = 0$ means that the collector voltage is fixed at its d.c. value. Similarly $i_b = 0$ indicates that the base current is fixed at the required d.c. level.

We shall now relate these parameters to the approximate characteristics as shown in *Figure 3.8*.

Now considering each of our definitions in turn:

$$h_{\text{1e}} = \left.\frac{v_{\text{be}}}{i_{\text{b}}}\right|_{v_{\text{ce}}=0}$$

If v_{ce} is zero the operating point can move only along the input characteristic on *Figure 3.8a*. h_{1e} is therefore the slope of the input

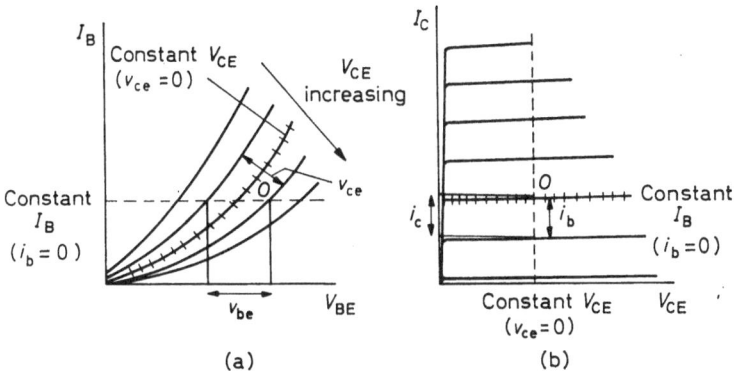

(a)   (b)

Figure 3.8. Transistor characteristics for determination of *h* parameters. (*a*) Input and (*b*) Output characteristics

characteristic with the output short circuit to a.c., i.e. it is the short circuit input resistance. Typical values are of the order of 1 000 Ω.

$$h_{\text{re}} = \left.\frac{v_{\text{be}}}{v_{\text{ce}}}\right|_{i_{\text{b}}=0}$$

If i_{b} is zero we can move only along the constant I_{B} line on *Figure 3.8a*. The relative v_{ce} and v_{be} under these conditions is shown and h_{re} is referred to as the reverse transfer parameter. This parameter is very small, typically 10^{-3}, and as we shall see in later chapters, may frequently be neglected.

$$h_{\text{fe}} = \left.\frac{i_{\text{c}}}{i_{\text{e}}}\right|_{v_{\text{ce}}=0}$$

Referring to *Figure 3.8b*, the output characteristic, we can see that h_{fe} is the direct ratio of i_{c} and i_{b} when v_{ce} is maintained at a constant level.

103

This is the most important parameter and is known as the short circuit current gain. Using conventional currents, both I_B and I_C flow out of the transistor (pnp), so changes in these directions are both negative when considering the four-terminal convention. The ratio of the two, h_{fe}, is therefore positive. In commercial transistors the value of h_{fe} will be in the range 10–500 depending upon type and application.

$$h_{oe} = \frac{i_c}{v_{ce}}\bigg|_{i_b=0}$$

Looking at *Figure 3.8b* we can see that this represents the slope of the output characteristic. h_{oe} is therefore an admittance and it is known as the open circuit output admittance. A typical value for h_{oe} is 100 μmhos, i.e. an output resistance of 10 kΩ.

To complete this section we should look at the resulting h parameter equivalent circuit as shown in *Figure 3.9*.

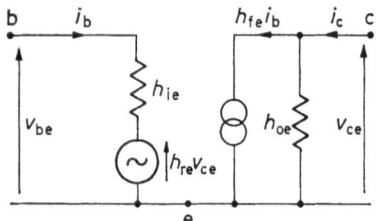

Figure 3.9. The hybrid parameter transistor equivalent circuit for common emitter connection

Common collector and common base parameters may be obtained in the same way from the appropriate sets of characteristics if these are available. But it is usually more convenient to convert, using network methods from the available set of parameters to those required. These manipulations will be shown in Chapter 5. In either case the resulting h parameter equivalent circuit will have the same form as that shown in *Figure 3.9* but with appropriate notation.

The T Equivalent Circuit

Another equivalent circuit often encountered is the equivalent T. This circuit is based on the common base configuration and is sometimes thought to represent the physical structure of the transistor. It is shown in *Figure 3.10*.

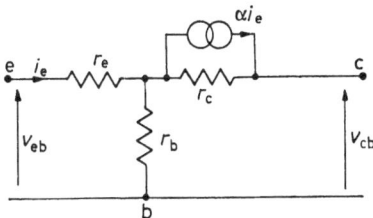

Figure 3.10. The T equivalent for a transistor in the common base configuration

Typical values for these parameters are r_e 5Ω, r_b 1 000 Ω, r_c 1 MΩ, and α0·99. Conversion to the common emitter form is achieved by the steps shown in *Figure 3.11*.

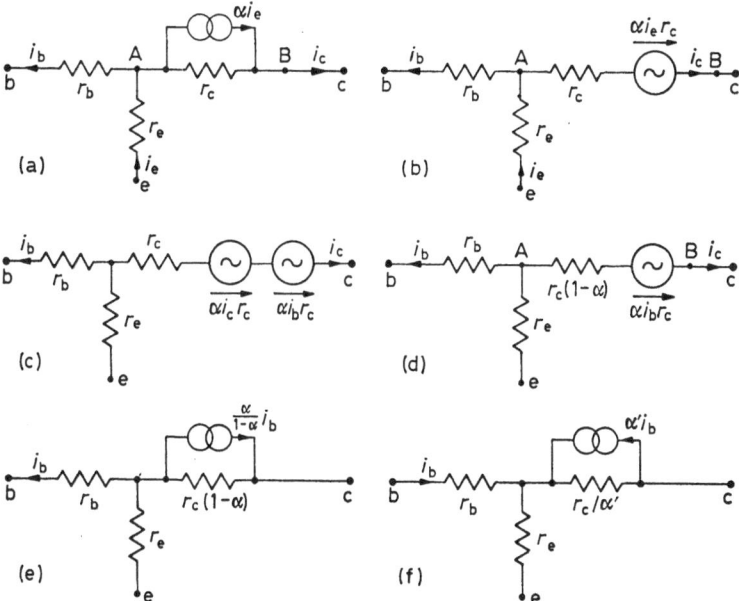

Figure 3.11. Conversion from common base T to common emitter T equivalent circuits

(*a*) Shows the T circuit turned so that the emitter is common.
(*b*) Thèvenin's theorem has been applied to branch AB.
(*c*) From fundamentals, i_c may be expressed as sum of i_c and i_b (equation 1.10). Thus $\alpha i_e r_c$ generator may be split into two components.

(*d*) Since i_c flows in branch AB, the $\alpha i_c r_c$ generator may be replaced by a resistor of $-\alpha r_c\ \Omega$. The minus sign gives the correct polarity for this e.m.f., and the two resulting series resistors have been added to make $r_c(1 - \alpha)$.

(*e*) Norton's theorem applied to branch AB.

(*f*) Since i_b is the input current, it is convenient to reverse its direction, and with it, the direction of the current generator. Also putting $\dfrac{\alpha}{1 - \alpha} = \alpha'$ and $(1 - \alpha) \simeq \dfrac{1}{\alpha'}$, the conversion is completed (equation 1.17). An example of this use of the circuit will be given in Chapter 5.

Conversion to the *h* parameter and vice versa may be achieved by applying the *h* parameter definitions to the T equivalent circuit. Knowledge of typical values allows valid approximations leading to simple conversion factors.

First for h_{1e} and h_{te} we must let v_{ce} be zero. The required circuit is shown in *Figure 3.12a*.

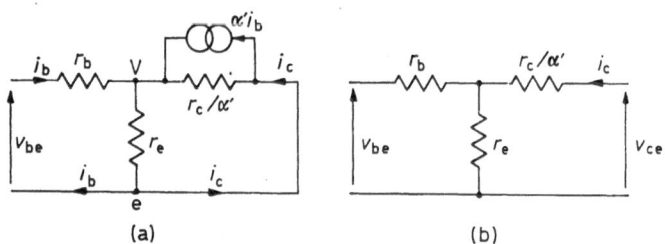

(a)   (b)

Figure 3.12. Circuits for determining *h* parameters from T parameters. (*a*) Short circuit output. (*b*) Open circuit input

By writing nodal equations we can solve for the voltage V in terms of i_b and thus put:

$$h_{1e} = r_b + \frac{V}{i_b}$$

$$i_b + \alpha' i_b = V\left(\frac{1}{r_e} + \frac{\alpha'}{r_c}\right) \text{ (Since the s/c puts } r_e \text{ and } r_c/\alpha' \text{ in parallel.)}$$

$$= \left(\frac{r_c + \alpha' r_e}{r_e r_c}\right)$$

$$\therefore \qquad \frac{V}{i_b} = \frac{r_e r_c(1 + \alpha')}{(r_c + \alpha' r_e)}$$

Now from the typical values we can see that $r_c \gg \alpha' r_e$.

So neglecting $\alpha' r_e$ and cancelling the r_c terms we obtain:

$$\frac{V}{i_b} \simeq r_e(1 + \alpha')$$

$$\therefore \qquad h_{1e} \simeq r_b + r_e(1 + \alpha') \qquad (3.9)$$

The current $(1 + \alpha')i_b$ flows into r_e and r_c/α' in parallel. But $r_c/\alpha' \gg r_e$ and we can assume that $(1 + \alpha')i_b$ flows in r_e. Since i_b must complete its circuit to the input terminals, the short circuit i_c must be the remainder.

$$i_c = \alpha' i_b$$

and

$$h_{fe} = \left. \frac{i_c}{i_b} \right|_{v_{ce}=0} = \alpha' \qquad (3.10)$$

For h_{oe} and h_{re} the condition is open circuit input. This makes i_b zero and therefore eliminates the current generator from our T equivalent. The result is shown in *Figure 3.12b*.

By inspection $\quad h_{oe} = \left. \frac{i_c}{v_{ce}} \right|_{i_b=0} = \dfrac{1}{r_e + \dfrac{r_c}{\alpha'}} \therefore h_{oe} \simeq \dfrac{\alpha'}{r_c} \qquad (3.11)$

Also $\quad h_{re} = \left. \dfrac{v_{be}}{v_{ce}} \right|_{i_b=0} = \dfrac{r_e}{r_e + \dfrac{r_c}{\alpha'}} \simeq \dfrac{\alpha' r_e}{r_c} \qquad (3.12)$

Equations 3.9, 3.10, 3.11 and 3.12 give the conversion factors for the h parameters in terms of the T parameters. The opposite conversions can be found using these results.

From equation 3.10,

$$\alpha' = h_{fe} \qquad (3.13)$$

From equation 3.11,

$$r_c = \frac{\alpha'}{h_{oe}} = \frac{h_{fe}}{h_{oe}} \qquad (3.14)$$

From equation 3.12,

$$r_e = \frac{h_{re} r_c}{\alpha'} = \frac{h_{re} h_{fe}}{h_{fe} h_{oe}} = \frac{h_{re}}{h_{oe}} \qquad (3.16)$$

To compare these values we shall consider a common general purpose transistor, the OC75. Manufacturer's published data quotes typical values for the h parameter as h_{1e} 1·3 kΩ, h_{re} 5 $\times$ 10^{-4}, h_{te} 90, and h_{oe} 125 μmhos.

Using the equations above:

$$\alpha' = 90$$

$$r_c = \frac{90}{125 \times 10^{-6}} = 720 \text{ k}\Omega$$

$$r_e = \frac{5 \times 10^{-4}}{125 \times 10^{-6}} = 4 \ \Omega$$

$$r_b = 1\ 300 - \frac{5 \times 10^{-4}}{125 \times 10^{-6}} (1 + 90)$$

$$= 936\ \Omega$$

Other transistor small signal equivalent circuits are used, but only for high frequency applications and these will be considered in a later chapter.

At some time in the future devices other than valves or transistors may come into general use. Where such devices are used under small signal conditions equivalent circuits will be used to represent them. These will be obtained in the same way by reference to the characteristics. If they are fundamentally four-terminal networks it may be convenient to measure the h parameters but it is quite possible that the z, y or g parameters will be more convenient. One such device is the field effect transistor for which manufacturers are quoting the y parameters. In all such cases, the general methods outlined in this and subsequent chapters will be applicable.

The applications of equivalent circuits to complete circuit arrangements will be discussed in Chapters 4 and 5 covering valve and transistor work respectively, and further examples appear in the remaining chapters of the book.

EXAMPLES

Example 3.1. A triode valve has the characteristics given in the table below. It is to be operated with an H.T. of 450 V and (a) R_L 30 kΩ, V_g -2 V or (b) R_L 60 kΩ, V_{gk} -12 V. For each case, determine the operating point, the components of the small signal equivalent circuit and hence the voltage amplification.

	V_{AK}	0	50	100	150	200	250	300	350	400	450
I_A (mA) for											
V_{GK}	0	0	5·0	13·7							
	−2		0	3·0	9·5	17·0					
	−4			0	2·2	6·5	14·0				
	−6				0	1·5	4·5	11·8			
	−8					0	1·3	4·0	10·0		
	−10						0	1·2	3·6	8·3	
	−12							0	1·1	3·1	7·2
	−14								0	1·4	3·5
	−16									0·1	1·3

Ans. (a) 153 V, 9·8 mA, r_a 6·6 kΩ, 36, Av − 29·5.
(b) 361 V, 1·5 mA, r_a 31 kΩ, 24, Av − 15·8.

Example 3.2. A transistor connected in the common emitter configuration has the input and output characteristics given in the table (p. 110). The d.c. operating point is given by (a) V_{CE} −4·5 V, I_C 8·8 mA, or (b) V_{BE} −130 mV, I_B 40 A. In each case, determine the small signal h parameters. (With respect to h_{re} these characteristics are not typical as they have been exaggerated to simplify graphical measurements.)

$-V_{BE}$ (mV)	0	50	100	150	200	250	300
I_B (μA) for $-V_{CE}$ 0	0	22	90	200	345		
1·5	0	4	18	60	135	235	360
3·0	0	2	14	50	117	214	331
4·5	0	0	10	43	105	200	313
6·0	0		8	38	98	190	200

V_{CE} (V)	0	0·2	0·4	1·0	6·0
I_c (mA) for I_B (μA) 0	0	0·2	0·2	0·2	0·2
40	0	1·2	1·4	1·5	1·85
80	0	2·4	2·8	2·9	3·9
120	0	3·6	4·1	4·3	5·7
160	0	4·8	5·6	5·85	7·6
200	0	6·0	6·9	7·25	9·45
240	0	7·6	8·4	8·65	11·5

Ans. (*a*) 470 Ω, 5×10^{-3}, 45, 450 μmho.
(*b*) 1 200 Ω, $6 \cdot 7 \times 10^{-3}$, 35, 97 μmho.

Example 3.3. The published data for a transistor connected in the common emitter configuration includes the following h parameters at the desired operating point.

$$h_{11} \ 1\ 000\ \Omega, \quad h_{12}\ 5 \times 10^{-4}, \quad h_{21}\ 120, \quad h_{22}\ 80\ \mu\text{mho}$$

Determine the components of (*a*) the common base and (*b*) the common emitter equivalent T circuits.

Ans. (*a*) 6·25 Ω, 244 Ω, 1·5 MΩ, 0·993.
 (*b*) 6·25 Ω, 244 Ω, 12·5 kΩ, 120.

Example 3.4. The T parameters for a transistor are given as r_e 15 Ω, r_b 500 Ω, r_c 1 MΩ, and α0·97. Determine the h parameters for the transistor connected in (*a*) the common base configuration and (*b*) the common collector configuration.

Ans. (*a*) h_{ib} 30 Ω, h_{rb} 5 × 10^{-4}, h_{fb} −0·97, h_{ob} 1 μmho.
 (*b*) h_{ic} 550 Ω, h_{rc} + 1, h_{fc} −33·3, h_{oc} 33 μmho.

4

USE OF THE VALVE EQUIVALENT CIRCUIT

In Chapter 1, simple valve circuits were solved by graphical means. This procedure is not only time consuming, but becomes exceedingly complex when more than one valve is used in the circuit. The problem is further complicated by the presence of reactive components which may affect the performance at certain operating frequencies. Chapter 3 showed that valves could be represented by equivalent circuits. The circuits developed were suitable for small changes of

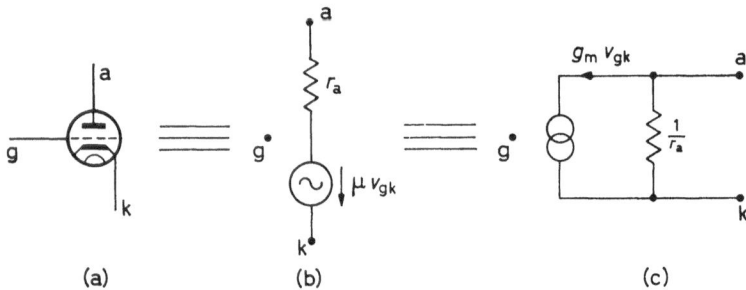

Figure 4.0. Triode valve and small signal equivalent circuits

electrode potentials and are hence known as small signal a.c. equivalent circuits. These circuits represent the valve only, at the particular d.c. operating point. Any change in operating point will modify the values of the constants r_a, g_m, and μ. In this chapter, we shall consider an exact method for solving any valve circuit operating within the limits of the equivalent circuits. Then, using this method, we shall analyse a range of amplifier circuits commonly found in electronic systems.

The procedure for using the equivalent circuits is always the same, the steps being as follows:

(1) For each valve in the complete circuit, draw the appropriate equivalent circuit. Either the constant voltage form *Figure 4.0b*

or the constant current form *Figure 4.0c* may be used and experience will show which will lead to the simplest solution. In general, circuits having most elements in series are best solved by using the constant voltage form, while those having more parallel components are more suited to the use of the constant current equivalent. Examples showing both forms are given later in this and other chapters. The reader should include all the details shown until he is familiar with the method.

(2) Taking each electrode in turn, connect it to earth through any component, source of e.m.f. or current source shown in the complete circuit. The d.c. positive H.T. line is taken as earth, since the a.c. resistance of a battery or power pack is negligible and may be regarded as a short circuit.

(3) If the constant voltage circuit is being used, indicate the sense of the unknown currents. These will normally be Maxwell's circulating currents in the clockwise direction, but in some instances other current definitions lead to simpler equations.

If the constant current circuit is to be used, the sense of the unknown node voltages must be chosen.

(4) In either type of circuit the generator is given in terms of v_{gk}. This means the a.c. potential at the grid, measured with respect to cathode. v_{gk} should now be expressed in terms of any applied voltages or signal sources together with any potentials existing across components placed between cathode and grid. In the nodal form such potentials will simply be v_a or $v_a - v_b$, but in the mesh form they will be in terms of the unknown currents as in i_aR_k or $(i_1 - i_2)Z_1$. The technique is to indicate the sense or direction of all such potential differences and applied e.m.f.s and proceeding from cathode to grid, add such p.d.s and e.m.f.s algebraically. Remember the potential across an impedance is positive if measured in the opposite sense to the direction of current flow. Any path between cathode and grid will lead to the correct result, but the shortest path usually leads to the quickest solution.

(5) Write the mesh or nodal equations for the circuit and solve for the unknown currents or voltages using substitution or determinant methods where necessary.

These results will normally lead to the solution of the problem for information concerning the voltage amplification, the output voltage or power, the frequency or phase response, and the input impedance. If the amplifier output impedance is required, the equivalent circuit is again used, but the different technique involved will be discussed later in the chapter. If a numerical solution only is required, the reader should insert values in the equations before solution. This will usually result in a simpler solution. The derivation of standard formulae may involve particular steps or even approximations which have to be memorized.

Example 4.1. Consider a single stage anode loaded amplifier with cathode bias and no decoupling capacitor. The valve has amplification factor μ and anode resistance r_a. The full circuit is given in *Figure 4.1a* and the steps in drawing the equivalent circuit in *b*, *c*, *d*, and *e*.

c, *d* and *e* show the connection of grid, cathode and anode respectively to earth. *f* shows the unknown current i_a and the positive sense of the potentials $i_a r_a$, $i_a R_k$, and $i_a R_L$ due to a positive i_a. Note that the output voltage v_0 is measured at the anode with respect to earth and that this is given by $+i_a R_L$. Step 4 in the procedure is to obtain v_{gk}. From the diagram, proceeding from cathode to grid we obtain:

$$v_{gk} = +i_a R_k + e_s \tag{4.1}$$

(Note, if i_a had been chosen to circulate in an anticlockwise direction; this would have led to $v_{gk} = -i_a R_k + e_s$ and $v_0 = -i_a R_L$.)

Continuing with step 5:

$$-\mu v_{gk} = i_a r_a + i_a R_L + i_a R_k \tag{4.2}$$

Substituting for v_{gk}

$$-\mu(i_a R_k + e_s) = i_a r_a + i_a R_L + i_a R_k \tag{4.3}$$

Collecting terms in i_a on the right hand side of the equation

$$-\mu e_s = i_a[r_a + R_L + R_k(1 + \mu)] \tag{4.4}$$

and

$$i_a = \frac{-\mu e_s}{r_a + R_L + R_k(1 + \mu)} \tag{4.5}$$

If the output voltage is required, $v_0 = i_a R_L$, and

$$v_0 = \frac{-\mu e_s R_L}{r_a + R_L + R_k(1 + \mu)} \tag{4.6}$$

and the voltage gain,

$$A_v = \frac{v_0}{e_s} = \frac{-\mu R_L}{r_a + R_L + R_k(1 + \mu)} \tag{4.7}$$

The minus sign indicates that the output voltage will be 180° out of phase with the input signal e_s.

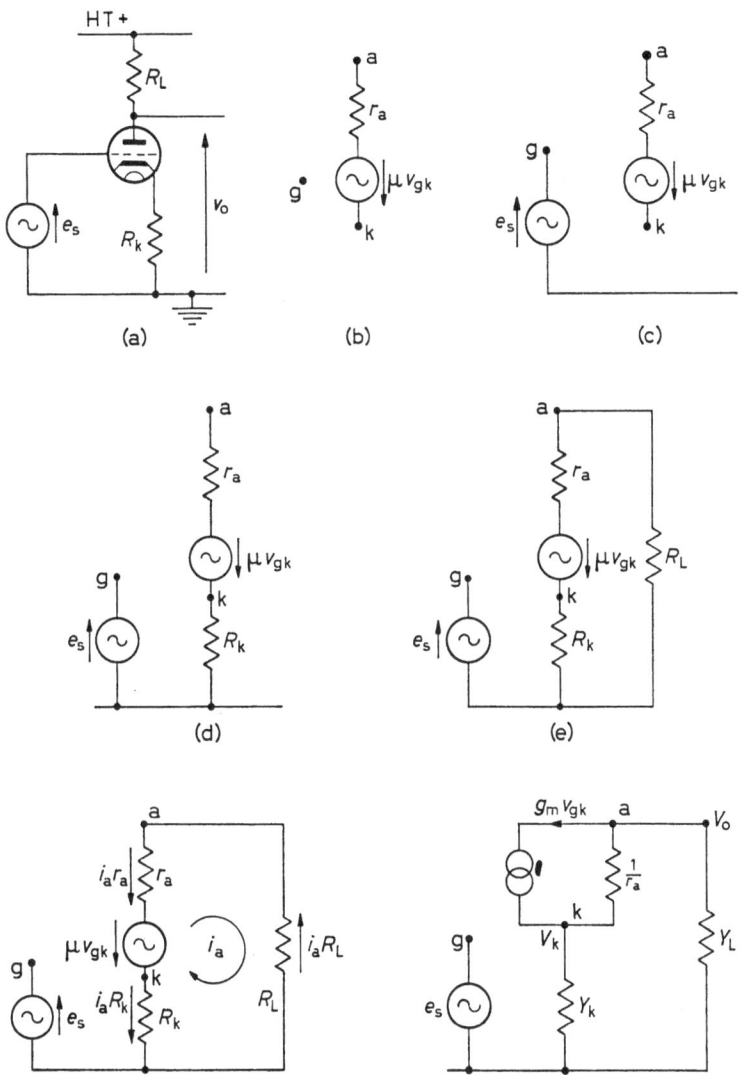

Figure 4.1. (*a*) Triode amplifier with un-decoupled cathode resistor. (*b*) to (*f*) Steps in the drawing of the equivalent circuit for (*a*) using the constant voltage form. (*g*) The constant current form of equivalent circuit for (*a*)

115

This circuit is essentially a series circuit. Mesh analysis therefore resulted in the simplest solution. If the constant current form is used, the equivalent circuit obtained is that shown in *Figure 4.1g*. There are now two unknowns and v_{gk} is given by:

$$v_{gk} = -V_2 + e_s$$

The reader should solve this for practice and obtain

$$v_o = V_1 = \frac{-g_m e_s Y_k}{y_a(Y_k + Y_L) + Y_L(Y_k + g_m)} \tag{4.8}$$

Then by putting

$$y_a = \frac{1}{r_a}, \quad Y_k = \frac{1}{R_k}, \quad Y_L = \frac{1}{R_L} \quad \text{and} \quad g_m = \frac{\mu}{r_a}$$

re-arrangement will lead to the result obtained using the constant voltage generator.

Two other series forms will be given as examples. These are the simple cathode follower or grounded anode amplifier and the grounded grid amplifier.

THE CATHODE FOLLOWER

Example 4.2. A triode valve having μ of 20 and r_a 30 kΩ is to be used as a simple cathode follower with a cathode load of 5 kΩ.

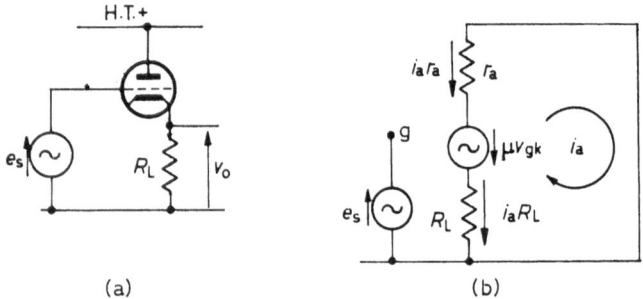

Figure 4.2. The simple cathode follower. (*a*) Full circuit and (*b*) the small signal equivalent circuit

Calculate the voltage gain and the load power if the input signal is 1 volt r.m.s.

Figure 4.2a shows the full circuit and *Figure 4.2b* the required equivalent circuit.

116

Using *Figure 4.2b*

$$v_{gk} = +i_a R_L + e_s \qquad (4.9)$$

or
$$v_{gk} = -\mu v_{gk} - i_a r_a + e_s \qquad (4.10)$$

and
$$v_{gk}(1 + \mu) = e_s - i_a r_a$$

giving
$$v_{gk} = \frac{e_s - i_a r_a}{1 + \mu} \qquad (4.11)$$

Both forms are correct, but since result 4.9 is simpler, this is the best to use.

Writing the circuit equation

$$-\mu(i_a R_L + e_s) = i_a R_L + i_a r_a \qquad (4.12)$$

$$-\mu e_s = i_a[r_a + R_L(1 + \mu)] \qquad (4.13)$$

Inserting values and working in mA, volts, and $k\Omega$

$$-20e_s = i_a[30 + 5(1 + 20)]$$

$$\therefore \qquad i_a = \frac{-20e_s}{135} \, \text{mA}$$

Now, $v_0 = -i_a R_L$ since $+i_a R_L$ is measured with respect to cathode and v_0 is measured with respect to earth.

$$\therefore \qquad v_0 = + \frac{20}{135} \, 5e_s = 0.74e_s \, \text{V}$$

$$\therefore \qquad A_v, \text{ the voltage gain} = \frac{v_0}{e_s} = 0.74$$

and the power dissipated in R_L is found from

$$i_a{}^2 R_L = \left(-\frac{20}{135} \times 10^{-3}\right)^2 \times 5 \times 10^3 \, \text{W}$$

$$= 0.11 \, \text{mW}$$

This result gives only the a.c. power in the load. In addition there will be d.c. power due to the direct anode current. Since this may be of the order of for example 5 mA resulting in 125 mW dissipation the load resistor should be rated at $\frac{1}{4}$ W to allow a margin of safety.

THE GROUNDED GRID AMPLIFIER

Example 4.3. Derive expressions for the voltage gain, the terminal input impedance, and the output impedance of a grounded grid amplifier driven by a source of open circuit voltage e_s, and internal resistance R_s. The anode is loaded with resistance R_L and the valve has amplification factor and anode resistance of μ and r_a respectively.

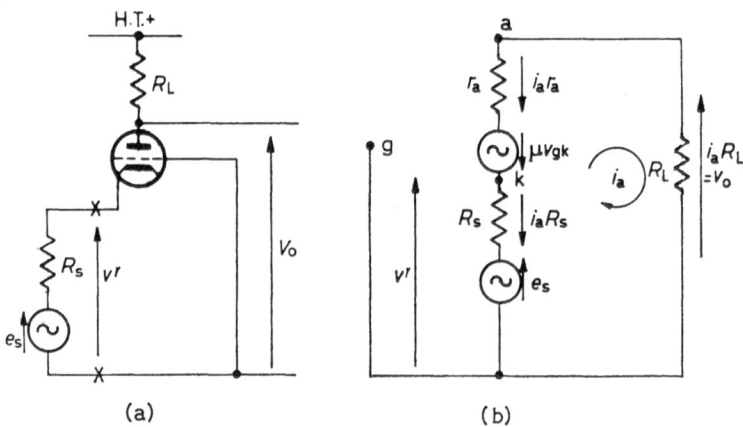

(a)   (b)

Figure 4.3. The grounded grid amplifier. (*a*) Full circuit and (*b*) the small signal equivalent circuit

Figure 4.3b shows the complete equivalent circuit for the grounded grid amplifier shown in *Figure 4.3a.* This may be used directly to determine the voltage amplfication v_0/e_s or the terminal voltage amplification v_0/v' and the terminal input impedance v'/i_a. The output impedance may be deduced from the first of these results, but a general method of calculation of output impedance will also be introduced.

Considering *Figure 4.3b*

$$v_{gk} = +i_a R_s - e_s \tag{4.14}$$

writing mesh equation

$$e_s - \mu(+i_a R_s - e_s) = i_a(R_L + R_s + r_a) \tag{4.15}$$

rearranging

$$e_s(1 + \mu) = i_a[r_a + R_L + R_s(1 + \mu)] \tag{4.16}$$

$$i_a = \frac{e_s(1 + \mu)}{r_a + R_L + R_s(1 + \mu)} \tag{4.17}$$

118

and
$$v_0 = i_a R_L = \frac{e_s(1 + \mu)R_L}{r_a + R_L + R_s(1 + \mu)} \qquad (4.18)$$

giving the overall voltage amplification

$$\frac{v_0}{e_s} = \frac{(1 + \mu)R_L}{r_a + R_L + R_s(1 + \mu)} \qquad (4.19)$$

The terminal input impedance

$$Z_{1n} = \frac{v'}{i_a}$$

and since
$$v' = e_s - i_a R_s$$

$$Z_{1n} = \frac{e_s}{i_a} - R_s = \frac{r_a + R_L + R_s(1 + \mu)}{(1 + \mu)} - R_s \qquad (4.20)$$

$$Z_{1n} = \frac{r_a + R_L}{(1 + \mu)} \qquad (4.20)$$

If it is required the terminal voltage gain A_{vt} may be obtained by calculating v' in terms of e_s using $v' = i_a Z_{1n}$, and putting

$$A_{vt} = \frac{v_0}{v'}$$

This results in a value for the terminal voltage amplification

$$A_{vt} = \frac{(1 + \mu)R_L}{r_a + R_L} \qquad (4.21)$$

CALCULATION OF OUTPUT IMPEDANCE

The output impedance of an amplifier is the effective internal impedance of the equivalent voltage generator. Consider a simple

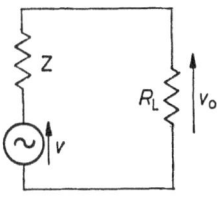

Figure 4.4. Circuit for discussion on output impedance

generator of v volts and internal impedance Z, connected to a load R_L (*Figure 4.4*). The output voltage is given by $vR_L/(Z + R_L)$. Now consider the expression for the output voltage of the grounded grid amplifier,

$$v_0 = \frac{e_s(1 + \mu)R_L}{r_a + R_s(1 + \mu) + R_L} \qquad (4.22)$$

which would be the result for a generator of $e_s(1 + \mu)$ volts and internal impedance

$$r_a + R_s(1 + \mu) \tag{4.23}$$

The output impedance of the grounded grid amplifier is thus $r_a + R_s(1 + \mu)\Omega$. This logical approach is not always so convenient and an alternative method may be adopted. The procedure for determining the output impedance is then as follows:

(1) Redraw the equivalent circuit with any external generators or e.m.f. sources suppressed (i.e. replaced by their internal impedances).

(2) Connect a generator of E volts and zero internal impedance to the output terminals. (See *Figure 4.5*.)

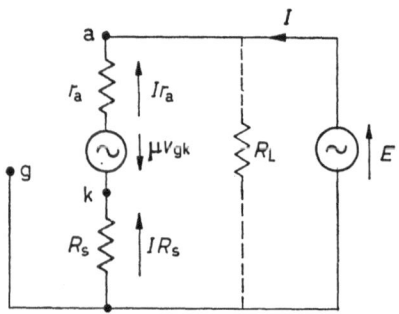

Figure 4.5. Equivalent circuit for determining the output impedance of the grounded grid amplifier

(3) Calculate the current I due to this generator and find $E/I = Z_{out}$. In practice it is often convenient to omit R_L in the calculation, and to add it in parallel to the calculated result.

Following this procedure for the grounded grid amplifier:

Since the external generator has been suppressed

$$v_{gk} = -IR_s$$

and writing the mesh equation

$$E + \mu(-IR_s) = I(r_a + R_s) \tag{4.24}$$

rearranging $\qquad E = I[r_a + R_s(1 + \mu)] \tag{4.25}$

and $\qquad Z_{out} = \dfrac{E}{I} = r_a + R_s(1 + \mu) \tag{4.26}$

which is the same result (4.23) as was obtained using the logical approach. The overall output impedance including the effect of R_L will therefore be

$$Z_{out} = \frac{R_L[r_a + R_s(1 + \mu)]}{R_L + r_a + R_s(1 + \mu)} \qquad (4.27)$$

This method of calculating the output impedance of a circuit is completely general and will be used in subsequent chapters.

Examples involving the solution of two mesh currents will now be considered. These are a form of cathode follower, designed to give a high resistive input impedance, and a two valve circuit known as a long tailed pair.

THE MODIFIED CATHODE FOLLOWER

Example 4.4. The circuit shown in *Figure 4.6a* is that of a cathode follower. The 1 kΩ resistor provides the correct d.c. bias and the

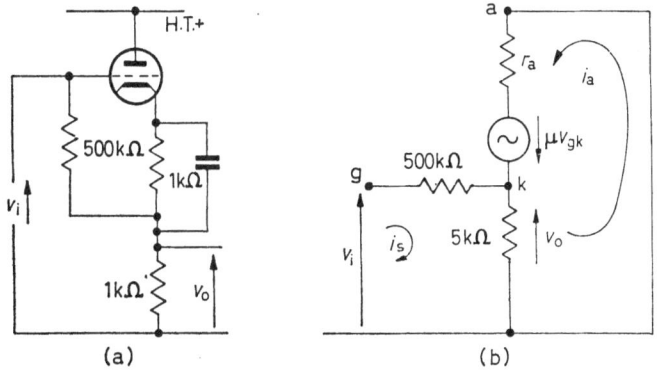

Figure 4.6. Full and equivalent circuits of the modified cathode follower circuit in Example 4.4

bypass capacitor may be assumed to have negligible reactance at all signal frequencies. The valve has μ 50 and r_a 10 kΩ. Calculate the input impedance and the output voltage if an r.m.s. signal of two volts is applied between grid and earth.

The equivalent circuit for the problem is shown in *Figure 4.6b*. The direction of the unknown mesh currents are chosen to make the output voltage $v_0 = +(i_a + i_s)R_k$. Taking all resistors in kΩ, all currents in mA, and all potential differences and e.m.f.s in volts, we have:

$$v_{gk} = 500i_s \quad [\text{or } -5(i_a + i_s) + v_1]$$

writing the mesh equations:

$$v_1 = 500i_s + 5(i_a + i_s) \tag{4.28}$$

$$500\mu i_s = 10i_a + 5(i_a + i_s) \tag{4.29}$$

Rearranging and collecting terms,

$$v_1 = 505i_s + 5i_a \tag{4.30}$$

$$0 = -25\,000i_s + 15i_a \tag{4.31}$$

(in equation 4.31, $+5i_s$ has been neglected.)

Using substitution methods, from equation 4.31

$$i_a = \frac{25\,000}{15} i_s = \frac{5\,000}{3} i_s \tag{4.32}$$

substituting in equation 4.30

$$v_1 = 500i_s + 5 \left(\frac{5\,000}{3} i_s + i_s\right) = 8\,834 I_s$$

$$Z_{1n} = \frac{v_1}{i_s} = 8 \cdot 834 \text{ M}\Omega$$

Also

$$i_s = \frac{v_1}{8\,834} \text{ mA}$$

Substitute in equation 4.30

$$v_1 = \frac{505}{8\,834} v_1 + 5i_a$$

therefore

$$i_a = \frac{v_1}{5} \left(1 - \frac{505}{8\,834}\right) = 0 \cdot 189v_1$$

but $i_a \gg i_s$, therefore

$$v_0 = 0 \cdot 189v_1 \times 5 = 0 \cdot 943v_1$$

and since v_1 is 2 volts,

$$v_0 = 1 \cdot 886 \text{ volts}$$

THE LONG TAILED PAIR

Example 4.5. The long tailed pair or cathode coupled amplifier shown in *Figure 4.7a* is to be used as a difference amplifier. Show that the voltage between the two anodes is proportional to the difference between the two input signals e_1 and e_2. Assume that the two valves and their loads are identical.

Taking i_1 and i_2 in the directions shown in *Figure 4.7b*, the equivalent circuit, we find

$$v_{gk1} = (i_1 + i_2)R_k + e_1 \qquad (4.33)$$

and
$$v_{gk2} = (i_1 + i_2)R_k + e_2 \qquad (4.34)$$

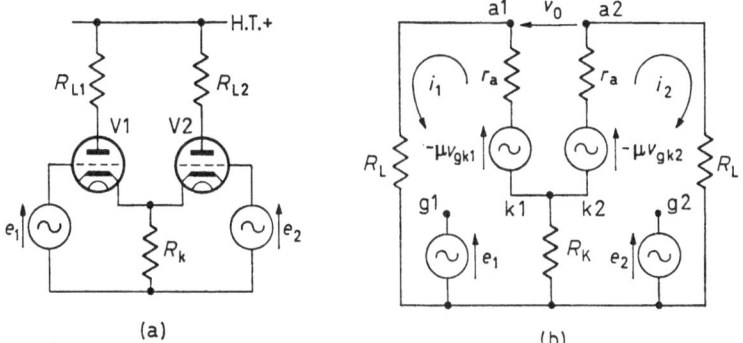

(a)

(b)

Figure 4.7. Full and equivalent circuits for the long tailed pair amplifier discussed in Example 4.5

The mesh equations may then be written:

$$-\mu e_1 - \mu R_k(i_1 + i_2) = i_1(r_a + R_L + R_k) + i_2 R_k \qquad (4.35)$$
$$-\mu e_2 - \mu R_k(i_1 + i_2) = i_1 R_k + i_2(r_a + R_L + R_k) \qquad (4.36)$$

rearranging:

$$-\mu e_1 = i_1[r_a + R_L + R_k(1 + \mu)] + i_2 R_k(1 + \mu) \qquad (4.37)$$
$$-\mu e_2 = i_1 R_k(1 + \mu) + i_2[r_a + R_L + R_k(1 + \mu)] \qquad (4.38)$$

The solution requires an expression for the voltage between the two anodes. By inspection this is given by

$$v_0 = i_1 R_L - i_2 R_L = R_L(i_1 - i_2)$$

Thus the equations above must be solved for i_1 and i_2. If substitution methods were used, very unwieldy expressions would appear. Solution is best achieved by the use of determinants.

$$i_1 = \frac{-\mu e_1[r_a + R_L + R_k(1 + \mu)] + \mu e_2 R_k(1 + \mu)}{[r_a + R_L + R_k(1 + \mu)]^2 - R_k^2(1 + \mu)^2} \qquad (4.39)$$

and
$$i_2 = \frac{-\mu e_2[r_a + R_L + R_k(1 + \mu)] + \mu e_1 R_k(1 + \mu)}{[r_a + R_L + R_k(1 + \mu)]^2 - R_k^2(1 + \mu)^2} \qquad (4.40)$$

These two expressions have the same denominator, and if the first term of this is expanded, a term $+R_k^2(1 + \mu)^2$ will appear. On collecting terms, this, and the second term in the denominator will cancel.

Therefore

$$i_1 - i_2 = \mu \frac{[r_a + R_L + R_k(1 + \mu)](e_2 - e_1) + \mu R_k(e_2 - e_1)(1 + \mu)}{(r_a + R_L)^2 + 2(r_a + R_L)R_k(1 + \mu)}$$

rearranging numerator and denominator

$$i_1 - i_2 = \frac{\mu(e_2 - e_1)[r_a + R_L + 2R_k(1 + \mu)]}{(r_a + R_L)[r_a + R_L + 2R_k(1 + \mu)]}$$

Thus
$$v_0 = (i_1 - i_2)R_L = \frac{-\mu(e_1 - e_2)R_L}{r_a + R_L} \tag{4.41}$$

From this result, it can be seen that this circuit behaves as a single stage amplifier with an input of $(e_1 - e_2)$ volts, and a load R_L.

This problem would have been much less cumbersome if it had been solved numerically. The student should try this for himself using typical values for the components and valve constants.

CIRCUITS CONTAINING REACTANCE

All the circuits discussed so far in this chapter have been non-frequency-conscious. In practice most circuits contain those reactive components necessary for interstage coupling, those for minimizing negative feedback, and stray reactances due to wiring etc. Usually the effect of these can be neglected over certain frequency ranges, but over other ranges both gain and phase shift will be modified. Initially the full equivalent circuit should be drawn; then at low frequencies shunt or parallel capacitors can be ignored, and at high frequencies series capacitors can be ignored. Similarly, at low frequencies, series inductors may be neglected, and at high frequencies shunt inductors are ignored. The question of whether a frequency is low, high or medium, depends on the magnitude of the particular reactance relative to its series or parallel resistive component.

Two examples involving reactive components will be given. The first demonstrates the effect of the bias decoupling capacitor. Detailed working will show how the decoupling capacitor may be regarded as a short-circuit at all frequencies above a certain level. The second example will give a detailed analysis of resistance capacity inter-stage coupling.

Effect of the Bias Decoupling Capacitor

Example 4.6. A triode valve having μ 39 and r_a 10 kΩ is loaded with R_L 20 kΩ. The cathode bias circuit consists of a 1 kΩ resistor in parallel with a 1 μF capacitor. Calculate the voltage gain at 159 Hz, 1 590 Hz and 15·9 kHz. Sketch the gain and phase response over the frequency range 100 Hz to 20 kHz.

Figure 4.8a shows the complete circuit; R_g is necessary to provide a d.c. connection between grid and earth. The equivalent circuit in

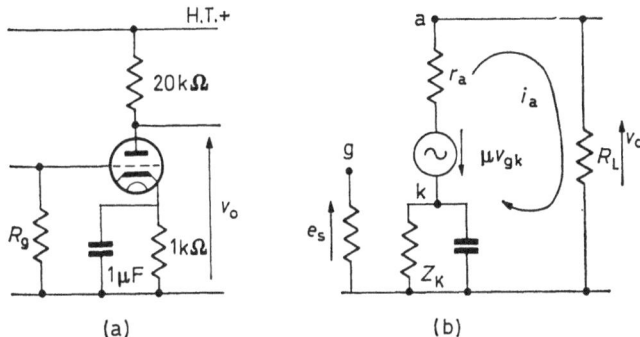

(a)   (b)

Figure 4.8. Circuits for Example 4.6

Figure 4.8b shows the bias circuit as Z_k, where Z_k is the parallel combination of 1 μF and 1 kΩ. This circuit is identical to that obtained in Example 4.1 except R_k has been replaced by Z_k. We can therefore use the result obtained in Example 4.1, equation 4.7, i.e.

$$A_v = \frac{-\mu R_L}{r_a + R_L + Z_k(1 + \mu)} \qquad (4.42)$$

At 159 Hz, $\qquad\qquad X_c = \dfrac{1}{\omega C} = 1$ kΩ

So, working in kΩ:

$$Z_k = \frac{-j \times 1}{1 - j} = \frac{-j(1 + j)}{2} = \frac{1 - j}{2}$$

Therefore from equation 4.42

$$A_v = \frac{-39 \times 20}{10 + 20 + \frac{40}{2}(1 - j)} = \frac{-39 \times 2}{5 - j2} = \frac{-78(5 + j2)}{25 + 4}$$

$$= -2·69(5 + j2) = -13·45 - j5·38$$

$$= \sqrt{(13·45^2 + 5·38^2)}\Big/ \tan^{-1}\frac{5·38}{13·45} \text{ in the third quadrant}$$

$$\underline{A_v = 14·53 \angle 202°}$$

At 1 590 Hz

$$X_c = 100 \ \Omega$$

Therefore

$$Z_k = \frac{-j0\cdot1}{1 - j0\cdot1} = \frac{-j0\cdot1(1 + j0\cdot1)}{1\cdot01} \simeq 0\cdot01 - j0\cdot1$$

and

$$A_v = \frac{-39 \times 20}{10 + 20 + 40(0\cdot01 + j0\cdot1)} = \frac{-78}{3\cdot04 - j0\cdot4}$$

$$= \frac{-78(3\cdot04 + j0\cdot4)}{9\cdot25 + 0\cdot16} = -8\cdot3(3\cdot04 + j0\cdot4)$$

$$= \underline{-25\cdot2 - j3\cdot3} = \underline{25\cdot4 \ \angle\ 187° \ 36'}$$

It can be seen that as frequency increases, the voltage gain tends rapidly towards a value where $Z_k(1 + \mu) \ll r_a + R_L$, and

$$A_v = \frac{-39 \times 20}{20 + 10} = 26 \ \angle 180°$$

Thus for this circuit, low frequencies would be less than say 2 kHz and high frequencies those above 2 kHz. In general the effect of the bias circuit may be neglected if $X_c < R_k/10$.

To sketch the required frequency response curves, the magnitude of A_v and the phase shift are best displayed when plotted to a base of log frequency as shown in *Figure 4.9a* and *b*.

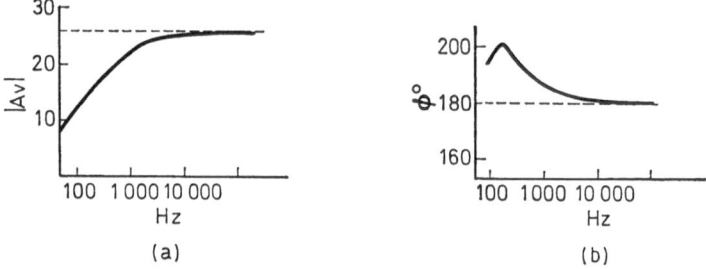

Figure 4.9. Gain and phase responses for circuits in Figure 4.8

In this problem, the reactive component is effectively multiplied by $(1 + \mu)$ due to the valve action. In coupling circuits this does not necessarily apply and in practice a capacitor may be neglected when $X_c > 5R$ for parallel combinations $(Z \simeq R)$, and to $X_c < R/5$ for series combinations $(Z \simeq R)$.

THE RESISTANCE CAPACITANCE COUPLED AMPLIFIER

Example 4.7. Derive general expressions for the voltage gain of an audio frequency amplifier using resistance capacity coupling to

126

a second stage. Sketch the gain and phase responses and calculate the 3 db frequencies, given that: $\mu = 20$, $r_a = 10$ kΩ, $R_L = 20$ kΩ, $R_g = 50$ kΩ, $C_c = 0{\cdot}1$ μF, and $C_s = 100$ pF.

The complete circuit is shown in *Figure 4.10a*. If the output voltage v_0 is to be applied to the grid of a second valve, the high positive voltage at the anode of the first valve must be blocked or isolated. C_c the coupling capacitor is included for this purpose. The second valve will normally have cathode bias and the grid must have a d.c. connection to earth. This is provided by R_g across which

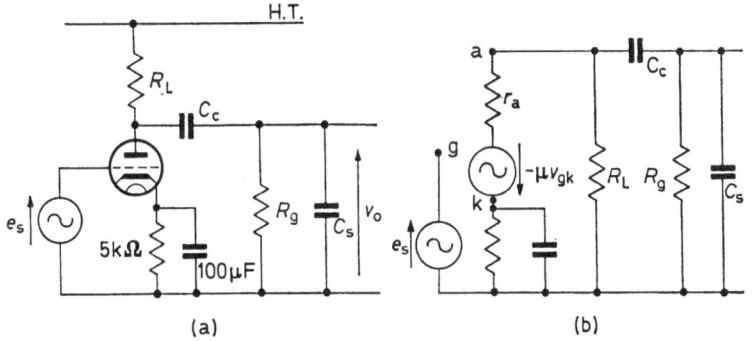

Figure 4.10. Full and equivalent circuits for valve amplifier with RC coupling network

the a.c. output voltage will be developed. C_s represents the stray capacitance which is principally caused by the inter-electrode capacitance of the second valve, but capacitance between connecting wires and earth is included in this component. The effects of inter-electrode capacitance will be considered in Chapter 9.

The full equivalent circuit shown in *Figure 4.10b* could be solved by mesh analysis, but the expressions obtained would be cumbersome and difficult to use. The first step then, is to simplify this circuit by examining the component values over certain frequency ranges. If the factor of 10 used in Example 4.5, is applied, certain components may be neglected as either open-circuit or short-circuit.

(1) The cathode bias circuit. If

$$X_c = \frac{R_k}{10} \quad \text{then} \quad \frac{10}{\omega C} = 5\,000\,\Omega$$

$$\omega = \frac{1}{500C} = \frac{10^6}{50\,000} = 20\,\text{rads/sec}$$

and $f = 3{\cdot}18$ Hz.

As shown in Example 4.5, for all frequencies greater than this the cathode bias circuit has negligible effect on the a.c. amplification.

(2) The coupling capacitor C_c is effectively in series with R_g the grid resistor. Therefore if X_{Cc} is less than one tenth of R_g it can be regarded as a short-circuit.

$$\frac{1}{\omega C_c} = \frac{R_g}{10}$$

$$\omega = \frac{10}{R_g C_c} = \frac{10^7}{50 \times 10^3 \times 0 \cdot 1} = 2\ 000\ \text{rad/sec}$$

Therefore $f = 318$ Hz

Now X_c is inversely proportional to frequency, so the coupling capacitor can be neglected at all frequencies above 318 Hz.

(3) The stray capacitance C_s is effectively in parallel with R_g. If the reactance is greater than $10R_g$, it may be regarded as open circuit. If

$$\frac{1}{\omega C_s} = 10R_g$$

$$\omega = \frac{1}{10R_g C_s} = \frac{10^{12}}{5 \times 10^5 \times 100} = 2 \times 10^4\ \text{rad/sec}$$

And $f = 3\ 180$ Hz

But capacitive reactance increases as frequency is reduced. C_s can therefore be regarded as open circuit for all frequencies less than 3 180 Hz.

If these results are examined, it can be seen that for frequencies in the range 318 Hz to 3 180 Hz, both capacitors and the bias circuit can be neglected. This range will be known as the medium frequency range and the voltage gain in this range as A_{vm}. At frequencies greater than 3 180 Hz only C_s need be included. This will be the high frequency range with a voltage gain of A_{vh}. The range 3·18 Hz to 318 Hz is the low frequency range with gain A_{vl}. At these low frequencies only C_c need be considered. The cathode bias circuit is an effective short circuit to a.c. at all frequencies above 3·18 Hz and since this is an audio frequency amplifier this applies to the whole range.

We can now draw equivalent circuits for medium frequencies, low frequencies, and high frequencies. The coupling circuit is in parallel with the valve which suggests the parallel or constant current form of equivalent circuit.

128

The required medium frequency equivalent circuit is shown in *Figure 4.11*. The three resistors in parallel may be combined to form one resistor R_e.

$$\frac{1}{R_e} = \frac{1}{r_a} + \frac{1}{R_L} + \frac{1}{R_g} \tag{4.43}$$

$$v_{gk} = e_s, \quad \text{and} \quad v_0 = -g_m v_{gk} R_e = -g_m e_s R_e$$

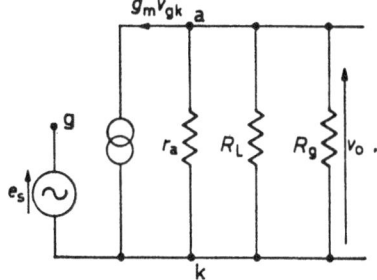

Figure 4.11. Equivalent circuit for RC coupled amplifier at medium frequencies

Therefore medium frequency gain

$$A_{vm} = -g_m R_e \tag{4.44}$$

Inserting numerical values in equations 4.43 and 4.44:

$$\frac{1}{R_e} = \frac{1}{10} + \frac{1}{20} + \frac{1}{50} = \frac{10 + 5 + 2}{100} \tag{4.45}$$

giving $\qquad R_e = 5 \cdot 89 \text{ k}\Omega$

$$g_m = \frac{\mu}{r_a} = \frac{20}{10} = 2 \text{ mA/V}$$

Therefore $\qquad A_{vm} = -2 \times 5 \cdot 89 = -11 \cdot 78$

In *Figure 4.12*, the high frequency equivalent circuit is shown with the three resistors combined as R_e. If the parallel combination

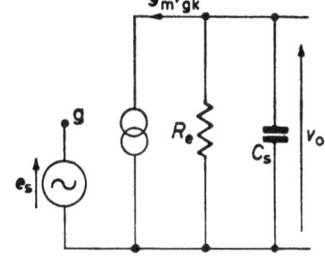

Figure 4.12. High frequency equivalent circuit for RC coupled amplifier

of R_e and C_s is denoted as Z_e, the high frequency gain A_{vh} is given by:

$$A_{vh} = -g_m Z_e \qquad (4.46)$$

$$= \frac{-g_m(-jX_{cs}R_e)}{R_e - jX_{cs}}$$

Dividing numerator and denominator by $-jX_{cs}$,

$$A_{vh} = \frac{-g_m R_e}{1 - \dfrac{R_e}{jX_{cs}}} = \frac{A_{vm}}{1 + \dfrac{jR_e}{X_{cs}}} \qquad (4.47)$$

Inserting values for X_{cs} at various frequencies would lead to the gain and phase shift at these frequencies. This approach is somewhat laborious and it is much easier to select values for R_e/X_{cs} such as 1, 0·2, 0·5, 2, and 5. From these values $1 + jR_e/X_{cs}$ can be readily calculated giving the modified gain and phase shift. The frequencies at which these values occur can then be found and the required frequency responses plotted. For example, let $R_e/X_{cs} = 1$, then

$$A_{vh} = \frac{A_{vm}}{1 + j} = \frac{A_{vm}}{\sqrt{2}\ \angle 45°} = \frac{11 \cdot 78\ \angle 180°}{\sqrt{2}\ \angle 45°} = 8 \cdot 33\ \angle 135°$$

This is known as the upper 3 db point since a voltage reduction by a factor of $1/\sqrt{2}$ is a reduction of 3 db in the logarithmic scale. (See Appendix 1.) The upper 3 db frequency occurs then when $R_e/X_{cs} = 1$, thus

$$R_e = X_{cs} = \frac{1}{\omega C_s} \quad \text{and} \quad \omega = \frac{1}{R_e C_s} = \frac{10^{12}}{100 \times 10^3 \times 5 \cdot 89}$$

But 5·89 was arrived at from $\dfrac{100}{17}$ (equation 4.45).

Therefore the frequency f is given by:

$$f = \frac{10^{12} \times 10^{-3} \times 17}{100 \times 10^3 \times 100 \times 2\pi} = \frac{1\ 700}{2\pi}\ \text{kHz}$$

$$f = 271\ \text{kHz}$$

This may be sufficient to complete the response curve, but if more accuracy is required, further values of R_e/X_{cs} may be taken, leading to the result shown in *Table 4.1*.

Table 4.1

$\dfrac{R_e}{X_{cs}}$	A_{vh}	f(kHz)
0·2	$11·53 \angle 169°$	54·2
0·5	$10·5 \angle 153° \, 30'$	135·5
1·0	$8·33 \angle 135°$	271
2·0	$5·25 \angle 116° \, 30'$	542
5·0	$2·3 \angle 101° \, 20'$	1 355

To obtain these points, the value of R_e/X_{cs} was changed. The frequencies are obtained by using the value for the 3 db frequency calculated above and then multiplying or dividing by 2 and 5. The voltage gains are obtained by dividing A_{vm} by $\sqrt{(1^2 + 0·2^2)}$, $\sqrt{(1^2 + 0·5^2)}$, $\sqrt{(1^2 + 2^2)}$, and $\sqrt{(1^2 + 5^2)}$. The new phase angle is obtained by subtracting $\tan^{-1} 0·2$, $\tan^{-1} 0·5$, $\tan^{-1} 2$, and $\tan^{-1} 5$ from 180°.

To obtain the low frequency equivalent circuit, C_s becomes open circuit but C_c must be included. This is shown in *Figure 4.13*.

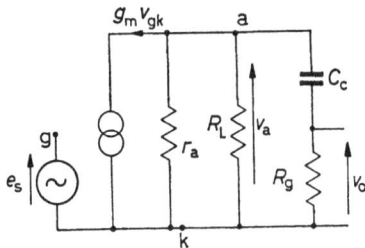

Figure 4.13. Low frequency equivalent circuit for RC coupled amplifier

First an expression for v_0 is obtained in terms of the a.c. anode voltage v_a.

$$v_0 = \frac{R_g v_a}{R_g - jX_{cc}} \quad \text{where} \quad \frac{1}{\omega C_c} = X_{cc} \qquad (4.48)$$

131

Now v_a is given by the product of the current $-g_m v_{gk}$ and the total impedance presented by r_a, R_L, and R_g. Therefore

$$v_0 = \frac{R_g}{R_g - jX_{cc}} \times \frac{-g_m v_{gk}\left(\dfrac{r_a R_L}{r_a + R_L}\right)(R_g - jX_{cc})}{\left(\dfrac{r_a R_L}{r_a + R_L}\right) + R_g - jX_{cc}} \quad (4.49)$$

The term $(R_g - jX_{cc})$ cancels and $v_{gk} = e_s$, and

$$A_{v1} = \frac{v_0}{e_s} = \frac{-g_m R_g\left(\dfrac{r_a R_L}{r_a + R_L}\right)}{\left(\dfrac{r_a R_L}{r_a + R_L}\right) + R_g - jX_{cc}} \quad (4.50)$$

Multiplying equation 4.50 by $(r_a + R_L)$

$$A_{v1} = \frac{-g_m R_g r_a R_L}{r_a R_L + r_a R_g + R_L R_g - jX_{cc}(r_a + R_L)} \quad (4.51)$$

The expression $\dfrac{R_g r_a R_L}{r_a R_L + r_a R_g + R_L R_g}$ is the parallel combination of r_a, R_L, and R_g; i.e. it is R_e. So if we divide the numerator and the denominator by $r_a R_L + r_a R_g + R_g R_L$ we obtain

$$A_{v1} = \frac{-g_m R_e}{1 - \dfrac{jX_{cc}(r_a + R_L)}{r_a R_L + r_a R_g + R_L R_g}}$$

But $\quad \dfrac{r_a R_L + r_a R_g + R_L R_g}{r_a + R_L} = R_g + \dfrac{r_a R_L}{r_a + R_L}$

Let $\quad R_g + \dfrac{r_a R_L}{r_a + R_L} = R' \quad (4.52)$

and since $-g_m R_e = A_{vm}$, from equation 4.44

$$A_{v1} = \frac{A_{vm}}{1 - \dfrac{jX_{cc}}{R'}} \quad (4.53)$$

This result can now be used in a similar manner to that applicable to the high frequency range. Values for X_{cc}/R' are chosen for simplicity of calculation and the appropriate frequencies subsequently determined.

As before $A_{vm} = 11.78 \angle 180°$, and

$$R' = \frac{r_a R_L}{r_a + R_L} + R_g$$

$$= \frac{200}{30} + 50 = 56.7 \text{ k}\Omega$$

$\dfrac{X_{cc}}{R'}$	A_{vl}	$f(\text{Hz})$
0·2	$11.53 \angle 191°$	140·5
0·5	$10.5 \angle 206° 30'$	56·2
1	$8.33 \angle 225°$	28·1
2	$5.25 \angle 243° 30'$	14
5	$2.3 \angle 258° 40'$	5·6

The values for X_{cc}/R' are the same as those chosen for R_e/X_{cs} in the high frequency case. The corresponding values for voltage gain will therefore be the same. The phase shift in this case must be greater than 180° since the imaginary term in the denominator is negative. This new phase is obtained by adding 180° to $\tan^{-1} 0.2$, $\tan^{-1} 0.5$ etc.

To calculate the frequencies, first consider the point where

$$\frac{X_{cc}}{R'} = 1$$

Therefore $\qquad R' = X_{cc} = \dfrac{1}{\omega C_c}$

and $\qquad \omega = \dfrac{1}{C_c R'} \text{ rad/sec}$

giving $\qquad f = \dfrac{1}{2\pi C_c R'} = \dfrac{10^6}{2\pi \times 0.1 \times 56.7 \times 10^3} \text{ Hz}$

$\qquad\qquad = 28.1 \text{ Hz}$

When the value for X_{cc}/R' is multiplied by a constant, the frequency must be divided by the same constant.

For example: put $X_{cc}/(R') = 2$, now

$$\frac{1}{\omega C_c} = 2R' \quad \text{and} \quad \omega = \frac{1}{2R_g C_c}$$

133

Thus the frequency must be divided by 2.

By definition, the 3 db frequencies are those at which the voltage gain falls to $1/\sqrt{2}$ of the minimum value.

From the tables:

$$\text{Upper 3 db frequency} = 271 \text{ kHz}$$
$$\text{Lower 3 db frequency} = 28\cdot1 \text{ Hz}$$

These give a measure of the useful frequency range or bandwidth of the amplifier. The gain and phase responses are given in *Figure 4.14* plotted in each case against frequency on a logarithmic scale.

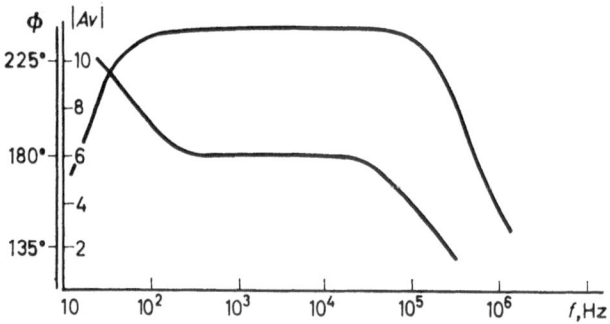

Figure 4.14. Gain and phase response curves for the RC coupled amplifier discussed in Example 4.7

This chapter has shown how the valve equivalent circuits may be used to solve a variety of simple amplifier circuits. Once the technique has been mastered, the reader will find that quite complicated circuits may quickly be reduced to a set of equations. Examples on more involved circuits will be found in later chapters.

EXAMPLES

Example 4.8. A triode valve has μ 40 and r_a 20 kΩ. Calculate the voltage gain when it is connected as a simple common cathode amplifier and the anode load is (*a*) 10 kΩ, (*b*) 20 kΩ and (*c*) 100 kΩ.

Ans. $-13\cdot3$, -20, $-33\cdot3$.

Example 4.9. An alternating voltage of 50 mV r.m.s. is applied between grid and cathode of a triode valve whose parameters are

134

g_m 3 mA/V, r_a 10 kΩ. Calculate the anode-cathode voltage when the anode is loaded with (a) 10 kΩ and (b) 50 kΩ.
Ans. 0·75 V, 1·25 V.

Example 4.10. A single stage amplifier is required to produce 20 V signal in a load of 33 kΩ or 15 V signal in a load of 20 kΩ. If the input signal is 0·8 V, determine the constants μ and r_a of a suitable valve.
Ans. 51·0, 34·3 kΩ.

Example 4.11. A triode valve employs cathode bias without decoupling to provide the desired operating point of V_{AK} 110 V, I_A 2 mA, and V_{GK} −3 V. If the anode load is 20 kΩ and the valve parameters μ and r_a are 85 and 12 kΩ respectively, calculate the voltage gain and the overall output impedance of the amplifier.
Ans. −10·5, 17·5 kΩ.

Example 4.12. A simple cathode follower (*Figure 4.2a*) operates with a cathode load of 2 kΩ. If μ and r_a are 25 and 10 kΩ respectively, calculate the voltage gain and the output impedance of the circuit. Hence find the voltage across a 500 Ω resistor, capacity coupled to the load. The capacitive reactance may be neglected and a signal of one volt is applied between the valve grid and earth.
Ans. 0·808, 323 Ω, 0·49 V.

Example 4.13. A grounded grid amplifier (*Figure 4.3a*) is driven by a source of e.m.f. 0·3 V having internal impedance 600 Ω. If the anode load is 27 kΩ, g_m 6 mA/V and r_a 10 kΩ, find the equivalent Thèvenin generator 'seen' between anode and earth.
Ans. 6·72 V, 17·1 kΩ.

Example 4.14. A modified cathode follower as shown in *Figure 4.6a* has a cathode load of 2 kΩ. Determine the value of grid leak resistor which will result in an input impedance of 5 MΩ. Find also the voltage gain and output impedance assuming the source impedance to be negligible. Take μ and r_a as 100 and 8 kΩ respectively.
Ans. 455 kΩ, 0·95, 80 Ω.

Example 4.15. Calculate the output impedance of the circuit described in Example 4.14 when the source impedance is 100 kΩ.
Ans. 97 Ω.

Example 4.16. The circuit shown in *Figure 4.15* is that of a long tailed pair connected as a difference amplifier. Assuming that the valves are identical with μ 40 and r_a 10 kΩ, calculate v_0 (*a*) if

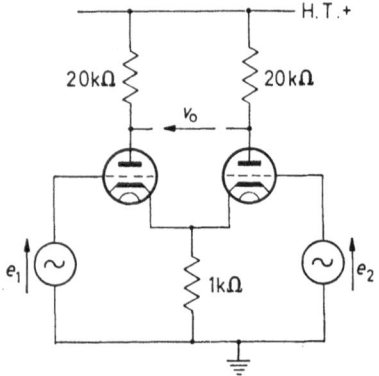

Figure 4.15. Circuit for Example 4.16

$e_1 = 1 \sin \omega t$ and $e_2 = 0.9 \sin \omega t$, or (*b*) if $e_1 = 0.5 \sin \omega t$ and $e_2 = 0.4 \sin (\omega t + \pi/4)$.

Ans. $1.01 \sin \omega t$, $4.3 \sin (\omega t + 41° 48')$.

Example 4.17. The d.c. amplifier circuit shown in *Figure 4.16* can be considered as a cathode follower and a grounded grid amplifier connected in cascade. Determine the voltage gain using this

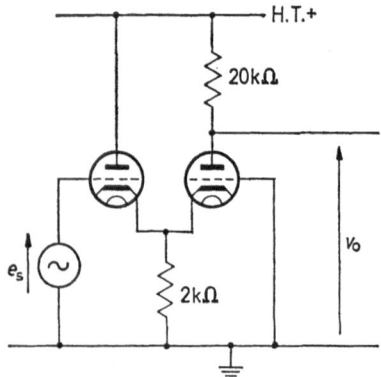

Figure 4.16. Circuit for Example 4.17

approach and check the result by solving the complete equivalent circuit. Take μ and r_a as 60 and 10 kΩ respectively, for each valve.

Ans. 28.3.

Example 4.18. The direct coupled amplifier shown in *Figure 4.17* employs identical valves with μ 36 and r_a 12 kΩ. Draw the complete equivalent circuit and calculate the small signal voltage gain v_0/e_s.

Ans. 35·6.

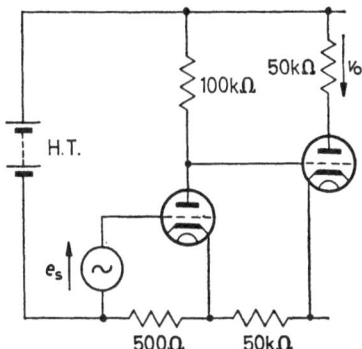

Figure 4.17. Circuit for Example 4.18

Example 4.19. If μ and r_a are 50 and 10 kΩ respectively, calculate the voltage gain and input impedance of the amplifier shown in *Figure 4.18*.

Ans. −9·62, 4·88 kΩ.

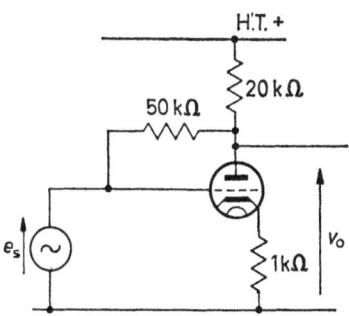

Figure 4.18. Circuit for Example 4.19

Example 4.20. A grounded cathode amplifier employs decoupled cathode bias using 2 kΩ in parallel with 0·1 μF for this purpose. The valve has μ 40 and r_a 15 kΩ and is loaded with 47 kΩ in series with the anode. Estimate the frequency range over which the voltage gain is frequency conscious and calculate the gain above and below this range.

Ans. 50 Hz to 10 kHz, −13, −30·4.

137

Example 4.21. Calculate the gain and phase shift of the amplifier shown in *Figure 4.19* if the signal frequency is 5 kHz. Take μ and r_a as 50 and 8 kΩ respectively.

Ans. 12·1 $\angle$ 241° 42'.

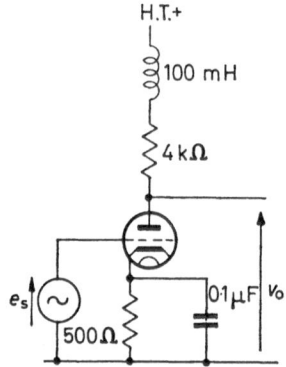

H.T.+

100 mH

4 kΩ

e_s

500 Ω

0·1 μF V_o

Figure 4.19. Circuit for Example 4.21

Example 4.22. An RC coupled amplifier having the circuit shown in *Figure 4.10a* has the following components; R_L 30 kΩ, R_g 100 kΩ, C_c 0·25 μF, C_s 330 pF. The valve parameters are μ 90 and r_a 30 kΩ and it may be assumed that R_k is effectively decoupled at all signal frequencies. Draw complete gain and phase frequency response curves and hence find the maximum gain and 3 db frequencies.

Ans. −39·15, 37 kHz, 5·5 Hz.

Example 4.23. An RC coupled amplifier is required to have a useful frequency range from 60 Hz to 500 kHz. The total shunt capacitance is 50 pF and the grid resistor for the next stage is 330 kΩ. Taking g_m as 5 mA/V and r_a as 20 kΩ calculate suitable values for R_L and C_c and the resulting maximum gain.

Ans. 9·6 kΩ, 0·01 μF, −31·9.

Example 4.24. An AF amplifier having three identical stages is required to have an overall gain of 1 000 and a 3 db bandwidth of 100 kHz. If pentodes having r_a 1 MΩ are used and the shunt capacitance per stage is 30 pF, calculate the minimum g_m for the valves and the common value of R_L. The reactance of the coupling capacitors can be neglected and the grid leak resistors will be much greater than the required R_L.

Ans. 0·37 mA/V, 27 kΩ.

138

Example 4.25. An RC coupled amplifier having two identical stages employs valves with r_a 40 kΩ. The coupling components are R_L 50 kΩ, C_c 0·1 μF, C_s 100 pF and R_g 120 kΩ. Calculate the frequency range over which the overall phase shift is less than $\pm10°$.

Ans. 128 Hz to 7·5 kHz.

5

USE OF TRANSISTOR EQUIVALENT CIRCUITS

In Chapter 3 small signal equivalent circuits were developed for valves and transistors. Both the hybrid or h parameter equivalent circuit and the equivalent T were shown to be suitable for representing the transistor. It must be stressed that the component values for such circuits will only apply at a particular d.c. operating point and that there may be considerable variation in these values for transistors of nominally the same type. It is appreciated that some readers will be principally interested in transistor circuits so the important ideas and techniques used for valve circuits in Chapter 4 will be repeated in this chapter where they apply.

Any form of equivalent circuit may be used for any problem, but the solution will be found more quickly if the appropriate circuit is used, i.e. if the transistor is to be used in the common base configuration, the best hybrid parameters to use are h_{1b}, h_{rb}, h_{fb}, and h_{ob}. Similarly the simple T equivalent involving only r_e, r_c, r_b and α is suitable for this circuit. If however, the available information includes the common emitter hybrid parameters, or even, for example, the components of the common collector equivalent T, there is no need to convert to the required form. The equivalent circuit method to be outlined, will, if correctly applied, produce the correct solution in all cases.

When the correct h parameters are available, the general solutions discussed in Chapter 2 may often be used with considerable time saving.

GENERAL METHODS

(1) Select the form of equivalent circuit to be used. This will usually depend upon the available information. Experience will show whether conversion is worthwhile.

(2) For each transistor in the full circuit, draw the chosen equivalent. Show all details, particularly the sense or direction of all voltages and currents, applied or generated, and mark clearly the points representing the emitter, base and collector connections.

(3) Taking each electrode in turn, connect it to earth through any components or generators shown in the complete circuit. Remember that batteries or other d.c. power supplies are short circuit to a.c. and that reactive components such as capacitors can sometimes be neglected at the applicable signal frequency.

(4) Indicate the sense of unknown mesh currents or node voltages. With hybrid parameters these unknowns will normally be I_1 and V_2, but with T equivalents, the two mesh currents are usually more convenient. In any case, if an equivalent circuit generator is expressed in terms of a circuit variable, ($h_{te} i_b$, $h_{re} V_{ce}$, etc.) the unknown mesh current or node voltage should correspond to the direction indicated on the equivalent circuit.

(5) Express any equivalent generator currents or voltages directly in terms of the unknown currents and voltages, or as iZ or vY products together with external generators appearing in the complete circuit.

(6) Write the circuit mesh or nodal equations and solve, using substitution methods or determinants.

These methods will normally lead directly to the calculation of voltage and current gain, and of the input impedance. If the output impedance is required the method outlined in Example 4.3 must be used. When the h parameters are given, the general solution derived in Chapter 2 may be used.

We shall now consider a simple example showing the complete application of the general method.

Example 5.1. The circuit shown in *Figure 5.1a* shows a single stage amplifier, the transistor having h parameters h_{ie} 900 Ω, h_{re} 5×10^{-4}, h_{fe} 90 and h_{oe} 125 μmho. Components R_1, R_2 and R_3 provide the stabilized d.c. operating point as discussed in Chapter 1. The output voltage is obtained across R_L and the amplifier is driven by a source e_s of internal resistance 500 Ω. Find the terminal voltage and current gain and the output voltage and current. Assume the reactance of the capacitor shown to be zero at the signal frequency.

To draw the correct equivalent circuit we must apply steps 1 to 3 in the general method. *Figure 5.1b* shows the h parameter equivalent circuit. Note that all details must be included if errors are to be avoided. *Figures 5.1c, d* and *e* show the connections of the emitter, collector and base respectively, to earth. Note that the negative supply line is at earth potential to a.c. and the coupling capacitor, having zero reactance, is shown as a short circuit. Note also that the h_{re} generator is expressed in terms of the voltage at the collector

141

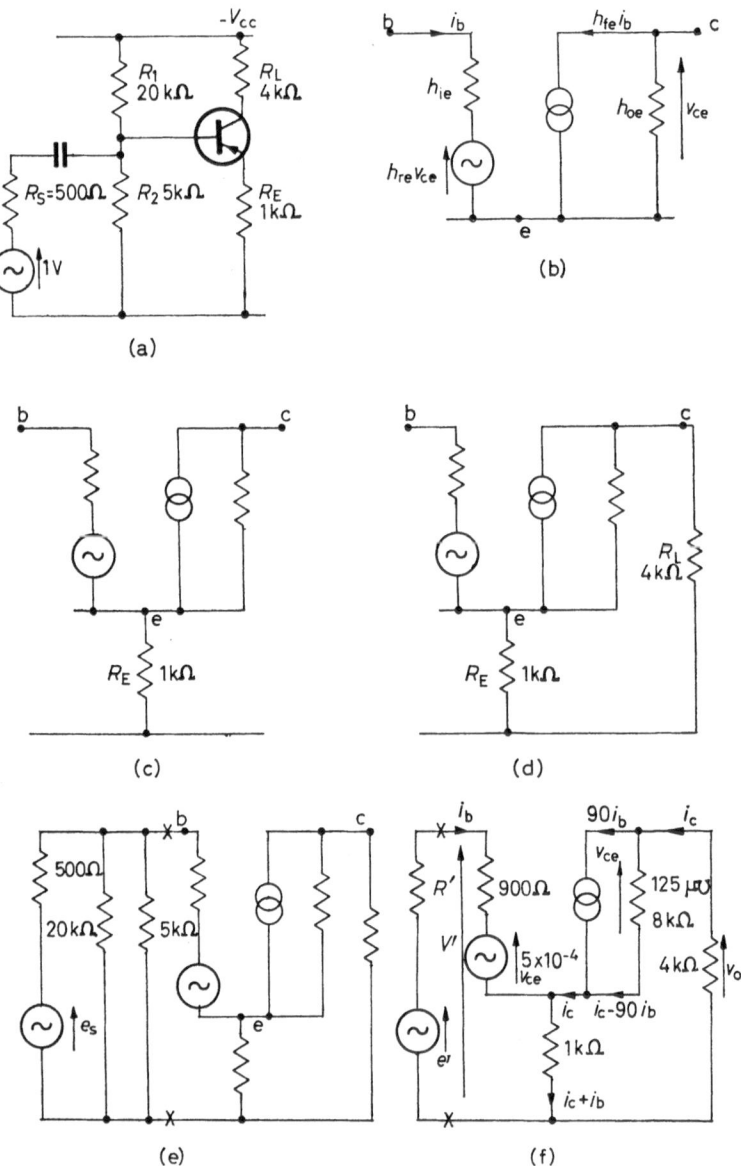

Figure 5.1. (*a*) Circuit for Example 5.1. (*b*) to (*f*) Stages in the equivalent circuit solution for the circuit in *Figure 5.1a*

142

with respect to emitter, not with respect to earth. The h_{fe} generator is expressed in terms of i_{b} flowing into the base.

Before continuing the solution, it is convenient to simplify the input section of the circuit. This includes the bias components R_1 and R_2 (in parallel to a.c.) and the signal source. This is achieved by the application of Thèvenin's theorem at points XX looking away from the transistor.

Putting R_1 and R_2 in parallel we obtain:

$$R_{\text{B}} = \frac{20 \times 5}{20 + 5} = 4 \text{ k}\Omega$$

Now applying Thèvenin's theorem,

$$R' = \frac{R_{\text{B}}R_{\text{S}}}{R_{\text{B}} + R_{\text{S}}} = \frac{0 \cdot 5 \times 4}{4 \cdot 5} \text{ k}\Omega$$

$$= 0 \cdot 445 \text{ k}\Omega$$

and

$$e' = \frac{e_{\text{s}}R_{\text{B}}}{R_{\text{B}} + R_{\text{S}}} = \frac{1 \times 4}{4 \cdot 5} \text{ V}$$

$$= 0 \cdot 89 \text{ V}$$

Figure 5.1f shows the complete equivalent circuit including this simplification. The unknown currents chosen for mesh analysis are i_{b} and i_{c}. Applying Kirchhoff's current law, we find the current in the emitter load is $(i_{\text{b}} + i_{\text{c}})$ and the current in the h_{oe} branch is $(i_{\text{c}} - 90i_{\text{b}})$. For mesh analysis the h_{oe} branch is expressed as an impedance from

$$Z = \frac{1}{Y} = \frac{1}{125 \times 10^{-6}} = 8 \text{ k}\Omega$$

For paragraph 5 of the general method, we must examine the equivalent circuit generators and express them in terms of our unknown currents or voltages. The h_{fe} current generator is already expressed in terms of i_{b} and the rule is satisfied. The h_{re} generator is expressed in terms of v_{ce}: this may be expressed in terms of i_{b} and i_{c} in one of three alternative ways. These may be found by proceeding from the emitter to the collector around the circuit by any route, and by adding any iZ products and generated e.m.f.s. If proceeding against the current arrow, take the product as negative, and if proceeding with the current arrow, take the product as positive. The three alternatives are:

(a) $v_{ce} = 8\,000(i_c - 90i_b)V$, or

(b) $v_{ce} = [-1\,000(i_c + i_b) - 4\,000i_c]V$, or

(c) $v_{ce} = [5 \times 10^{-4}v_{ce} + 900i_b + R'i_b - e' - 4\,000i_c]V$.

Each of these results would lead to the correct solution, but the shortest path is usually the most convenient. Taking (a) above and writing the mesh equations in the normal manner:

$$e' - 5 \times 10^{-4} \times 8\,000(i_c - 90i_b) = (900 + R')i_b + 1\,000(i_b + i_c) \tag{5.1}$$

$$0 = 4\,000i_c + 8\,000(i_c - 90i_b) + 1\,000(i_c + i_b) \tag{5.2}$$

Rearranging these equations, collecting terms and inserting the value of R',

$$e' = 4i_c - 360i_b + 1\,345i_b + 1\,000i_b + 1\,000i_c$$
$$= 1\,985i_b + 1\,004i_c \tag{5.3}$$

and $\quad 0 = 4\,000i_c + 8\,000i_c - 72 \times 10^4i_b + 1\,000i_c + 1\,000i_b \tag{5.3}$

$$= 13 \times 10^3i_c - 71 \cdot 9 \times 10^4i_b \tag{5.4}$$

From 5.4 $\qquad i_b = \dfrac{13 \times 10^3}{71 \cdot 9 \times 10^4} i_c = 1 \cdot 81 \times 10^{-2}i_c$

Substituting in 5.3,

$$e' = 19 \cdot 85 \times 1 \cdot 81i_c + 1\,004i_c$$

and $\qquad\qquad i_c = \dfrac{e'}{1\,040}$ A

$\therefore \qquad\qquad i_b = \dfrac{1 \cdot 81 \times 10^{-5}e'}{1 \cdot 04} = 1 \cdot 74 \times 10^{-5}e'$ A

The output current is normally expressed as flowing towards earth. In this case,

$$i_0 = -i_c = \frac{-e'}{1\,040}\ A\ .$$

Inserting the value found for e' and expressing the answer in μA

$$i_c = \frac{-0 \cdot 89 \times 10^6}{1\,040}\ \mu A$$

$$= -856\ \mu A$$

(Note the minus, indicating 180° phase shift.)

144

The terminal input current is i_b, and substituting for terminal current gain

$$A_1 = \frac{-i_c}{i_b} = \frac{\dfrac{-e'}{1\,040}}{1\cdot74 \times 10^{-5}e'}$$

$$\therefore \qquad A_1 = -\frac{10^5}{1\,040 \times 1\cdot74} = -55\cdot2$$

The output voltage v_0 is now expressed with respect to earth,

$$\therefore \qquad v_0 = -i_c R_L$$
$$= -856 \times 10^{-5} \times 4 \times 10^3 \text{ V}$$
$$= 3\cdot44 \ \angle 180° \text{ V}$$

To obtain the terminal voltage gain we require the terminal input voltage V' shown in *Figure 5.1f*. Applying Kirchhoff's voltage law:

$$V' = e' - i_b R'$$
$$= 0\cdot89 - 1\cdot74 \times 10^{-5} \times 445$$
$$= 0\cdot89 - 0\cdot00775 \simeq 0\cdot89 \text{ V}$$

$$\therefore \qquad A_v = \frac{v_0}{V'} = 3\cdot87 \ \angle 180°$$

Note that we could also obtain the terminal input impedance from:

$$Z_{in} = \frac{V'}{i_b}$$
$$= \frac{10^{-3}}{1\cdot74 \times 10^{-5}} \text{ k}\Omega$$
$$= 57\cdot5 \text{ k}\Omega$$

This high input impedance is to be expected, since the input voltage must not only drive i_b into the transistor, but must also produce $i_e R_E$ volts across the emitter resistor. Thus with an amplifier having an unbypassed emitter resistor we can say the input impedance will be of the order of $h_{te}R_E$ (since $i_e \simeq h_{te}i_b$).

This analysis may seem very long, but techniques to be introduced in Chapter 7 will reduce the solution of this problem to a few lines. On the other hand this technique, correctly applied, will always lead to the correct solution. Any circuit involving transistors with known h parameters may be solved in this way. If however the common emitter parameters are given and the transistors are connected

145

so that the emitters are earthed to a.c., the h parameter general solutions may be used. The next example showing a two stage amplifier is particularly suitable for this approach.

Example 5.2. Determine the overall voltage gain, current gain and input impedance for the amplifier shown in *Figure 5.2a*. Find also the output impedance assuming a source impedance of 500 Ω.

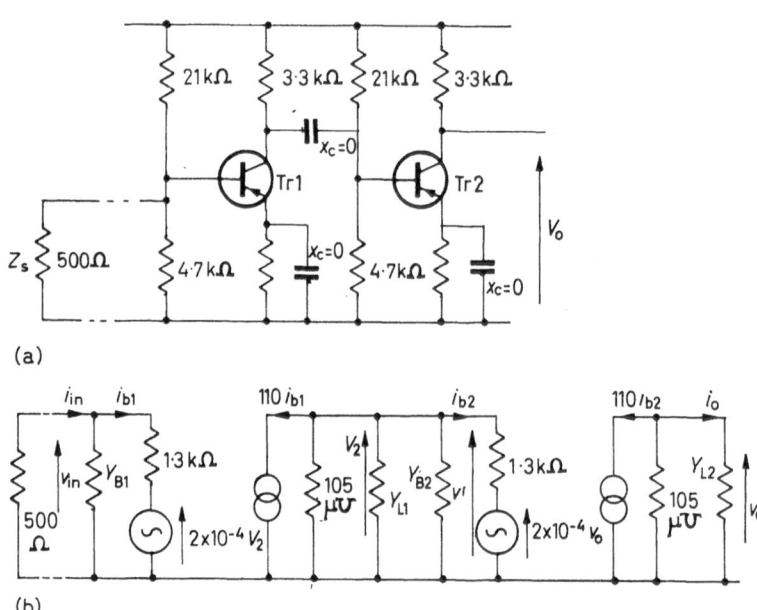

Figure 5.2. Circuit for the two stage amplifier in Example 5.2

The transistors are identical and have the following common emitter parameters, h_{ie} 1·3 kΩ, h_{re} 2 × 10^{-4}, h_{fe} 110, and h_{oe} 105 μmho. Assume all capacitive reactances to be zero at the signal frequency.

Figure 5.2b shows the complete equivalent circuit for the amplifier. The following points should be noted.

(1) Since capacitive reactances are zero, these components have been shown as short circuits. This also eliminates the emitter resistors.

(2) The bias resistors have been combined and expressed as admittances, as have the load resistors.

(3) The required input and output voltages and currents have been indicated v_{in}, v_o, i_{in} and i_o.

146

(4) For convenience the terminal input voltage to Tr2 has been shown as v'.

This circuit could be solved by converting the h_{re} voltage generators to current generators by Norton's theorem and using nodal analysis; a more convenient method for multistage amplifiers is as follows:

(1) Find the input admittance to Tr2.

(2) Calculate the resulting effective load to Tr1.

(3) Find the input admittance to Tr1 and hence the overall input impedance Z_{1n}.

(4) Using the effective load found in 2 find the voltage gain v'/v_{1n} for Tr1.

(5) Determine the voltage gain for Tr2 and hence the overall voltage gain v_0/v_{1n}.

(6) From $i_{1n} = v_{1n}/Z_{1n}$, and $i_0 = v_0 Y_{L2}$ find the current gain i_0/i_{1n}. Alternatively taking current division between parallel admittances and using the input admittances found in 1 and 3 write a single expression for the current gain.

Before applying these steps to the problem in hand, we must find the admittance values for the bias and load components shown in our equivalent circuit. A simple rule for this conversion is as follows. To obtain the admittance of a component in μmhos, divide 1 000 by the resistance in kilohms. The reverse conversion is: To obtain the impedance of a component in kilohms divide 1 000 by the number of μmhos. Applying this rule,

$$Y_{L1} = Y_{L2} = \frac{1\ 000}{3\cdot 3} = 303\ \mu\text{mho}$$

$$Y_{B1} = Y_{B2} = \frac{1\ 000}{21} + \frac{1\ 000}{4\cdot 7} = 47\cdot 6 + 212\cdot 7 \simeq 260\ \mu\text{mho}$$

Now writing a mesh equation for v', and a nodal equation at v_0.

$$v' = 1\ 300i_{b2} + 2 \times 10^{-4}v_0 \tag{5.5}$$

$$-110i_b = v_0(105 + 303)10^{-6} \tag{5.6}$$

Substituting for v_0 in 5·5,

$$v' = 1\ 300i_{b2} + \frac{2 \times 10^{-4}(-110i_{b2})}{408 \times 10^{-b}}$$

$$Z_{1n2} = \frac{v'}{i_{b2}} = 1\ 300 - \frac{220}{4\cdot 08} = 1\ 246\ \Omega$$

147

Note that this result could have been obtained directly, using the general solution found in Chapter 2.

$$Z_{1n2} = h_{1e} - \frac{h_{re}h_{te}}{h_{oe} + Y_L} = 1\ 300 - \frac{110 \times 2 \times 10^{-4}}{(303 + 105)10^{-6}} = 1\ 246\ \Omega$$

Proceeding to step 2,

$$Y_{1n2} = \frac{1\ 000}{1 \cdot 246} = 803\ \mu\text{mho}$$

The effective load Y_{L1}' for Tr1 is the parallel combination of Y_{L1}, Y_{B2} and Y_{1n2}.

$$\therefore \qquad Y_{L \cdot 1}' = 303 + 260 + 803 = 1\ 366\ \mu\text{mho}$$

Now from the general solution, for step 3,

$$Z_{1n1} = 1\ 300 - \frac{110 \times 2 \times 10^{-4}}{(1\ 366 + 105)10^{-6}} = 1\ 285\ \Omega$$

and $\qquad Y_{1n1} = \frac{1\ 000}{1 \cdot 285} = 778\ \mu\text{mho}$

Including Y_{B1} for overall input impedance

$$Z_{1n} = \frac{1\ 000}{778 + 260} = \underline{0 \cdot 964\ \text{k}\Omega}$$

For step 4, we require a mesh equation for V_{1n} and a nodal equation at V_2.

$$V_{1n} = 1\ 300 i_{b1} + 2 \times 10^{-4} V_2 \qquad (5.7)$$

$$-110 i_{b1} = V_2(105 + 1\ 366)10^{-6} \qquad (5.8)$$

Substituting for i_{b1} in 5.7

$$V_{1n} = \frac{1\ 300(1\ 471)10^{-6}}{-110} V_2 + 2 \times 10^{-4} V_2$$

Voltage gain $\quad A_{v1} = \dfrac{V_2}{V_{1n}} = \dfrac{-100}{1\ 300(1\ 471)10^{-6} - 100 \times 2 \times 10^{-4}}$

$$= \frac{-110}{1 \cdot 3 \times 1 \cdot 471 - 0 \cdot 022}$$

$$= -58 \cdot 2 \text{ or } 58 \cdot 2\ \angle\ 180°$$

This result could also have been obtained from the appropriate general solution.

$$A_v = \frac{-h_{fe}}{h_{1e}(h_{oe} + Y_L) - h_{te}h_{re}}$$

$$= \frac{-110}{1\,300(1\,471)10^{-6} - 110 \times 2 \times 10^{-4}} = -58\cdot2$$

Using this general solution for A_{v2}

$$A_{v2} = \frac{-110}{1\,300(105 + 303)10^{-6} - 110 \times 2 \times 10^{-4}}$$

$$= \frac{-110}{0\cdot53 - 0\cdot022}$$

$$= -216\cdot5 \text{ or } 216\cdot5 \angle 180°$$

The overall voltage gain $\dfrac{v_o}{v_{1n}} = A_{v1} \times A_{v2} = -216\cdot5 \times -58\cdot2$

$$= \underline{12\,600 \angle 0°}$$

For the current gain, refer to step 6.

$$A_1 = \frac{i_o}{i_{1n}} = \frac{v_o Y_{L2}}{\dfrac{v_{1n}}{Z_{1n}}} = A_v Y_{L2} Z_{1n}$$

$$\therefore \qquad A_1 = 12\,600 \times 303 \times 10^{-6} \times 0\cdot964 \times 10^3$$

$$= 12\cdot6 \times 30\cdot3 \times 9\cdot64$$

$$= \underline{3\,680 \angle 0°}$$

The alternative procedure, when voltage gain is not required is to find the input admittances as before and applying the current splitting rule write:

$$i_{b1} = \frac{i_{1n}Y_{1n1}}{Y_{1n1} + Y_{B1}} = \frac{i_{1n}778}{778 + 260} = \frac{7\cdot78}{10\cdot38}\,i_{1n}$$

$$i_{B2} = \frac{-110i_{b1} \times 803}{803 + 260 + 303 + 105} = -\frac{11 \times 8\cdot03}{1\cdot471}\,i_{b1}$$

$$i_o = \frac{-110i_{b2} \times 303}{303 + 105} = \frac{-110 \times 3\cdot03}{4\cdot08}\,i_{b2}$$

$$\therefore \qquad A_1 = \frac{i_o}{i_{1n}} = \frac{7\cdot78}{10\cdot38} \times \frac{-11 \times 8\cdot03}{1\cdot471} \times \frac{-11 \times 30\cdot3}{4\cdot08}$$

$$= \underline{3\,680 \angle 0°}$$

For the output impedance calculations we shall rely simply on the general solution.

$$Y_o = h_{oe} - \frac{h_{re}h_{te}}{h_{ie} + Z_s}$$

In this case we proceed from the input end taking the following steps:

(1) Find the equivalent Z_{s1} including the bias resistors.
(2) Find Y_{o1} and hence the equivalent Z_{s2}.
(3) Find Y_{o2} and including Y_{L2}, the overall output admittance and impedance.

For Step 1:

$$Y_{s1} = (260 + 2\,000)\ \mu\text{mho}$$

$$\therefore \qquad Z_{s1} = \frac{1\,000}{2\,260} = 0{\cdot}442\ \text{k}\Omega$$

For Step 2:

$$Y_{o1} = 105 \times 10^{-6} - \frac{110 \times 2 \times 10^{-4}}{1\,300 + 442}\ \mu\text{mho}$$

$$= 105 - \frac{220}{17{\cdot}42}\ \mu\text{mho}$$

$$= 92{\cdot}4\ \mu\text{mho}$$

$$Y_{s2} = (92{\cdot}4 + 303 + 260)\ \mu\text{mho}$$

$$= 655{\cdot}4\ \mu\text{mho}$$

$$\therefore \qquad Z_{s2} = \frac{1\,000}{655{\cdot}4} = 1{\cdot}53\ \text{k}\Omega$$

Now $\qquad Y_{o2} = 105 \times 10^{-6} - \dfrac{110 \times 2 \times 10^{-4}}{1\,300 + 1\,530}\ \mu\text{mho}$

$$= 105 - \frac{220}{28{\cdot}3}\ \mu\text{mho}$$

$$= 97{\cdot}2\ \mu\text{mho}$$

Including Y_{L2},

$$\text{Overall output admittance} = 97{\cdot}2 + 303\ \mu\text{mho}$$

$$\simeq 400\ \mu\text{mho}$$

$$\therefore \qquad \text{Output impedance} = \frac{1\,000}{400} = 2{\cdot}5\ \text{k}\Omega$$

Thus the complete amplifier can be reduced to the Thèvenin equivalent generator shown in *Figure 5.3* where e_s is the source

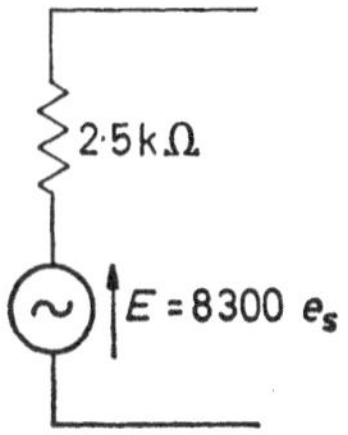

Figure 5.3. The equivalent generator solution for Example 5.2

$$2\cdot5\,k\Omega$$

$$E = 8300\ e_s$$

voltage. E, the open circuit output voltage is the overall gain multiplied by v_{in}. But

$$v_{in} = \frac{e_s Z_{in}}{Z_{in} + Z_s} = \frac{e_s \times 964}{500 \times 964}\ \text{V}$$

$$= 0\cdot658 e_s$$

$$\therefore \qquad E = 0\cdot648 \times 12\ 600 e_s = 8\ 300 e_s$$

This solution for Example 5.2 is accurate but not really practical. The first thing to notice, is that the effect of h_{re} on the results is very small. In the calculation of input impedance or output admittance for a stage, the neglecting of h_{re} might cause 10 per cent error. In each case however, further components are added in parallel, so the error in the effective load admittance or source impedance is very much less. In any case, the load and bias resistors will probably have a 20 per cent tolerance on the stated value.

A further valid approximation is to note that the effect of h_{oe} on A_{v1} and Z_{in1} is small and is in any case smaller than the effect of component tolerances. Thus for a multistage amplifier we can neglect h_{oe} for all transistors except the last.

Finally, the bias components, having an impedance much greater than h_{ie}, may be neglected or at least rounded off.

To show the effect of these approximations, we shall reconsider Example 5.2 as far as the calculation of voltage gain. First note the effect upon the general solutions if h_{re} is assumed to be zero, and h_{oe} is neglected.

$$Z_{in} = h_{ie} - \frac{0 \times h_{te}}{h_{oe} + Y_L} = h_{ie}$$

$$A_v = \frac{-h_{te}}{h_{ie}(h_{oe} + Y_L)} \simeq \frac{-h_{te}Z_L}{h_{ie}}$$

151

Now applying these results to the problem,

$$A_{v2} = \frac{-110}{1{\cdot}300(105 + 303)10^{-6}} = -208$$

$$Z_{in2} = 1\ 300$$

$\therefore$ $\quad Y_{in2} = 770\ \mu\text{mho}$

and $\quad Y_{L1\ eff} = (770 + 303 + 260)\ \mu\text{mho} = 1\ 333\ \mu\text{mho}$

$\therefore$ $\quad Z_{L1} = 0{\cdot}75\ \text{k}\Omega$

so $\quad A_{v1} = -90 \times \dfrac{750}{1\ 300} = -63{\cdot}5$

and $\quad$ Overall voltage gain $= -63{\cdot}5 \times -208 = 13\ 200$

This result must be compared with the accurate result of 12 600 and may be expressed as a percentage error.

$$\frac{600}{12\ 600} \times 100 \text{ per cent} = 4{\cdot}75 \text{ per cent}$$

This is much less than the probable error when component and parameter tolerances are allowed for.

Having considered the common emitter amplifier, a summary of the performance would be useful for comparison with the other configurations.

Table 5.1. Properties of Common Emitter Amplifiers

Input impedance	Medium	500 Ω–2 kΩ
Output impedance	Medium	5 kΩ–20 kΩ
Voltage gain	High	up to 500
Current gain	High	up to 250
Phase change	180°	

Common base amplifiers are only used in practice at very high frequencies. Low frequency analysis could be achieved by using either common emitter parameters and normal equivalent circuit methods or by use of the general solution in terms of the common base parameters h_{ib}, h_{rb}, h_{fb} and h_{ob}. The resulting properties are listed in *Table 5.2*.

Table 5.2. Properties of Common Base Amplifiers

Input impedance	Low	5–30 Ω
Output impedance	High	0·5–2 MΩ
Voltage gain	High	up to 500
Current gain	Low	up to 0·99
Phase change	0°	

The common collector or emitter follower is a very important circuit and it is used in many applications. Since common collector parameters are not normally quoted we shall have to consider the best approach to solving common collector circuit. Analysis may be achieved either by use of the common emitter parameters, or by determining the common collector parameters and using the general solutions. In the next example, these alternative methods will be compared.

Example 5.3. Investigate the performance of the common collector amplifier shown in *Figure 5.4a*, firstly, by direct use of common

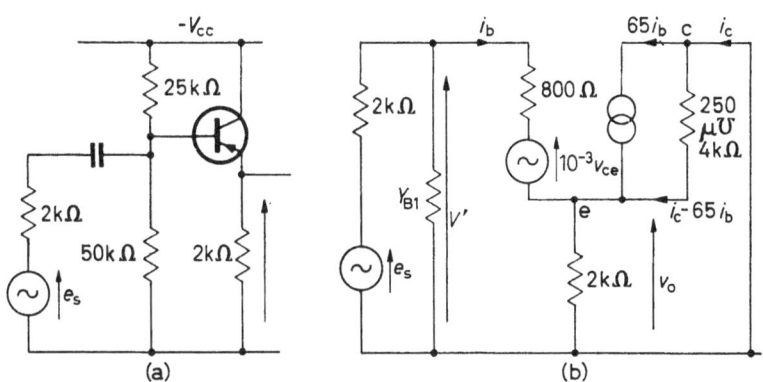

Figure 5.4. Circuits for the common collector amplifier in Example 5.3

emitter parameters, and secondly, by determining the common collector parameters and using the general solutions. Take h_{ie} 800 Ω, h_{re} 10^{-3}, h_{oe} 250 μmho and h_{fe} 65.

Figure 5.4b shows the complete equivalent circuit using common emitter parameters. Proceeding with the general method we must first find v_{ce} in terms of i_{c} and i_{b} our unknown currents.

Since the collector is connected to earth we can see that

$$v_{\text{ce}} = -v_0 = -2\,000(i_{\text{b}} + i_{\text{c}})$$

153

Now writing the mesh equations and working in mA and kΩ,

$$V' = 0 \cdot 8i_b - 10^{-3}(i_b + i_c)2 + 2(i_b + i_c)$$

$$0 = 4(i_c - 65i_b) + 2(i_b + i_c)$$

Collecting terms,

$$V' = 2 \cdot 8i_b + 2i_c \qquad \text{(neglecting the } 2 \times 10^{-3} \text{ terms)}$$

$$0 = -258i_b + 6i_c$$

From the second equation,

$$i_c = \frac{258}{6} i_b$$

Now
$$\text{Output current} = i_b + i_c$$

$$= i_b \left(1 + \frac{258}{6}\right)$$

$$= 44i_b$$

$\therefore$
$$\text{Current gain} \frac{i_e}{i_b} = 44$$

To find the input impedance we need to know i_b in terms of V', so substituting for i_c in the first equation,

$$V' = 2 \cdot 8i_b + 2 \times \frac{258}{6} i_b$$

$$Z_{in} = \frac{V'}{i_b} = 2 \cdot 8 + 86 = 88 \cdot 8 \text{ k}\Omega$$

$\therefore$
$$Y_{in} = 11 \cdot 25 \ \mu\text{mho}$$

Now for the overall current gain and input impedance we must include the effect of the bias components. From *Figure 5.4a*, Y_{B1} is given by the two bias resistors in parallel.

$\therefore$
$$Y_{B1} = \frac{1\,000}{50} + \frac{1\,000}{25} \ \mu\text{mho} = 60 \ \mu\text{mho}$$

and
$$Z_{B1} = \frac{1\,000}{60} = 16 \cdot 7 \text{ k}\Omega$$

$\therefore$ Overall input admittance $= Y_{B1} + Y_{in} = 71 \cdot 25 \ \mu\text{mho}$

and the overall input impedance $= 14$ kΩ.

By current splitting, the overall current gain is given by

$$A_1 = \frac{44 \times 11\cdot25}{11\cdot25 + 60} = 6\cdot95 \angle 0°$$

But   Input voltage $= i_{in}Z_{in}$

and   Output voltage $= i_o Z_L$

$$\therefore \quad\quad A_v = \frac{i_o Z_L}{i_{in} Z_{in}} = A_1 Z_L Y_{in}$$

$\therefore$ Voltage gain, $A_v = 6\cdot95 \times 2 \times 10^3 \times 71\cdot25 \times 10^{-6}$

$$= 0\cdot99 \angle 0°$$

To find the output admittance we must redraw the equivalent circuit, suppressing any external generators, and apply a generator of I amps to the output terminals. Then by calculating the resultant output voltage V we find

$$Y_o = \frac{I}{V}$$

This equivalent circuit is more easily understood if the branches are shown in parallel between the emitter terminal and earth, as shown in *Figure 5.5a.*

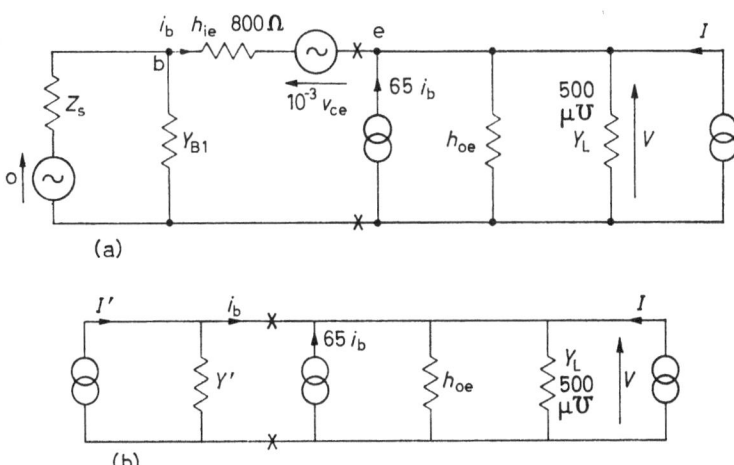

Figure 5.5. Equivalent circuit solution for the output admittance of the common collector amplifier in Example 5.3

This equivalent circuit (*Figure 5.5a*) is of mixed form and is best converted for nodal analysis by applying Norton's Theorem at XX. The result of this conversion is shown in *Figure 5.5b*.

First $v_{ce} = -V$, so the sense of the h_{re} generator can be reversed and the value changed to $10^{-3}V$.

$$\text{Now} \qquad I' = \frac{10^{-3}V}{\left(0{\cdot}8 + \dfrac{16{\cdot}7 \times 2}{16{\cdot}7 + 2}\right)} = 0{\cdot}384 \times 10^{-6}V$$

$$\text{and} \qquad Y' = \frac{1\ 000}{\left(0{\cdot}8 + \dfrac{16{\cdot}7 \times 2}{16{\cdot}7 + 2}\right)}\ \mu\text{mho}$$

$$= 384\ \mu\text{mhos}$$

At this stage, care must be taken not to lose i_b since this is required for the h_{te} generator.

Applying Kirchhoff's current law:

$$i_b = I' - VY'$$
$$= 0{\cdot}384 \times 10^{-6}V - 384 \times 10^{-6}V$$
$$\simeq -384 \times 10^{-6}V$$

(This is effectively neglecting h_{re}.)

Now writing the nodal equation,

$$0{\cdot}384 \times 10^{-6} - 65 \times 384 \times 10^{-6}V + I = V(500 + 250 + 384)10^{-6}$$

Neglecting the first term,

$$I = V(500 + 250 + 384 + 24\ 950)10^{-6}$$

The output admittance

$$Y_0 = \frac{I}{V} = 26{\cdot}1\ \text{mmho}$$

and the output impedance

$$Z_0 = 38{\cdot}4\ \Omega$$

These results may now be summarized for comparison with *Tables 5.1* and *5.2*.

Table 5.3. Properties of Common Collector Amplifiers

Current gain	Medium	5–20
	(limited by bias components)	
Voltage gain	Low	0·8–0·99
Input impedance	High	Bias components
Output impedance	Low	20–100 Ω
Phase shift	0°	

The alternative procedure is to find the common collector h parameters and to use the general solutions. The conversion from h_e parameters to h_c parameters can be simply achieved by redrawing the common emitter equivalent circuit with the collector as the common terminal as shown in *Figure 5.6*. The normal h parameter definitions may then be applied to find the h_c parameters.

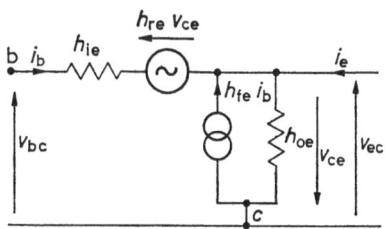

Figure 5.6. Circuit for determining the common collector h parameters from the common emitter h parameters

Working with the required voltages and currents, v_{bc}, v_{ec}, i_b and i_e as shown

$$h_{1c} = \frac{v_{bc}}{i_b}\bigg|_{v_{ec}=0} \qquad h_{rc} = \frac{v_{bc}}{v_{ec}}\bigg|_{i_b=0}$$

$$h_{fc} = \frac{i_e}{i_b}\bigg|_{v_{ec}=0} \qquad h_{oc} = \frac{i_e}{v_{ec}}\bigg|_{i_b=0}$$

First if v_{ec} is zero we have a short circuit across the emitter collector terminals. Thus v_{ce} is also zero and the h_{re} generator disappears. All the current from the h_{fe} generator, together with i_b flow into the short circuit. Thus

$$i_e = -(i_b + h_{fe}i_b)$$

and $$h_{fc} = -(h_{fe} + 1)$$

and $$h_{1c} = h_{1e}$$

Now for i_b zero, the base terminal is open circuit and the h_{fe} generator disappears.

This leaves $h_{oc} = h_{oe}$.

Also since $v_{ec} = -v_{ce}$, the h_{re} generator becomes $-h_{re}v_{ec}$. There is no volt drop across h_{1e} so:

$$v_{bc} = v_{ec} - h_{re}v_{ec} \quad \text{and} \quad h_{rc} = 1 - h_{re}$$

157

Now applying these to Example 5.3:

$$h_{1c} = 800 \ \Omega \qquad h_{rc} = 1 - 10^{-3} \simeq 1$$

$$h_{fc} = -66 \qquad h_{oc} = 250 \ \mu\text{mho}$$

Applying the general solutions

$$A_v = \frac{+66}{800(250 + 500)10^{-6} + 66} = 0.99 \ \angle 0°$$

$$Z_{1n} = 800 + \frac{66}{(250 + 500)10^{-6}} = 88.8 \ \text{k}\Omega$$

For the output admittance, we require the effective Z_s, but Y_s is given by

$$Y_s = 500 + 40 + 20 \ \mu\text{mho}$$

$$\therefore \qquad Z_s = \frac{1 \ 000}{560} \ \text{k}\Omega = 1.788 \ \text{k}\Omega$$

Now $$Y_0 = 250 + \frac{66 \times 10^6}{1 \ 788 + 800} \ \mu\text{mho}$$

$$= 250 + 25 \ 500 \ \mu\text{mho}$$

For overall Y_0, including load,

$$Y_0 = 25 \ 750 + 500 \ \mu\text{mho}$$

and $$Z_0 = 38.2 \ \Omega$$

This method is obviously very much quicker and is therefore preferable even if the necessary conversion factors have to be found.

ALTERNATIVE COUPLING METHODS

Two further examples of the application of equivalent circuit techniques to practical configurations will now be given. The first circuit is known as the emitter coupled or long tailed pair amplifier. With slight modifications it can be used as a difference amplifier, a phase splitter or a d.c. amplifier. The second circuit is one in which two transistors are interconnected and used as one. This combination is sometimes known as a Darlington connected, or super α pair. The properties of such a combination will become apparent from the example.

Example 5.4. Determine the voltage gain and input impedance of the emitter coupled amplifier shown in *Figure 5.7a*. The transistors have the following hybrid parameters at the d.c. operating points determined by the circuit, h_{ie} 1 300 Ω, h_{re} 5 × 10^{-4}, h_{oe} 125 μmho and h_{fe} 90.

The amplifier circuit shown in *Figure 5.7a* could be treated in a number of ways. The output of the first stage is taken from the

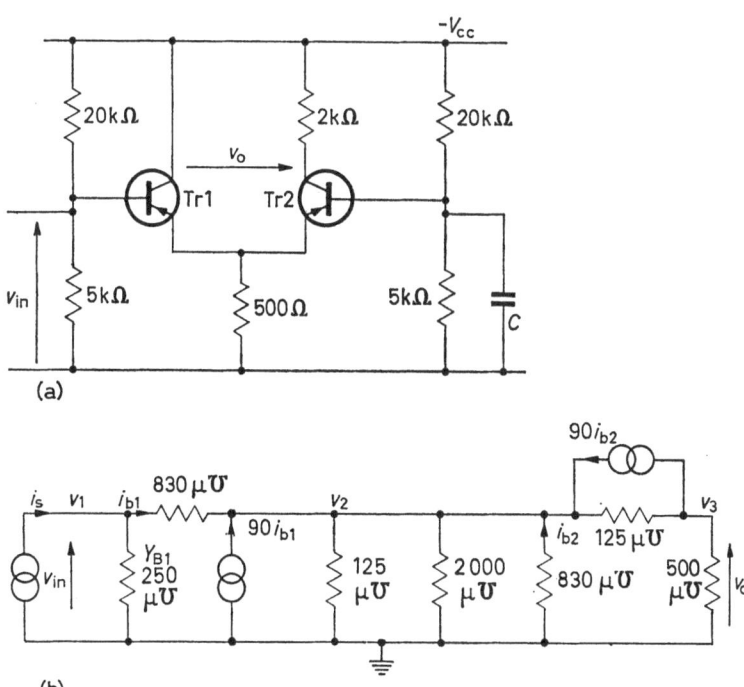

(a)

(b)

Figure 5.7. Circuits for Example 5.4

emitter and the collector is earthed through the d.c. supply. The base of Tr2 is also earthed to a.c. by the capacitor shown (assuming $X_C \ll 4$ kΩ). One possible approach is to take Tr2 as a common base amplifier and Tr1 as a common collector amplifier. Thus if the h_c and h_b parameters are known, the general solutions can be applied in the normal manner. In this example only the h_e parameters are available, so we shall draw the equivalent circuit neglecting h_{re} and solve using nodal analysis.

The required equivalent circuit is shown in *Figure 5.7b*. Note the components representing h_{1e} have been expressed as admittances, as have the combined bias components on Tr1 and the loads on both stages. The input is shown as being supplied with a current i_s. This is necessary, since for nodal analysis the only permissible generators are current generators. Note also that i_{b1} and i_{b2} are clearly indicated as are the unknown node voltages v_1, v_2, and v_3. The solutions to the nodal equations will express v_1, v_2 and v_3 in terms of i_s. The ratio of v_3 to v_1 will give the required terminal voltage gain and that of v_1 to i_s the input impedance.

The next step is to express any equivalent generators in terms of the unknown voltages, circuit admittances and external generators.

For i_{b1}, we may write

$$i_{b1} = (v_1 - v_2)830 \ \mu A$$

and $$i_{b2} = -v_2 \ 830 \ \mu A$$

Now for node 1, equating currents entering the node from generators to currents leaving the node as voltage admittance products:

$$i_s = (v_1 - v_2)830 + v_1 \ 250$$

The admittances are expressed in μmho, therefore if v_1, v_2 and v_3 are measured in volts the currents will be given in μA.

Now following the general method for writing nodal equations as stated in Chapter 2 on page 59.

For node 2:

$$90(v_1 - v_2)830 + 90(-830v_2)$$
$$= -v_1 830 + v_2(830 + 125 + 2\,000 + 125 + 830) - 125v_3$$

and for node 3:

$$-90(-830v_2) = -125v_2 + 625v_3$$

Now collecting terms and dividing by 1 000, which changes the units to volts, kΩ and mA:

$$i_s = 1{\cdot}08v_1 - 0{\cdot}83v_2$$
$$0 = -75{\cdot}5v_1 + 153v_2 - 0{\cdot}125v_3$$
$$0 = \qquad\qquad -75v_2 + 0{\cdot}625v_3$$

160

Using determinants, the input impedance Z_{1n} is given by

$$Z_{1n} = \frac{v_1}{i_s} = \frac{1}{i_s} \frac{\begin{vmatrix} i_s & -0\cdot83 & 0 \\ 0 & +153 & -0\cdot125 \\ 0 & -75 & +0\cdot625 \end{vmatrix}}{\begin{vmatrix} 1\cdot08 & -0\cdot83 & 0 \\ -75\cdot5 & +153 & -0\cdot125 \\ 0 & -75 & +0\cdot625 \end{vmatrix}}$$

$$\therefore \quad Z_{1n} = \frac{1}{i_s} \times$$

$$\times \frac{i_s(153 \times 0\cdot625 - 0\cdot125 \times 75) + 0\cdot83 \times 0 + 0}{1\cdot08(153 \times 0\cdot625 - 0\cdot125 \times 75) + 0\cdot83(-75\cdot5 \times 0\cdot625) + 0}$$

$$= \frac{87}{54\cdot8} = 1\cdot59 \text{ k}\Omega$$

This answer might seem low for a common collector stage, but the effective load on the stage is the input impedance to a common base stage which is also very low.

For the voltage gain we require the ratio v_3/v_1 and in determinant form this is given by

$$A_v = \frac{v_3}{v_1} = \frac{\begin{vmatrix} 1\cdot08 & -0\cdot83 & i_s \\ -75\cdot5 & +153 & 0 \\ 0 & -75 & 0 \end{vmatrix}}{\begin{vmatrix} i_s & -0\cdot83 & 0 \\ 0 & +153 & -0\cdot125 \\ 0 & -75 & +0\cdot625 \end{vmatrix}}$$

The denominator has already been found in the numerator of the expression for Z_{1n}.

$$\therefore \quad A_v = \frac{1\cdot08(0) + 0\cdot83(0) + i_s(-75\cdot5 \times \times 75 - 0)}{87 i_s}$$

$$= \frac{75\cdot5 \times 75}{87} = 65$$

This expression for voltage gain is positive and there is therefore no phase reversal. This is to be expected, since each stage of a cascaded common base, common collector amplifier has no phase reversal.

Example 5.5. An electronic circuit shows two transistors connected as a Darlington pair. Determine the *h* parameters for the composite unit used in the common emitter configuration, and hence find the conditions leading to high current gain and high input impedance. Assume the transistors to have h_{ie} 2 000Ω, h_{fe} 120, h_{oe} 150 μmho and negligible h_{re}.

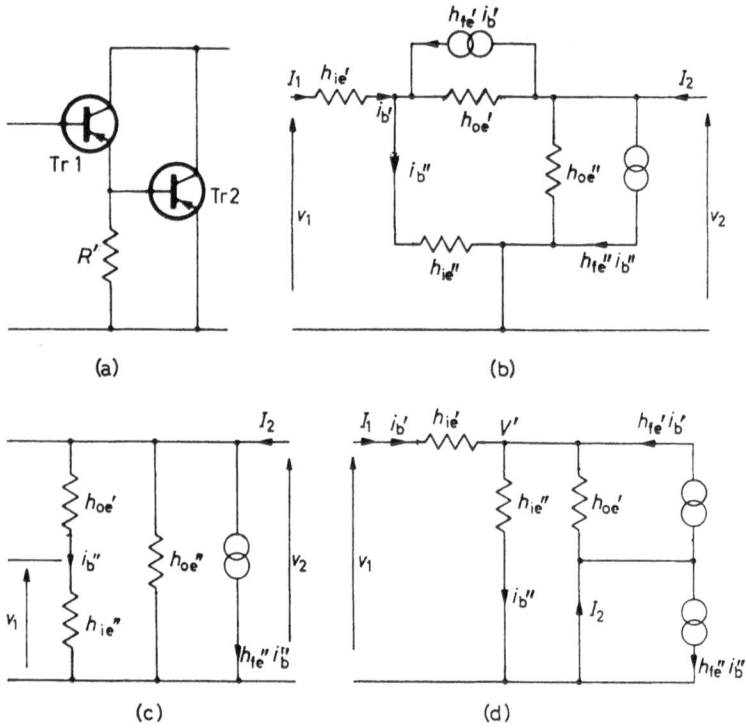

(a)   (b)   (c)   (d)

Figure 5.8. Compound connected transistors and circuits for determining the overall *h* parameters

The required Darlington connection is shown in *Figure 5.8a*. A resistor R' is normally included to provide the required d.c. conditions and for calculation it may be included in the h_{ie} for Tr2. Assuming Tr1 to have parameters h_{ie}', h_{oe}', h_{fe}' and Tr2 to have parameters h_{ie}'', h_{oe}'' and h_{fe}'', the equivalent circuit is shown in *Figure 5.8b*.

Rewriting the h parameter equations, where h_{11}, h_{12}, h_{21}, h_{22} are the parameters for the overall circuit,

$$V_1 = I_1 h_{11} + V_2 h_{12}$$

$$I_2 = I_1 h_{21} + V_2 h_{22}$$

We can see that the required conditions for finding the parameters, are open circuit input, for h_{12} and h_{22}, and short circuit output, for h_{21} and h_{11}.

Figure 5.8c shows the modified equivalent circuit with the input open circuit. Under these conditions I_1 and i_b' are zero. Thus the h_{fe}' i_b' generator is eliminated, and since there is no p.d. across h_{1e}', this can also be deleted.

Now
$$h_{12} = \frac{V_1}{V_2}\bigg|_{I_1=0}$$

From *Figure 5.8c* we can see that h_{oe}' and h_{1e}'' form a potential divider across v_2, so h_{12} may be written:

$$h_{12} = \frac{h_{1e}''}{h_{1e}'' + \dfrac{1}{h_{oe}'}} = \frac{h_{1e}'' h_{oe}'}{h_{1e}'' h_{oe}' + 1} \tag{5.9}$$

To find h_{22}, we can write a nodal equation, but first we must find i_b'' in terms of the node voltage. Here

$$i_b'' = \frac{V_2}{h_{1e}'' + \dfrac{1}{h_{oe}'}}$$

Now,
$$I_2 = h_{fe}'' i_b'' + V_2 Y$$

where Y is the combination of h_{1e}'', h_{oe}' and h_{oe}''.

$$\therefore \quad I_2 = \frac{h_{fe}'' V_2}{h_{1e}'' + \dfrac{1}{h_{oe}'}} + V_2 \left(h_{oe}'' + \frac{\dfrac{1}{h_{oe}'} + h_{1e}''}{\dfrac{h_{1e}''}{h_{oe}'}} \right)$$

and
$$h_{22} = \frac{I_2}{V_2} = \frac{h_{fe}'' h_{oe}'}{1 + h_{oe}' h_{1e}''} + h_{oe} + \frac{1 + h_{1e}'' h_{oe}'}{h_{1e}''}$$

$$= h_{oe}'' + h_{oe}' + \frac{1}{h_{1e}''} + \frac{h_{fe}'' h_{oe}'}{1 + h_{oe}' h_{1e}''} \tag{5.10}$$

When practical values are inserted, we shall see that the last term of this result is the predominant part.

Proceeding to h_{11} and h_{21} we require the short circuit output condition. The redrawn equivalent circuit is shown in *Figure 5.8d*. The $h_{te}''i_b''$ generator can have no effect on I_1 since the entire generator current flows in the short circuit.

By inspection, the input impedance

$$\frac{V_1}{I_1} = h_{11} = h_{1e}' + \frac{V'}{i_b'}$$

Writing a nodal equation for V',

$$i_b' + h_{te}'i_b' = V'\left(\frac{1}{h_{1e}'} + h_{oe}'\right)$$

$$\therefore \qquad \frac{V'}{i_b'} = \frac{1 + h_{te}'}{\dfrac{1}{h_{1e}''} + h_{oe}'} \qquad (5.11)$$

$$\therefore \qquad h_{11} = h_{1e}' + \frac{(1 + h_{te}')h_{1e}''}{1 + h_{1e}''h_{oe}'} \qquad (5.12)$$

Once again, numerical values will prove the last term to predominate.

For h_{21}, we must find the value of the current in the short circuit output, I_2, in terms of I_1. In this case, $I_1 = i_b'$. Now applying the superposition theorem, I_2 is given by the sum of the currents flowing in the short circuit due to the two current generators and i_b' taken separately.

$$I_2 \text{ due to } i_b' \text{ alone} \quad = \frac{-i_b'h_{oe}'}{h_{oe}' + \dfrac{1}{h_{1e}''}} = \frac{-i_b'h_{oe}'h_{1e}''}{1 + h_{oe}'h_{1e}''} \quad (5.13)$$

$$I_2 \text{ due to } h_{te}'i_b' \text{ alone} = \frac{+h_{te}'i_b'\dfrac{1}{h_{1e}''}}{h_{oe}' + \dfrac{1}{h_{1e}''}} = \frac{+h_{te}'i_b'}{1 + h_{oe}'h_{1e}''} \quad (5.14)$$

I_2 due to $h_{te}''i_b''$ alone $= h_{te}''i_b''$

But from equation 5.11,

$$V' = \frac{i_b'(1 + h_{te}')h_{1e}''}{1 + h_{1e}''h_{oe}'} \qquad (5.15)$$

and $$i_b'' = \frac{V'}{h_{1e}''} \qquad (5.16)$$

Substituting from equations 5.15 and 5.16,

$$h_{te}{}''i_b{}'' = \frac{h_{te}{}''(1 + h_{te}{}')h_{1e}{}''i_b{}'}{(1 + h_{1e}{}''h_{oe}{}')h_{1e}{}''}$$

$$= \frac{h_{te}{}''(1 + h_{te}{}')i_b{}'}{1 + h_{1e}{}''h_{oe}{}'} \qquad (5.17)$$

Now adding equations 5.13, 5.14, and 5.17,

$$\frac{I_2}{i_b{}'} = h_{21} = \frac{h_{te}{}' - h_{oe}{}'h_{1e}{}'' + h_{te}{}''(1 + h_{te}{}')}{1 + h_{1e}{}''h_{oe}{}'} \qquad (5.18)$$

In this case, practical values will approximate h_{21} to $h_{te}{}''(1 + h_{te}{}')$.
Taking the values for the parameters given in the question:

$$h_{12} = \frac{2\,000 \times 150 \times 10^{-6}}{1 + 2\,000 \times 150 \times 10^{-6}} = \frac{0 \cdot 3}{1 \cdot 3} = 0 \cdot 231$$

$$h_{22} = (150 + 150)10^{-6} + \frac{1}{2\,000} + \frac{120 \times 150 \times 10^{-6}}{1 + 2\,000 \times 150 \times 10^{-6}} \text{ mho}$$

$$= 800 \times 10^{-6} + \frac{0 \cdot 018}{1 \cdot 3} \text{ mho} = 0 \cdot 8 + 13 \cdot 8 \text{ mmho} = 14 \cdot 6 \text{ mmho}$$

Note, this result is equivalent to an output impedance of $68 \cdot 5\ \Omega$.
From equation 5.12,

$$h_{11} = 2\,000 + \frac{121 \times 2\,000}{1 \cdot 3} = 188 \text{ k}\Omega$$

From equation 5.18,

$$h_{21} = \frac{120 - 0 \cdot 3 + 120 \times 121}{1 \cdot 3} = 11\,200$$

Thus from the general solutions, since $A_1 = \dfrac{h_{21}Y_L}{Y_L + h_{22}}$ a high current
gain will be obtained if Y_L is of the same order as, or greater than
h_{22}. In this case a suitable value of load would be 100 Ω or less.
Using this value, the input impedance is given by:

$$Z_{1n} = h_{11} - \frac{h_{12}h_{21}}{h_{22} + Y_L}$$

$$Z_{1n} = 188 \times 10^3 - \frac{11\,200 \times 0 \cdot 231}{(14 \cdot 6 + 10) \times 10^{-3}} \Omega = 83 \text{ k}\Omega$$

This of course will be modified by the bias components in parallel with the input. These may have far higher values than are necessary for a single transistor, as the large h_{21} permits a very low d.c. base current for the first transistor. An even higher input impedance can be achieved, without loss of current gain, by using the composite transistor in the common collector configuration. The conversion of parameters is obtained by using the results found in Example 5.3.

$$h_{11c} = 188 \text{ k}\Omega \qquad h_{12c} = 1 - 0{\cdot}231 = 0{\cdot}769$$

$$h_{21c} = -11\ 200 \qquad h_{22c} = 14{\cdot}6 \text{ mmho}$$

The current gain is unchanged, except in phase, and the input impedance becomes

$$188 \times 10^3 + \frac{11\ 200 \times 0{\cdot}769}{24{\cdot}6 \times 10^{-3}} = 538 \text{ k}\Omega$$

All the examples so far considered, have neglected the effect of reactances. Capacitors, where shown, have been assumed to have negligible reactance at all signal frequencies. Stray capacitance due to wiring, and that due to transistor properties, have not been shown, since at low frequencies their reactance is very much greater than shunt resistive components. Detailed analysis of high frequency performance will be considered in a later chapter, but the effect of an overall shunt capacitance C_s will be included in the next example. The effect of capacitors used for decoupling emitter resistors is exactly the same as those used in the cathode circuit of valve amplifiers. At this point then, we shall investigate the effects of C_s and C_c, the coupling capacitor. This analysis is similar to that used for the RC coupled valve amplifier in Chapter 4.

Example 5.6. The audio voltage amplifier shown in *Figure 5.9a* is to be used at frequencies from 100 Hz upwards. By means of a general analysis, determine, (a) a suitable value for the coupling capacitor C_c, (b) the high frequency above which the gain is more than 3 db below the maximum value. Investigate methods by which the gain and phase responses can most easily be recorded. The transistor h_e parameters are h_{1e} 1 000 Ω, h_{fe} 110, h_{oe} 100 μmho, and h_{re} negligible. The total shunt capacitance is 1 200 pF.

Figure 5.9b shows those parts of the equivalent circuit essential to the general analysis. The output section of the second transistor is not required since the absence of reactive components in the load make the gain, A_{v2}, independent of frequency. Tr1 bias components have no effect on the voltage gain of the stage and only modify

166

input impedance. Y_{in2} is the total input admittance to the second stage and will include the bias components Y_{B1} and Y_{B2}. Since h_{re} is zero, the terminal input impedance to Tr2 becomes h_{1e}.

$$\therefore \qquad Y_{in2} = Y_{B1} + Y_{B2} + \frac{1}{h_{1e}} \qquad (5.19)$$

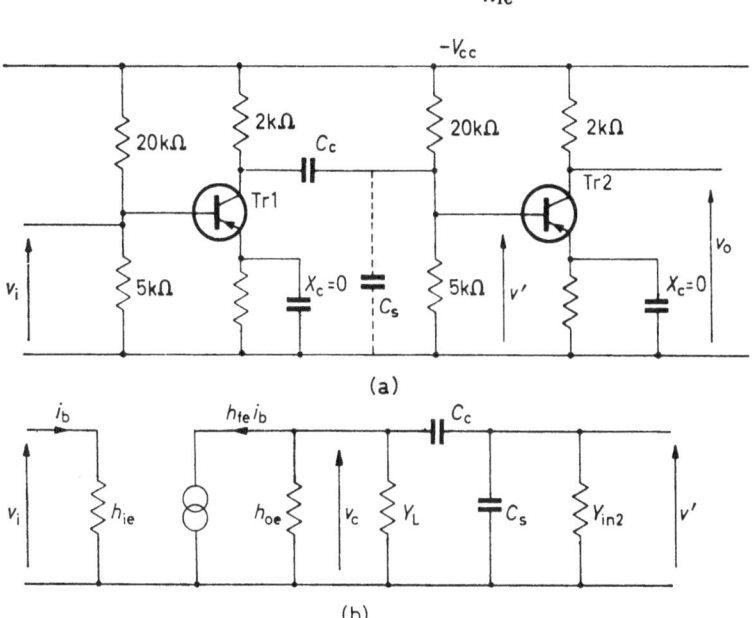

(a)

(b)

Figure 5.9. The RC coupled amplifier for Example 5.6

For simplicity, certain components may be combined for the analysis. By definition, let

$$R_e = \frac{1}{h_{oe} + Y_L + Y_{in2}} \qquad (5.20)$$

and let

$$R' = \frac{1}{h_{oe} + Y_L} + Z_{in2} \qquad (5.21)$$

where

$$Z_{in2} = \frac{1}{Y_{in2}}$$

and let

$$R_L' = \frac{1}{h_{oe} + Y_L} \qquad (5.22)$$

167

Also, frequencies at which both C_c and C_s may be neglected, will be referred to as medium frequencies; those at which C_s only may be neglected, as low frequencies, and those at which C_c only may be neglected as high frequencies. The corresponding voltage gain for these frequency ranges will be referred to as A_{vm}, A_{vl} and A_{vh} respectively.

First, considering *Figure 5.9b* at medium frequencies:

$$X_{Cc} = \frac{1}{\omega C_c} \ll Z_{1n2}$$

and

$$X_{Cs} = \frac{1}{\omega C_s} \gg R_e$$

Thus

$$v' = -h_{fe}i_b \times R_e$$

but

$$i_b = \frac{v_1}{h_{1e}}$$

$\therefore$

$$A_{vm} = \frac{v'}{v_1} = \frac{-h_{fe}}{h_{1e}} R_e \qquad (5.23)$$

At high frequencies, X_{Cs} is in parallel with R_e,

$\therefore$

$$v' = -h_{fe}i_b \times \frac{-jX_{Cs}R_e}{R_e - jX_{Cs}}$$

Dividing numerator and denominator by $-jX_{Cs}$,

$$A_{vh} = \frac{-h_{fe}}{h_{1e}} R_e \left(\frac{1}{1 - \dfrac{R_e}{jX_{Cs}}} \right)$$

$$= \frac{A_{vm}}{1 + \dfrac{jR_e}{X_{Cs}}} \qquad (5.24)$$

At medium and high frequencies, v_c shown in *Figure 5.9b* is the same as v'. At low frequencies, X_{Cc} and Z_{1n2} act as a potential divider circuit

$\therefore$

$$v' = \frac{v_c Z_{1n2}}{Z_{1n2} - jX_{Cc}}$$

v_c may be found in exactly the same way as v' was found at medium and high frequencies.

$\therefore$

$$v' = -h_{fe}i_b \times \frac{R_L'(Z_{1n2} - jX_{Cc})}{R_L' + Z_{1n2} - jX_{Cc}} \times \frac{Z_{1n2}}{(Z_{1n2} - jX_{Cc})}$$

The terms in brackets cancel and i_b is the same as that for medium frequencies.

$$\therefore \qquad A_{v1} = \frac{-h_{fe}}{h_{ie}} \times \frac{R_L' Z_{in2}}{R_L' + Z_{in2} - jX_{Cc}}$$

$$= \frac{-h_{fe}}{h_{ie}} \frac{R_L' Z_{in2}}{R_L' + Z_{in2}} \times \frac{1}{1 - \dfrac{jX_{Cc}}{R_L' + Z_{in2}}}$$

But $\qquad \dfrac{R_L' Z_{in2}}{R_L' + Z_{in2}} = R_e \quad \text{and} \quad R_L' + Z_{in2} = R'$

(Equations 5.19, 5.20, 5.21, and 5.22.)

Thus from equation 5.23,

$$A_{v1} = \frac{A_{vm}}{1 - \dfrac{jX_{Cc}}{R'}} \tag{5.25}$$

Equations 5.23, 5.24 and 5.25 provide a convenient means for plotting graphs of the variation of gain and phase shift of the amplifier as the frequency is changed. At any particular frequency, the value of X_{Cs} or X_{Cc} could be calculated and used to determine the complex gain in the polar form $A \angle \theta$. For a general investigation, it is more convenient to choose a simple numerical value for X_{Cc}/R' or R_e/X_{Cs} and then to determine the complex gain and the frequency at which it applies. In this numerical example, from equation 5.20,

$$R_e = \frac{10^6}{100 + 250 + 1\,000 + 500} \, \Omega = 540 \, \Omega$$

and from equation 5.23,

$$A_{vm} = \frac{-110 \times 540}{1\,000} = -59 \cdot 3$$

Now considering equation 5.24, let ω_h be the frequency at which $R_e/X_{Cs} = 1$.

This results in $\quad A_{vh} = \dfrac{-59 \cdot 3}{1 + j} = \dfrac{59 \cdot 3 \angle 180°}{\sqrt{2} \angle 45°} = 42 \angle 135°$

Now if $\qquad \dfrac{R_e}{X_{Cs}} = 1 = \omega_h C_s R_e \tag{5.26}$

Then $\quad \omega_\mathrm{h} = \dfrac{1}{C_\mathrm{s}R_\mathrm{e}} = \dfrac{10^{12}}{1\,200 \times 540} = 1\cdot54 \times 10^6$ rad/sec

$\therefore \qquad\qquad\qquad f_\mathrm{h} = \dfrac{\omega_\mathrm{h}}{2\pi} = 245$ kHz

Other values for the complex gain at particular frequencies may be obtained by putting $R_\mathrm{e}/X_{C_\mathrm{s}}$ as $\frac{1}{2}$, 2, 3, 5, etc. Inspection of equation 5.26 shows that the corresponding angular frequencies will be given by $\omega_\mathrm{h}/2$, $2\omega_\mathrm{h}$, $3\omega_\mathrm{h}$, and $5\omega_\mathrm{h}$ respectively. At these frequencies, the gain may be obtained by dividing A_vm by $\sqrt{1\cdot25}$, $\sqrt{5}$, $\sqrt{10}$ and $\sqrt{26}$ respectively with phase shifts changed from 180° by $\tan^{-1} 0\cdot25$, $\tan^{-1} 2$, $\tan^{-1} 3$ and $\tan^{-1} 5$ respectively. These results are shown in *Table 5.4* on page 171.

At low frequencies, the problem requires that the amplifier should be suitable for use down to 100 Hz. In practice, this implies that the gain shall not fall by more than 3 db (see Appendix 1).

Thus at 100 Hz, $\qquad 3 = 20 \log_{10} \left| \dfrac{A_\mathrm{vm}}{A_\mathrm{vl}} \right|$

$\therefore \qquad\qquad \left| \dfrac{A_\mathrm{vm}}{A_\mathrm{vl}} \right| = \text{antilog}_{10}\, 0\cdot15 = \sqrt{2}$

$\therefore \qquad\qquad |A_\mathrm{vl}| = \dfrac{|A_\mathrm{vm}|}{\sqrt{2}}$

With reference to equation 5.24, this corresponds to the frequency at which $X_{C_\mathrm{c}}/R' = 1$.

$$\dfrac{1}{2\pi f C_\mathrm{c}} = R'$$

and $\qquad\qquad\qquad C_\mathrm{c} = \dfrac{1}{2\pi f R'}$ Farad

From equations 5.19 and 5.21,

$$R' = \dfrac{10^6}{500 + 100} + \dfrac{10^6}{200 + 50 + 1\,000}\, \Omega$$

$$= 1\,670 + 800 = 2\,470\, \Omega$$

$\therefore \qquad C_\mathrm{c} = \dfrac{10^6}{2\pi \times 100 \times 2\,470}\, \mu\mathrm{F} = 0\cdot63\ \mu\mathrm{F}$

Thus in practice a 1 μF capacitor would be suitable.

Table 5.4. $A_{vm} = 59 \cdot 3 \angle 180°$

$\dfrac{X_{Cc}}{R'}$	A_{v1}	Phase shift = $180° + \theta$ where θ is	$f = \dfrac{\omega}{2\pi}$ where ω is	f
1·0	$\dfrac{\|A_{vm}\|}{\sqrt{2}} = 42 \cdot 0$	$\tan^{-1} 1 \quad 45°$	$\omega_1 = \dfrac{1}{R'C_c}$	100 Hz
0·5	$\dfrac{\|A_{vm}\|}{\sqrt{0 \cdot 25}} = 53 \cdot 1$	$\tan^{-1} 0 \cdot 5 \quad 26° \, 30'$	$2\omega_1$	200 Hz
0·25	$\dfrac{\|A_{vm}\|}{\sqrt{1 \cdot 0625}} = 57 \cdot 4$	$\tan^{-1} 0 \cdot 25 \quad 14°$	$4\omega_1$	400 Hz
2·0	$\dfrac{\|A_{vm}\|}{\sqrt{5}} = 26 \cdot 5$	$\tan^{-1} 2 \quad 63° \, 30'$	$\dfrac{\omega_1}{2}$	50 Hz
3·0	$\dfrac{\|A_{vm}\|}{\sqrt{10}} = 18 \cdot 8$	$\tan^{-1} 3 \quad 71° \, 30'$	$\dfrac{\omega_1}{3}$	33 Hz
5·0	$\dfrac{\|A_{vm}\|}{\sqrt{26}} = 11 \cdot 6$	$\tan^{-1} 5 \quad 78° \, 30'$	$\dfrac{\omega_1}{5}$	20 Hz
10	$\dfrac{\|A_{vm}\|}{\sqrt{101}} = 5 \cdot 93$	$\tan^{-1} 10 \quad 84° \, 18'$	$\dfrac{\omega_1}{10}$	10 Hz
20	$\dfrac{\|A_{vm}\|}{\sqrt{401}} = 2 \cdot 96$	$\tan^{-1} 20 \quad 87° \, 10'$	$\dfrac{\omega_1}{20}$	5 Hz
50	$\dfrac{\|A_{vm}\|}{\sqrt{2501}} = 1 \cdot 19$	$\tan^{-1} 50 \quad 88° \, 49'$	$\dfrac{\omega_1}{50}$	2 Hz

$\dfrac{R_e}{X_{Cs}}$	A_{vh}	where $-\theta$ is		
1·0	$\dfrac{\|A_{vm}\|}{\sqrt{2}} = 42 \cdot 0$	$\tan^{-1} \quad 45°$	$\omega_h = \dfrac{1}{R_e C_s}$	245 kHz
0·5	$\dfrac{\|A_{vm}\|}{\sqrt{1 \cdot 25}} = 53 \cdot 1$	$\tan^{-1} 0 \cdot 5 \quad 26° \, 30'$	$\dfrac{\omega_h}{2}$	122 kHz
0·25	$\dfrac{\|A_{vm}\|}{\sqrt{1 \cdot 0625}} = 57 \cdot 4$	$\tan^{-1} 0 \cdot 25 \quad 14°$	$\dfrac{\omega_h}{4}$	61 kHz
2·0	$\dfrac{\|A_{vm}\|}{\sqrt{5}} = 26 \cdot 5$	$\tan^{-1} 2 \quad 63° \, 30'$	$2\omega_h$	490 kHz
3·0	$\dfrac{\|A_{vm}\|}{\sqrt{10}} = 18 \cdot 8$	$\tan^{-1} 3 \quad 71° \, 30'$	$3\omega_h$	735 kHz
5·0	$\dfrac{\|A_{vm}\|}{\sqrt{26}} = 11 \cdot 6$	$\tan^{-1} 5 \quad 78° \, 30'$	$5\omega_h$	1·015 MHz
10	$\dfrac{\|A_{vm}\|}{\sqrt{101}} = 5 \cdot 93$	$\tan^{-1} 10 \quad 84° \, 18'$	$10\omega_h$	2·45 MHz
20	$\dfrac{\|A_{vm}\|}{\sqrt{401}} = 2 \cdot 96$	$\tan^{-1} 20 \quad 87° \, 10'$	$20\omega_h$	4·9 MHz
50	$\dfrac{\|A_{vm}\|}{\sqrt{2501}} = 1 \cdot 19$	$\tan^{-1} 50 \quad 88° \, 49'$	$50\omega_h$	12·2 MHz

Further points on the frequency response graphs may be obtained in the same way as that used for the high frequency range, i.e. values of $\frac{1}{2}$, 2, 3, etc. may be assigned to X_{Cc}/R' and the corresponding complex gain and frequencies determined. The resulting gains will have the same values as those found at high frequencies but in this case, the additional phase shift will be added to 180° since the j term in the denominator is negative instead of positive.

Also, since we are putting $\dfrac{X_{Cc}}{R'} = \dfrac{1}{\omega C_c R'} = \frac{1}{2}$, 2, 3, etc. the

corresponding frequencies will be found from $2\omega_1, \dfrac{\omega_1}{2}, \dfrac{\omega_1}{3}$ respectively

where $\omega_1 = \dfrac{1}{C_c R'}$. These results are also shown in *Table 5.4*.

To calculate the overall gain v_0/v_1, of the two stage amplifier, the gain of the first stage, at each frequency, must be multiplied by A_{v2}, the gain of the second stage. This may be found by application of the general solution in terms of the h parameters.

$$A_{v2} = \frac{-h_{te}}{h_{ie}(h_{oe} + Y_L) - h_{te}h_{re}}$$

$$= \frac{-110}{1\,000(100 + 500)10^{-6}} = 184 \angle 180°$$

Thus the overall gain at any frequency is given by the result shown on *Table 5.4* multiplied by 184 and the phase shift may be found by adding 180° to the angle given in the same table.

We must now consider the question of displaying the information obtained from the above analysis. Considering the single stage first, we could plot graphs of $|A_v|$ and phase shift against frequency, but since most of the change in gain occurs between 1 and 400 Hz and between 100 kHz and 10 MHz, a linear frequency scale hides much of the information at low frequencies. This is shown in *Figure 5.10a* where we can see that all detail is lost at frequencies below 1 MHz. An improved display is obtained by using $\log_{10}$ of frequency as a base. *Figure 5.10b* shows both gain and phase shift plotted in this manner. The same graphs are correct for the two stage amplifier under consideration if the vertical scales are changed. The scale for the phase shift graph would range from $-90°$ (at the top) to $+90°$, while the gain scale would be multiplied by 184.

The gain and phase variations may be shown simultaneously by considering $|A_v| \angle \phi$ as a vector and drawing the appropriate vectors for a number of frequencies. This is shown, for the single

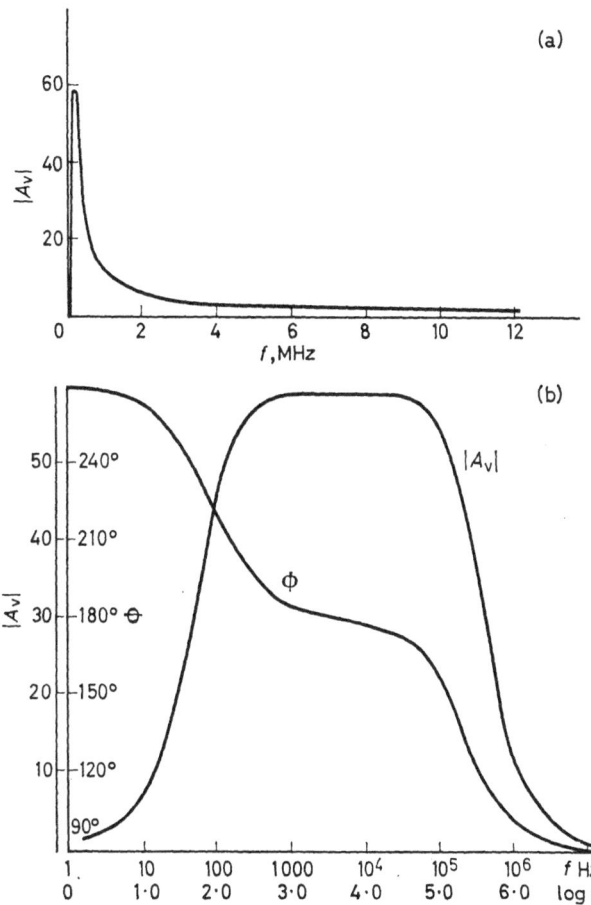

Figure 5.10. Graphs of the variation of gain and phase shift with frequency for Example 5.6. (*a*) Frequency on a linear scale and (*b*) frequency on a log scale

stage, in *Figure 5.11a*. If the ends of all these vectors are joined, we obtain the locus of the gain vector, plotted in the complex plane. In this case, the locus is a circle, with the origin on the circumference. The locus for the two stage amplifier will be obtained by multiplying each vector by 184 and adding 180° to the phase shift. This locus is shown in *Figure 5.11b*. Both methods of presenting the information are useful and examples of their application will appear in later chapters.

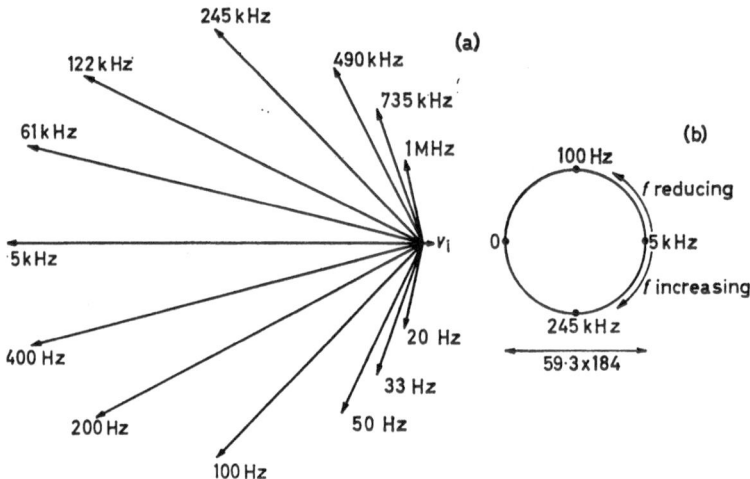

Figure 5.11. Gain vectors for the amplifier in Example 5.6

ALTERNATIVE EQUIVALENT CIRCUITS

The examples so far considered in this chapter have made use of the *h* parameter equivalent circuit. In Chapter 3 other possible circuits were mentioned. Examples of the use of the *y* parameter circuit and the hybrid *π* circuit will appear in Chapter 9, but to show that the general method is applicable to all equivalent circuits, we shall complete this chapter with a solution using the common emitter T equivalent circuit.

Example 5.7. An alternative form of bias circuit for a common emitter amplifier is shown in *Figure 5.12a*. If the transistor employed

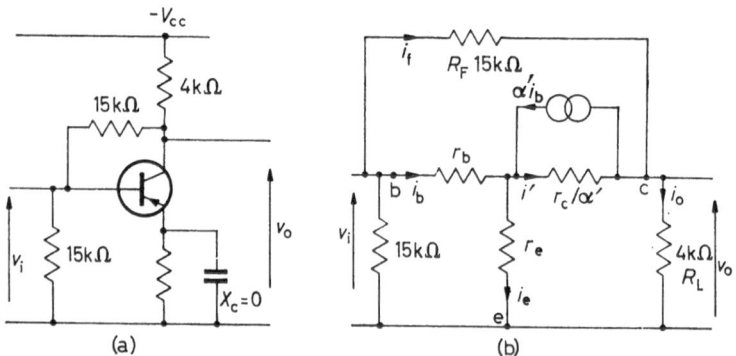

Figure 5.12. Circuits for Example 5.7 using the transistor T equivalent circuit

174

has r_e 30 Ω, r_b 800 kΩ and α' 80, determine the amplifier input impedance and voltage gain.

Figure 5.12b shows the amplifier equivalent circuit drawn using the common emitter T equivalent for the transistor. The unknown currents have been selected as i_b, i_e, and i_t. In terms of these three, the other circuit currents shown are given by:

$$i_o = i_t + i_b - i_e$$
$$i' = i_b - i_e + \alpha'i_b = i_b(1 + \alpha') - i_e$$

Writing the three mesh equations,

$$v_1 = i_b r_b + i_e r_e$$
$$0 = -i_e r_e + i'\frac{r_c}{\alpha'} + i_o R_L$$
$$0 = i_t R_F - i'\frac{r_c}{\alpha'} - i_b r_b$$

Putting $r_c/\alpha' = 10$ kΩ, substituting for i_o and i', inserting values and working in kΩ and mA,

$$v_1 = 0{\cdot}7i_b + 0{\cdot}3i_e \tag{5.27}$$
$$0 = -0{\cdot}03i_e + 10(81i_b - i_e) + 4(i_t + i_b - i_e) \tag{5.28}$$
$$0 = 15i_t - 10(81i_b - i_e) - 0{\cdot}7i_b \tag{5.29}$$

Rearranging and collecting terms,

$$v_1 = 0{\cdot}03i_e + 0{\cdot}7i_b$$
$$0 = -14i_e + 814i_b + 4i_t$$
$$0 = 10i_e - 810i_b + 15i_t$$

Solving by determinants for i_e,

$$i_e = \frac{\begin{vmatrix} v_1 & +0{\cdot}7 & 0 \\ 0 & +814 & +4 \\ 0 & -810 & +15 \end{vmatrix}}{\begin{vmatrix} 0{\cdot}03 & +0{\cdot}7 & 0 \\ -14 & +814 & +4 \\ +10 & -810 & +15 \end{vmatrix}} \text{ mA}$$

$$i_e = \frac{v_1(814 \times 15 + 810 \times 4)}{0{\cdot}03(814 \times 15 + 810 \times 4) - 0{\cdot}7(-14 \times 15 - 4 \times 10) + 0} \text{ mA}$$

$$= \frac{15\,450v_1}{463 + 175} = 24{\cdot}2v_1 \text{ mA}$$

Substituting in equation 5.27,

$$i_b = \frac{v_1 - 0\cdot03v_1 \times 24\cdot2}{0\cdot7} = 0\cdot392v_1 \text{ mA}$$

Substituting in equation 5.28,

$$i_f = \frac{14i_e - 814i_b}{4} = 3\cdot5 \times 24\cdot2v_1 - 203\cdot5 \times 0\cdot392v_1$$

$$= 1\cdot3v_1 \text{ mA}$$

$$\therefore \quad v_0 = i_0R_L = (i_f - i_b - i_e)4 \text{ V} = -22\cdot5 \times 4v_1$$

$$\therefore \quad A_v = \frac{v_0}{v_1} = -90$$

For the input impedance, the input current is $i_b + i_f$,

$$\therefore \qquad Z_{1n} = \frac{v_1}{i_{1n}} = \frac{v_1}{0\cdot392v_1 + 1\cdot3v_1} = 590 \ \Omega$$

But the overall input impedance must include the 15 kΩ bias resistor, it is therefore given by:

$$Z_{1n} = \frac{15 \times 0\cdot59}{15\cdot59} \text{ k}\Omega = 568 \ \Omega$$

SUMMARY

In this chapter, we have investigated the use of small signal equivalent circuits for the solution of a wide range of transistor amplifiers. In the majority of the examples, the h parameters have been used since these are the parameters that are usually available. The methods used, however, are equally applicable to any other form of equivalent circuit. In fact, circuits using devices other than transistors may be analysed in the same way provided the appropriate equivalent circuit parameters are available.

It should be stressed that the parameters quoted in each example are those for a particular transistor at a particular operating point. In the design of a practical amplifier, the typical values quoted in the manufacturers' published data should be used. However, component tolerances and spread of transistor parameters will lead to a possibly large degree of error. This should be allowed for and if a tight specification is required, feedback methods should be used. These methods will be discussed in the next two chapters.

EXAMPLES

Example 5.8. The common emitter amplifier shown in *Figure 5.13* employs a transistor having h_{ie} 1 100 Ω, h_{fe} 75, h_{re} 10^{-3}, h_{oe} 90 μmho.

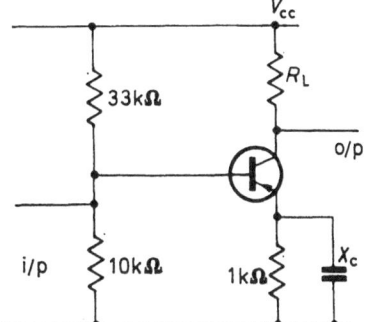

Figure 5.13. Circuit for Example 5.8

If R_L is 2 kΩ calculate the amplifier input impedance and voltage and current gain. Assume $X_c = 0$.

Ans. 862 Ω, -130, $-56\cdot5$.

Example 5.9. Repeat the calculations performed in Example 5.8 using R_L0, 100 Ω, 10 kΩ, and ∞. Hence sketch graphs showing the variation of Z_{in}, A_v, and A_i against R_L.

Ans. 960 Ω, 960 Ω, 645 Ω, 258 Ω. 0, $-0\cdot68$, -560, -3 130. $-65\cdot6$, $-65\cdot6$, $-36\cdot2$, 0.

Example 5.10. Repeat Example 5.8 taking X_c as being infinite.

Ans. $6\cdot83$ kΩ, $-1\cdot97$, $-6\cdot72$.

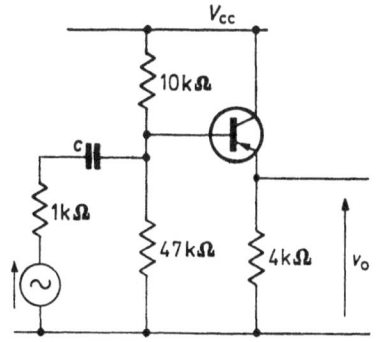

Figure 5.14. Circuit for Example 5.11

Example 5.11. The common collector amplifier shown in *Figure 5.14* employs a transistor having the same parameters as that in Example 5.8. Calculate the output voltage and the output impedance.

Ans. $8\cdot83$ mV, $25\cdot7$ Ω.

177

Example 5.12. Repeat Example 5.11 by finding the common collector *h* parameters and using the general *h* parameter solutions.

Ans. 1 100 Ω, 90 μmho, -76, 1, 8·83 mV, 25·7 Ω.

Example 5.13. The transistor shown in the amplifier circuit in *Figure 5.15* has h_{ie} 1 300 Ω, h_{re} 5 $\times$ 10^{-4}, h_{fe} 90, h_{oe} 120 μmho.

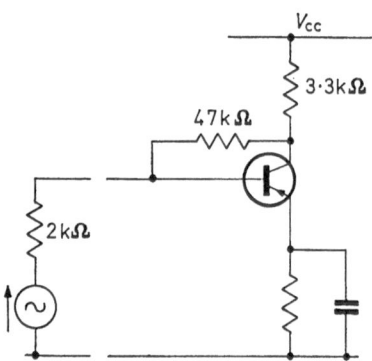

Figure 5.15. Circuit for Example 5.13

Calculate the current gain and the amplifier input and output impedances.

Ans. $-11\cdot35$, 264 Ω, 656 Ω.

Example 5.14. A long tailed pair amplifier has the circuit shown in *Figure 5.7a*. The load on Tr2 is 3 kΩ, the emitter resistor is 1 kΩ and the shunt bias components 47 kΩ and 68 kΩ. If the transistor parameters are h_{ie} 1 kΩ, h_{fe} 110, h_{oe} 80 μmho, h_{re} 0, calculate the voltage gain and the input impedance. Assume that the output is taken between Tr2 collector and earth and that $X_{\text{c}} = 0$.

Ans. 147, 2·22 kΩ.

Example 5.15. A three stage common emitter amplifier has identical transistors for each stage. The corresponding *h* parameters are h_{ie} 1·8 kΩ, h_{fe} 150, h_{oe} 80 μmho and negligible h_{re}. Each stage has a collector load of 2 kΩ and the bias components effectively shunt the input of each stage with 10 kΩ. Calculate the overall voltage and current gain and the input impedance.

Ans. 650 $\times$ 10^3, 496 $\times$ 10^3, 1 528 Ω.

Example 5.16. A two stage RC coupled amplifier has the following components. Collector loads, 3·3 kΩ, shunt bias resistors 8 kΩ

effective, per stage, coupling capacitor 0·5 μF, effective inter-stage shunt capacitance 500 pF. If the transistors have h_{ie} 1·2 kΩ, h_{fe} 80, h_{oe} 130 μmho, and h_{re} 0, calculate A_{vm} and the 3 db frequencies.

Ans. 7 370, 444 kHz, 105 Hz.

Example 5.17. A single stage common emitter amplifier is loaded with a 2 kΩ resistor in parallel with a 0·01 μF capacitor. The bias components place 15 kΩ in parallel with the input and the transistor parameters are h_{ie} 1·3 kΩ, h_{re} 10^{-3}, h_{fe} 125, and h_{oe} 130 μmho. Calculate the voltage gain and input impedance at a frequency of 4 kHz.

Ans. 162 $\angle$ 168°, 1 130 $\angle$ 3° 30′.

Example 5.18. The equivalent T parameters of a transistor are r_{e} 20 Ω, r_{b} 350 Ω, r_{c} 750 kΩ, and α 0·992. It is connected as a common emitter amplifier and loaded with 1·5 kΩ. If the shunt bias components total 20 kΩ, calculate the voltage and current gain and the input impedance.

Ans. −63·3, −89, 2·10 kΩ.

Example 5.19. The amplifier shown in *Figure 5.16* employs a matched pair of transistors having h_{ie} 1 300 Ω, h_{fe} 90, h_{re} 5 × 10^{-4},

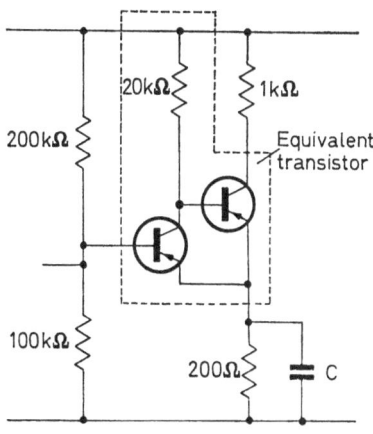

Figure 5.16. Circuit for Example 5.19

and h_{oe} 125 μmho. Find the h parameters of the equivalent transistor (inside the broken line) and hence determine A_{v} and Z_{in} with C considered as either short circuit or open circuit.

Ans. 1 123 Ω, −6 600, 20·4 × 10^{-8}, 119 μmho.

5 260, 1 122 Ω, 5, 63 kΩ.

Example 5.20. A Darlington pair of transistors is connected as shown in *Figure 5.8a*. The transistor parameters are Tr1 (small signal), h_{1e} 1·5 kΩ, h_{re} 4 × 10⁻⁴, h_{oe} 110 μmho, h_{fe} 130 and Tr2 (power) h_{1e} 200 Ω, h_{re} 10⁻³, h_{oe} 500 μmho, h_{fe} 70. Calculate the voltage gain, the current gain and the input and output impedance if (*a*) a load of 400 Ω is connected in the combined collector or (*b*) if the same load is connected in the emitter load of Tr2. In each case, take the combined shunt bias components to be 100 kΩ. Assume Z_8 to be 10 kΩ.

Ans. (*a*) −41·0, −1 270, 14·6 kΩ, 88 Ω. (*b*) 0·973, 218, 90 kΩ, 4·0 Ω.

6

THE THEORY OF FEEDBACK
AMPLIFIERS

In the preceding chapters, a number of properties of electronic amplifiers have become apparent. Before defining and investigating feedback it will be useful to review these properties.

THE PROPERTIES OF AMPLIFIERS

Amplification

An amplifier will have voltage gain, current gain or both. In this context we mean that the alternating voltage across the load will be greater in magnitude than the alternating voltage at the input terminals, or that the alternating current in the load will be greater in magnitude than the alternating current flowing into the input terminals. In either case the gain will be a function of the passive amplifier components, and of the device parameters. The passive components may vary with temperature and time, and if one is replaced by another of nominally the same value, manufacturing tolerances may result in a considerable change in exact value. The device parameters may also vary with time and temperature, and in the event of a replacement, the probability of obtaining identical parameters is most unlikely. In addition the parameters may be very sensitive to d.c. operating conditions, and a change of d.c. supply voltage may result in a considerable change in gain.

Impedance

Amplifiers also have both input and output impedance. The input impedance is the ratio of input voltage to input current. The output impedance is best compared with the internal resistance of a signal generator. The input impedance determines the suitability of the amplifier for use with a particular source. If the source internal impedance is much greater than the amplifier input impedance, the terminal input voltage may be so small that the amplifier output is less than the original open circuit source voltage. Similarly, the output impedance determines the suitability of a particular load. In this context the load may well be the input impedance of another amplifier or electronic circuit.

These impedances will be sensitive to changes in both passive components and device parameters in the same way as the gain.

Phase Shift

Amplifiers introduce a phase shift between input and output current or voltage. At medium frequencies this will be either 180° or 360° depending upon the number of stages in the amplifier, and their configuration.

Frequency Response

All the properties discussed above may vary with frequency. In general the gain will be smaller at very low and very high frequencies. The impedance will usually increase at very low frequencies and be reduced at very high frequencies. The phase shift will rise with lower frequencies and be reduced at higher frequencies. These variations are principally due to changes in the reactance of capacitors in the amplifier circuits, but device parameters may also be frequency sensitive, particularly at high frequencies.

Distortion and Noise

Since all active devices are basically non-linear, some degree of distortion will always be introduced. This takes the form of additional alternating voltages known as harmonics in the output. These will occur at frequencies which are multiples of the desired signal frequency. Other unwanted signals may also be introduced within the amplifier due to residual mains variation or hum in the d.c. supply or due to electrical noise occurring in either active or passive circuit components.

Definition of Feedback

In this chapter we shall see how all these amplifier properties may be modified by the use of feedback. In general a feedback amplifier or system is one in which the terminal input signal is the sum of an external signal and a feedback signal proportional to the output signal. At this stage it is convenient to work in general quantities or signals rather than in voltages or currents. The basic feedback amplifier is shown in *Figure 6.1*.

The large box represents the amplifier having a gain A, where in general $A = |A| \angle \theta$.

Thus $\left|\dfrac{S_o}{S'}\right| = |A|$ and S_o leads S' by a phase angle θ. The small box represents a feedback network having an attenuation β, where in general $\beta = |\beta| \angle \phi$.

Thus $\left|\dfrac{\beta S_o}{S_o}\right| = |\beta|$ and βS_o leads S_o by a phase angle ϕ. Usually, but not essentially, $|\beta|$ is less than one.

S is the input signal applied from an outside source, S' is the terminal input signal, and S_o is the output signal.

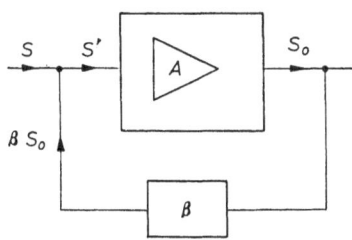

Figure 6.1. Block diagram for an amplifier with feedback

From the diagram:

$$S' = S + \beta S_o \qquad (6.1)$$

But $$S_o = AS'$$

$\therefore$ $$S' = S + \beta AS'$$

and $$S'(1 - \beta A) = S$$

so $$S' = \frac{S}{1 - \beta A}$$

Finally $$S_o = \frac{AS}{1 - \beta A}$$

and the overall gain with feedback

$$A_f = \frac{S_o}{S} = \frac{A}{1 - \beta A}$$

Remembering that both β and A will be complex, the complete expression for gain with feedback

$$A_f = \frac{|A| \angle \theta}{1 - |\beta A| \angle \theta + \phi} \qquad (6.2)$$

The modulus of $(1 - |\beta A| \angle \theta + \phi)$ may be either greater than 1 or less than 1, so application of feedback can either increase or reduce the overall gain. In this respect the feedback is said to be positive if the overall gain is increased and negative if it is reduced, i.e.

$$|A_f| > |A|, \text{ Feedback positive}$$
$$|A_f| < |A|, \text{ Feedback negative}$$

It is important to realise that the same feedback system may result in negative feedback at some frequencies and positive feedback at other frequencies, so in general it is not correct to refer to a 'negative feedback amplifier'.

Simple Negative Feedback

When the term is used, the implication is that over the normal operating frequency ranges, the feedback is negative. These would normally be the 'medium frequencies' at which the amplifier phase shift is 180° or 360°. Under these conditions the β factor is usually provided by a simple resistive network, the connection of which makes $\theta + \phi = 180°$.

Thus equation 6.2 becomes

$$|A_f| = \frac{|A|}{1 + |A\beta|} \tag{6.3}$$

This condition will be referred to as *simple negative feedback*.

We shall now consider the effects of feedback, and in particular negative feedback, on the properties of amplifiers discussed above.

Amplification with Feedback

Application of simple negative feedback may be shown to improve the gain stability of an amplifier, i.e. the gain becomes less susceptible to the changes of parameters etc. discussed above. With reference to equation 6.3 suppose β is such that $|\beta A| \gg 1$.

Now
$$|A_f| \simeq \frac{A}{|\beta A|} = \frac{1}{|\beta|}$$

But β is normally a fraction resulting from a simple resistive network and will not be subject to changes resulting from device ageing or replacement, d.c. supply changes, or temperature changes. (Since all resistors will change in approximately the same proportion.)

Example 6.1. A two stage transistor amplifier is constructed using transistors with a nominal h_{fe} of 125 resulting in an overall current gain of 3 000. Simple negative feedback is applied using a β of 1/600.

If a change in the direct supply voltage reduces h_{te} to 80, determine the percentage change in overall gain with and without feedback.

We shall first find the change in gain without feedback. Since gain per stage is proportional to h_{te}, we can say that the overall gain will be proportional to h_{te}^2

$$\therefore \qquad 3\ 000 = K(125)^2$$

where K is a constant.

$$\therefore \qquad K = \frac{3\ 000}{125^2} = 0 \cdot 191$$

$\therefore$ the new gain $A_2 = 0 \cdot 191(80)^2 = 1\ 220$.

Percentage change in terms of the original gain

$$= \frac{3\ 000 - 1\ 220}{3\ 000} \times 100 \text{ per cent}$$

Percentage change $= 59 \cdot 3$ per cent

Now applying simple negative feedback to the original amplifier

$$A_{1f} = \frac{3\ 000}{1 + \dfrac{3\ 000}{600}} = 500$$

and

$$A_{2f} = \frac{1\ 200}{1 + \dfrac{1\ 220}{600}} = 402$$

Percentage change in gain with feedback $= \dfrac{500 - 402}{500} \times 100$ per cent

$$= 19 \cdot 5 \text{ per cent}$$

Thus the application of feedback has reduced the percentage change in gain from approximately 60 per cent to approximately 20 per cent, which is a considerable improvement.

Before we can consider in detail the values of β and A and the effects of feedback on gain, input and output impedance, we must look back at the definition of a feedback amplifier.

SERIES AND PARALLEL CONNECTED FEEDBACK

From the definition of a feedback system, the terminal input signal is the *sum* of an input signal and a feedback signal. With electronic amplifiers, the feedback signal may be added in *series* or in *parallel* with the input signal. If two electrical quantities are to be added in series, the two quantities must be voltages.

Alternatively, if they are to be added in parallel, the quantities must be currents. In general amplifiers will amplify both current and voltage, but depending on the method of application of the feedback signal they must be treated as either voltage amplifiers or current amplifiers.

Thus if the feedback signal is added in series, the circuit will be treated as a voltage amplifier and the terminal current gain will be unchanged. The overall current gain however will be modified by changes in input and output impedance due to the voltage feedback. Similarly current feedback will not change the terminal voltage gain.

VOLTAGE AMPLIFIERS WITH FEEDBACK

To investigate these effects, it is convenient to introduce a general equivalent circuit for firstly, a voltage amplifier. This is shown in *Figure 6.2*, the equivalent for the amplifier being that part of the diagram within the box.

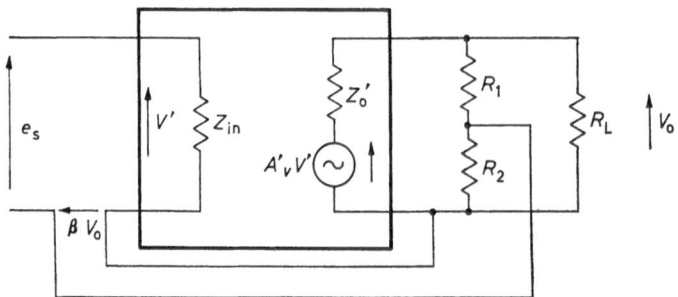

Figure 6.2. Equivalent circuit for a voltage amplifier with voltage derived feedback

Our equivalent voltage amplifier has three properties, an input impedance Z_{1n}, an open circuit output impedance Z_0' and an open circuit voltage gain A_v'. If this were to represent a multi-stage transistor amplifier, these properties would be given approximately by h_{1e} in parallel with the bias components for Z_{1n}, h_{oe}^{-1} for Z_0', and the terminal voltage gain with infinite final load for A_v'.

The feedback voltage may be obtained from the output in one of two ways. *Figure 6.2* shows the first of these; the feedback circuit, R_1, R_2, is in parallel with the output and the feedback voltage is said to be derived from the output voltage. The alternative will be

186

referred to as current derived feedback and is obtained when the feedback circuit is in series with the amplifier load. This is shown in *Figure 6.4*.

Voltage Derived, Series Applied Feedback

Referring again to *Figure 6.2* we shall investigate the effects of voltage derived series applied feedback upon the overall voltage gain A_{vf} and input impedances Z_{inf}. β is defined as that fraction of the output voltage that is fed back and added to the input signal voltage. Inspection of the circuit shows:

$$\beta V_0 = \frac{R_2}{R_1 + R_2} V_0$$

$\therefore$
$$\beta = \frac{R_2}{R_1 + R_2} \qquad (6.4)$$

Now the gain without feedback A_{vo} is given by

$$A_{vo} = \frac{A_v' R_L'}{R_L' + Z_0'} \qquad (6.5)$$

where
$$R_L' = \frac{R_L(R_1 + R_2)}{R_L + R_1 + R_2} \simeq R_L \qquad (6.5a)$$

since in practice $R_1 + R_2 \gg R_L$. (This is not always the case and should be checked.)

Now summing voltages at the input,

$$V' = e_s + \beta V_0 = e_s + A_{vo}\beta V' \qquad (6.6)$$

$\therefore$
$$V' = \frac{e_s}{1 - A_{vo}\beta}$$

but
$$V_0 = A_{vo}V' = \frac{A_{vo}e_s}{1 - A_{vo}\beta}$$

$\therefore$
$$\text{Gain with feedback} = \frac{V_0}{e_s} = \frac{A_{vo}}{1 - \beta A_{vo}} \qquad (6.7)$$

From this result, since β is positive (equation 6.4) simple negative feedback will occur if A_{vo} is negative. If A_{vo} is positive, an alternative connection may be used to make β negative.

Input Impedance With Feedback

From equation 6.6
$$e_s = V'(1 - \beta A_{vo})$$

If this equation is divided by i the input current,

$$\frac{e_\mathrm{s}}{i} = \frac{V'}{i}(1 - \beta A_\mathrm{vo})$$

But inspection of *Figure 6.2* shows that V'/i is the amplifier input impedance Z_in. Also e_s/i must be the input impedance with feedback Z_inf. The previous equation therefore becomes:

$$Z_\mathrm{inf} = Z_\mathrm{in}(1 - \beta A_\mathrm{vo}) \tag{6.8}$$

Thus with simple negative feedback the amplifier input impedance is increased.

Output Impedance with Feedback

To determine the output impedance, we must redraw the equivalent circuit with any external generators suppressed, and apply a generator of E volts to the output terminals. By calculating the resulting current I the output impedance is given by E/I. The redrawn equivalent circuit is shown in *Figure 6.3*.

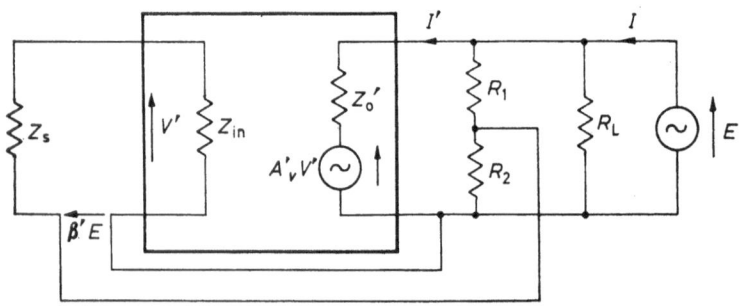

Figure 6.3. Circuit for determining the output impedance of a voltage amplifier with voltage derived feedback

We shall first determine the terminal output impedance with feedback Z_ot as given by E/I'. The overall resulting output impedance can then be found by adding R_L and $R_1 + R_2$ in parallel with Z_ot. If $Z_\mathrm{s} + Z_\mathrm{in} \gg R_2$ and if $Z_\mathrm{s} \ll Z_\mathrm{in}$ we can say

$$V' = \beta V_\mathrm{o} = \beta E \tag{6.9}$$

If not, R_2' should be used in determining β'. Where

$$R_2' = \frac{R_2(Z_\mathrm{s} + Z_\mathrm{in})}{R_2 + Z_\mathrm{s} + Z_\mathrm{in}}$$

188

and
$$V' = \beta' E \frac{Z_{in}}{Z_s + Z_{in}}$$

For most practical purposes equation 6.9 is satisfactory.
  Writing a mesh equation:

$$E - A_v'V' = I'Z_0'$$

From equation 6.9

∴
$$E(1 - \beta A_v') = I'Z_0'$$

$$Z_{of} = \frac{E}{I'} = \frac{Z_0'}{1 - \beta A_v'} \tag{6.10}$$

The overall output impedance can thus be found from

$$\frac{1}{Z} = \frac{1}{Z_{of}} + \frac{1}{R_L} + \frac{1}{R_1 + R_2}$$

Alternatively the loaded output impedance without feedback, and the loaded voltage gain A_{vo} may be used in equation 6.10 and the same result will be achieved. In practice however where simple negative feedback has been used $Z_{of} \ll R_L < R_1 + R_2$ and the shunting effect of R_L and $R_1 + R_2$ is negligible.

Current Derived, Series Applied Feedback

We must now consider the effects of current derived feedback. The required circuit is shown in *Figure 6.4*.

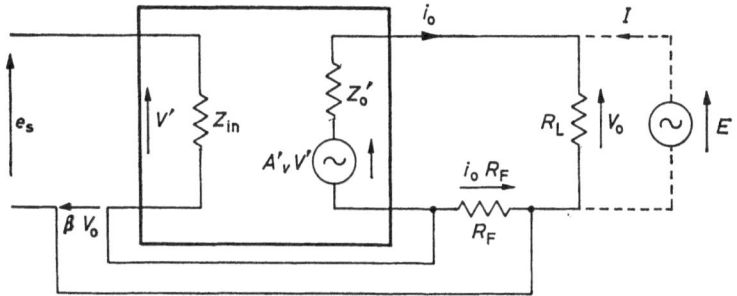

Figure 6.4. Equivalent circuit for a voltage amplifier with current derived feedback

In this case the output current flows in both the load and a resistor R_F. The resulting voltage i_0R_F is applied in series with the signal as βV_0.

189

Thus if $R_F \ll Z_s + Z_{1n}$ (as is usual)

$$\frac{\beta V_0}{V_0} = \frac{i_0 R_F}{i_0 R_L}$$

$$\therefore \qquad \beta = \frac{R_F}{R_L} \qquad (6.11)$$

In addition, A_{vo} is modified from equation 6.5 to,

$$A_{vo} = \frac{A_v' R_L}{R_L + R_F + Z_0'} \qquad (6.12)$$

With these new values for A_{vo} and β the results for gain (equation 6.7) and input impedance (equation 6.8) are unchanged.

Output Impedance with Feedback

For the output impedance calculation the generator E (shown in *Figure 6.4*) is applied and the source e_s suppressed. As with the voltage derived case R_L can be included afterwards if required.

Now since I is in the opposite direction to i_0,

$$V' = -IR_F \text{ (if } Z_{1n} + Z_s \gg R_F \text{ and } Z_s \ll Z_{1n})$$

and writing the mesh equation

$$E - A_v'(-IR_F) = I(Z_0' + R_F)$$

arranging and collecting terms,

$$E = IZ_0' + IR_F(1 - A_v')$$

or $\qquad Z_{ot} = \frac{E}{I} = Z_0' + R_F(1 - A_v') \qquad (6.13)$

With simple negative feedback this can result in a considerable increase in terminal output impedance. However the shunting effect of R_L on the overall output impedance normally makes this effect unimportant.

Example 6.2. An amplifier having input impedance 1 kΩ and open circuit output impedance 10 kΩ has a voltage gain of 800 when loaded with a 4 kΩ resistor.

A feedback voltage is derived from (*a*) a potential divider chain in parallel with the load having R_1 9·9 kΩ and R_2 100 Ω, or (*b*) a 100 Ω resistor in series with the load. In each case the series connection at the input ensures simple negative feedback.

Determine, for both methods, the overall terminal voltage gain, the modified input and output impedance, and the change in output current if the amplifier is driven from a source of 1 mV e.m.f. and internal impedance 1 kΩ.

For the amplifier without feedback, the input current i_{in} is given by:

$$i_{in} = \frac{1 \text{ mV}}{1 \text{ k}\Omega + 1 \text{ k}\Omega} = 0\cdot5 \text{ }\mu\text{A}$$

and $\qquad V' = i_{in}Z_{in} = 0\cdot5 \text{ }\mu\text{A} \times 1 \text{ k}\Omega = 0\cdot5 \text{ mV}$

Since $\qquad\qquad$ Voltage gain $= 800$

$$\text{Output voltage } v_o = 800 \times 0\cdot5 \text{ mV} = 0\cdot4 \text{ V}$$

$$\text{Output current } i_o = \frac{0\cdot4}{4\,000} \text{ A} = 100 \text{ }\mu\text{A} \qquad (6.14)$$

To investigate the effects of the feedback circuits, we must first find the open circuit voltage gain A_v'.

Applying equation 6.5

$$800 = \frac{A_v'4}{4 + 10}$$

$\therefore \qquad\qquad A_v' = \frac{800 \times 14}{4} = 2\,800$

Now consider case (a), the voltage derived feedback.

The effect of $R_1 + R_2$ cannot be neglected, so we must find R_L' and A_{vo} from equations 6.5 and 6.5a.

Since $\qquad R_1 + R_2 = 10 \text{ k}\Omega, \quad R_L' = \frac{10 \times 4}{14} = 2\cdot86$

$\therefore \qquad\qquad A_{vo} = \frac{2\,800 \times 2\cdot86}{10 + 2\cdot86} = 623$

Since $R_2 \ll (Z_{in} + Z_s)$, from equation 6.4,

$$\beta = \frac{100}{10\,000} = 0\cdot01$$

As the feedback is negative, from equations 6.7 and 6.8,

$$A_{vf} = \frac{623}{1 + 623 \times 0\cdot01} = 86\cdot2$$

and $\qquad Z_{in} = 1(1 + 623 \times 0\cdot01) \text{ k}\Omega = 7\cdot23 \text{ k}\Omega$

Now   New input current $= \dfrac{1 \text{ mV}}{8\cdot23 \text{ k}\Omega} = 0\cdot122 \ \mu\text{A}$

and   Overall input voltage $= 0\cdot122 \times 7\cdot23 \text{ mV} = 0\cdot88 \text{ mV}$

∴ $v_0 = 0\cdot88 \times 86\cdot2 \text{ mV} = 75\cdot8 \text{ mV}$

New output current $= \dfrac{75\cdot8 \text{ mV}}{4 \text{ k}\Omega} = 19 \ \mu\text{A}$

Note that although this current is considerably less than that found for the amplifier without feedback (equation 6.14), the current gain is unchanged.

If the original input current $0\cdot5 \ \mu\text{A}$ was applied, the overall input voltage is given by:

$$v_{\text{in}} = 0\cdot5 \ \mu\text{A} \times 1 \text{ k}\Omega(1 + 6\cdot23)$$

and $V_0 = A_{\text{vf}}V_{\text{in}}$

$$= 0\cdot5 \ \mu\text{A} \times 1 \text{ k}\Omega(1 + 6\cdot23) \times \frac{623}{1 + 6\cdot23}$$

$$= 311 \text{ mV}$$

The output current is given by v_0/R_{L}',

∴ $i_0 = \dfrac{311 \text{ mV}}{2\cdot68 \text{ k}\Omega} = 109 \ \mu\text{A}$

With reference to equation 6.14, it would appear that the current gain has increased! This is in fact so, and the reason is simply that the load has been changed from an R_{L} of 4 kΩ to the R_{L}' of 2·86 kΩ.

To return to the problem, we must find the modified output impedance using equation 6.10.

$$Z_{\text{of}} = \frac{10 \text{ k}\Omega}{1 + 6\cdot23} = 1\cdot38 \text{ k}\Omega$$

The overall output impedance including the load is found by shunting Z_{of} with R_{L}'

∴   Overall $Z_0 = \dfrac{1\cdot38 \times 2\cdot86}{4\cdot24} \text{ k}\Omega = 930 \ \Omega$

In part (b) of the example we have current derived feedback, and using equations 6.11, 6.12, 6.7 and 6.8 we find,

$$\beta = \frac{R_{\text{F}}}{R_{\text{L}}} = \frac{100 \ \Omega}{4 \text{ k}\Omega} = 0\cdot025$$

$$A_{\text{vo}} = \frac{2\,800 \times 4}{14\cdot1} = 795$$

192

$$\therefore \qquad A_{vf} = \frac{795}{1 + 795 \times 0.025} = 38.1$$

and $\qquad Z_{inf} = 1(1 + 795 \times 0.025)\, k\Omega = 20.8\, k\Omega$

For the output current calculation

$$i_{in} = \frac{1\, mV}{21.8\, k\Omega}$$

and $\qquad v_{in} = i_{in}Z_{inf} = \frac{20.8}{21.8}\, mV$

$$v_o = 38.1 \times \frac{20.8}{21.8}\, mV$$

$$i_o = \frac{v_o}{R_L} = \frac{38.1 \times 20.8}{4 \times 21.8}\, \mu A = 9.1\, \mu A$$

As before, the current gain has not been reduced, as has the input current, as a result of the increase in input impedance.

Finally for the output impedance with current derived feedback, we refer to equation 6.13. From which

$$Z_{of} = 10\, k\Omega + 100\, \Omega(1 + 2\,800)$$
$$= 290\, k\Omega$$
$$\text{Overall output impedance} = \frac{290 \times 4}{294}\, k\Omega$$
$$= 3.95\, k\Omega$$

This result should be compared with the loaded output impedance of the original amplifier without feedback.

This is given by $\qquad \dfrac{4 \times 10}{14} = 2.86\, k\Omega$

Summarizing these results:
  No feedback:

$\qquad A_v$ 800, $\qquad Z_{in}$ 1 000 Ω, $\qquad Z_o$ 2 860 Ω

  (a) Voltage derived feedback,

$\qquad A_v$ 86.2 $\qquad Z_{in}$ 7.23 kΩ $\qquad Z_o$ 930 Ω

  (b) Current derived feedback,

$\qquad A_v$ 38.1 $\qquad Z_{in}$ 20.8 kΩ $\qquad Z_o$ 3 950 Ω

Thus in each case negative feedback reduces the voltage gain and increases the input impedance. With voltage derived feedback, the output impedance is reduced, but with current derived feedback the output impedance is increased.

CURRENT AMPLIFIERS WITH FEEDBACK

If the feedback signal is applied in parallel with the input signal, the two quantities to be added must be currents, and the amplifier must be treated as a current amplifier. The equivalent circuit for a current amplifier is shown in *Figure 6.5*.

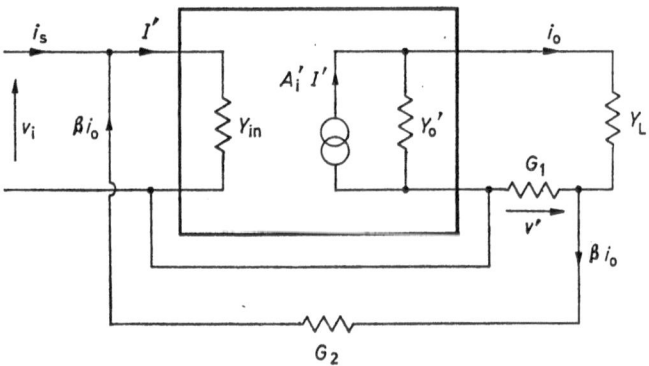

Figure 6.5. Circuit for a current amplifier with current derived feedback

As with the equivalent voltage amplifier shown in *Figure 6.2* the amplifier is represented by that part of the diagram enclosed in the box. The equivalent circuit properties are an input admittance Y_{in}, a terminal output admittance (excluding final load) of Y_o', and a short circuit current gain A_1'. For a multi-stage transistor amplifier, these would be given by:

$$Y_{\text{in}} \simeq \frac{1}{h_{1e}}$$

in parallel with the bias components.

$$Y_o' \simeq h_{oe}$$

for the final transistor.

$A_1' \simeq h_{fe}$ of the last stage multiplied by the current gain of all preceding stages.

194

Current Derived Current Feedback

In *Figure 6.5*, Y_L is the final load, and G_1 and G_2 form the current derived feedback network.

The feedback current

$$\beta i_0 = \frac{G_2}{G_1 + G_2} i_0$$

$$\therefore \qquad \beta = \frac{G_2'}{G_1 + G_2'} \qquad (6.15)$$

Note: The input admittance Y_{in} is negligible in the determination of β since v_1 is very much less than v', the voltage across G_1

$$\therefore \qquad Bi_c = (v' - v_1)G_2 \simeq v'G_2 \qquad (6.16)$$

Next we require A_{1o} the current gain without feedback. This is given by:

$$A_{1o} = \frac{A_1' Y_L'}{Y_o' + Y_L'} \qquad (6.17)$$

where

$$Y_L' = \frac{Y_L(G_1 + G_2)}{Y_L + G_1 + G_2} \qquad (6.18)$$

which in practice may often be approximated to Y_L.

Adding the currents at the input we obtain

$$I' = i_s + \beta i_0$$
$$= i_s + \beta A_{1o} I'$$

$$\therefore \qquad i_s = I'(1 - \beta A_{1o}) \qquad (6.19)$$

and

$$I' = \frac{i_s}{1 - \beta A_{1o}}$$

But

$$i_0 = \frac{A_{1o} i_s}{1 - \beta A_{1o}}$$

$$\therefore \qquad \text{Current gain with feedback } A_{1f} = \frac{A_{1o}}{1 - \beta A_{1o}} \qquad (6.20)$$

To find the input admittance with feedback we divide equation 6.19 by the terminal input voltage v_1.

$$\frac{i_s}{v_1} = \frac{I'}{v_1}(1 - \beta A_{1o})$$

But

$$\frac{I'}{v_1} = Y_{in} \quad \text{and} \quad \frac{i_s}{v_1} = Y_{inf}$$

the input admittance with feedback

$$\therefore \qquad Y_{inf} = Y_{in}(1 - \beta A_{1o}) \qquad (6.21)$$

195

Note from equations 6.20 and 6.24, with simple negative feedback, the current gain is reduced, and the input admittance is increased (i.e. the input impedance is reduced).

The reader may have noticed by this stage, that this derivation is proceeding along identical lines to that for the voltage amplifier with voltage derived feedback. Since admittances and currents are being used *Figure 6.5* can be regarded as the dual of *Figure 6.2*.

Output Admittance with Feedback

For the output admittance with feedback, the current source must be replaced by its internal admittance, and a generator of I amps connected to the output terminals. Calculation of the resulting output voltage V leads to the output admittance Y_{ot} being calculated from $Y_{ot} = I/V$.

This circuit arrangement is shown in *Figure 6.6*.

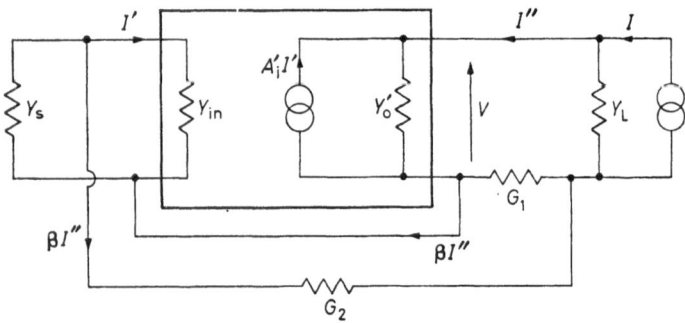

Figure 6.6. Circuit for determining the output admittance of a current amplifier with current derived feedback

First we shall calculate the terminal output admittance given by I''/V. The effect of Y_L, G_1 and G_2 may be included afterwards to give the overall output admittance if required.

First note that if $Y_s \gg Y_{in}$ where Y_s is the source admittance

$$I' = \beta' I'' \quad \text{where} \quad \beta' = \frac{G_2'}{G_1 + G_2'} \quad \text{and} \quad G_2' = \frac{G_2 + Y_{in} + Y_s}{G_2(Y_{in} + Y_s)}$$

(6.22)

if not use β'' where

$$\beta'' = \frac{\beta' Y_{in}}{Y_s + Y_{in}}$$

196

Now writing a nodal equation for V:

$$I'' + A_1'I' = VY_0'$$
$$\therefore \qquad I'' - \beta''A'I'' = VY_0'$$

and the output admittance with feedback Y_{of} is given by

$$Y_{of} = \frac{I''}{V} = \frac{Y_0'}{1 - \beta''A_1'} \qquad (6.23)$$

The overall output admittance can then be obtained by adding $G_1 + G_2'$ in series with Y_{of} and Y_L in parallel with the result.

$$\therefore \quad \text{Overall output admittance} = \frac{Y_{of}(G_1 + G_2')}{Y_{of} + G_1 + G_2'} + Y_L \quad (6.24)$$

With typical values, Y_L will be the predominent term and equation 6.24 will approximate to Y_L.

Voltage Derived Current Feedback

The alternative connection for current amplifiers, resulting in voltage derived feedback, is shown in *Figure 6.7*.

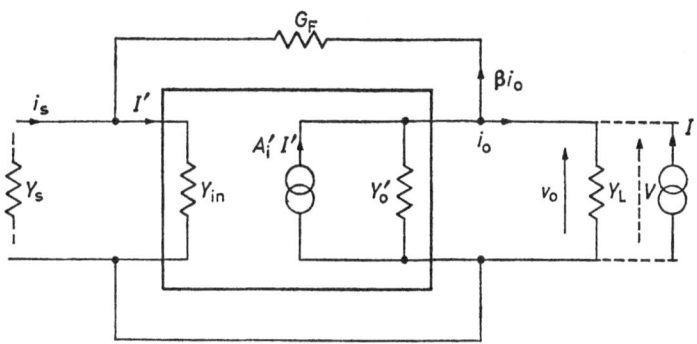

Figure 6.7. Circuit for a current amplifier with voltage derived feedback

As with the current derived feedback, we must first define β and A_{1o}. Examination of the circuit shows that

$$i_o = v_o Y_L$$
and
$$\beta i_o = v_o G_F$$
since
$$v_1 \ll v_o.$$

197

Thus
$$\beta = \frac{G_F}{Y_L} \qquad (6.25)$$

Also,
$$A_{1o} = \frac{A_1' Y_L}{Y_0' + Y_L + G_F'} \simeq \frac{A_1' Y_L}{Y_0' + Y_L} \qquad (6.26)$$

Using these values of β and A_{1o}, the current gain and input admittance with feedback may be determined from equations 6.20 and 6.21 respectively.

Output Admittance

The output admittance is calculated in the usual manner by connecting the generator of I amps to the output terminals, suppressing the source generator, and calculating the resulting V.

Under these conditions

$$I' = VG_F' \times \frac{Y_{1n},}{Y_{1n} + Y_s} \quad \text{where} \quad G_F' = \frac{G_F(Y_{1n} + Y_s)}{G_F + Y_{1n} + Y_s}$$

and putting

$$A_1'' = A_1' \frac{Y_{1n}}{Y_{1n} + Y_s}$$

Writing a nodal equation

$$I + A_1'' VG_F' = V(Y_0' + G_F' + Y_L)$$

and

Overall output admittance $= \dfrac{I}{V} = Y_0' + Y_L + G_F'(1 - A_1'')$ (6.27)

Once again the last term will usually predominate and

$$Y_{of} \simeq G_F(1 - A_1')$$

which with negative feedback makes the output impedance very low.

Example 6.3. The amplifier described in Example 6.2 is to have feedback connected in shunt with the i/p by one of two alternative methods. In each case the connection of the feedback network ensures that simple negative feedback is applied. The two networks are (a) a 200 Ω resistor (G_1) connected in series with the load and a 50 kΩ resistor (G_2) connected in parallel with G_1 to the input terminals and (b) a 100 kΩ resistor (G_F) connected directly from the output terminal to the input terminal. Assuming $Y_s \ll Y_{1n}$ determine in each case, the overall current gain, input admittance, and output admittance.

We must first find the parameters of our equivalent current amplifier.

$$Y_{in} = \frac{1}{Z_{in}} = 1\,000\ \mu mho$$

$$Y_o' = \frac{1}{Z_o'} = 100\ \mu mho$$

$$A_1' = \frac{i_{sc}}{i_{in}} = \frac{\dfrac{2\,800v'}{10\,000}}{\dfrac{v'}{1\,000}} = 280$$

where i_{sc} is the current that would flow in a short circuit connected across the output terminals, and i_{in} is the current flowing into the input terminals without feedback. Also

$$Y_L = \frac{1}{Z_L} = 250\ \mu mho$$

For part (a) applying equations 6.15, 6.16, 6.17 and 6.18

$$G_2 = \frac{1}{50\ k\Omega} = 20\ \mu mho \quad and \quad G_1 = \frac{1}{200\ \Omega} = 5\,000\ \mu mho$$

$$\beta = \frac{20}{20 + 5\,000} = \frac{1}{251}$$

and

$$Y_L' = \frac{250\,(5\,020)}{250 + 5\,020} = 238\ \mu mho$$

∴

$$A_{1o} = \frac{280 \times 238}{238 + 100} = 197$$

Now applying equations 6.20 and 6.21, remembering that βA_{1o} will be negative.

$$A_{1f} = \frac{197}{1 + \dfrac{197}{251}} = 110$$

and

$$Y_{inf} = 1\,000 \left(1 + \frac{197}{251}\right)\ \mu mho = 1\,786\ \mu mho$$

(representing an input impedance of 560 Ω).

For the terminal output admittance, applying equations 6.22 and 6.23

$$\beta' = \frac{19\cdot6}{5\,000 + 19\cdot6} \simeq \frac{1}{251}$$

$$\therefore \qquad Y_{ot} = \frac{100}{1 + \dfrac{197}{251}}\,\mu\text{mho} = 56\,\mu\text{mho}$$

and the overall output admittance from equation 6.24:

$$Y_0 = \frac{56 \times 5\,020}{5\,020 + 56} + 250 = 305\,\mu\text{mho}$$

(representing an output impedance of $3\cdot28$ kΩ).

This should be compared with the original overall output admittance of $100 + 250 = 350\,\mu$mho.

For the voltage derived feedback in case (b), β and A_{1o} are obtained from equations 6.25 and 6.26.

$$G_F' = \frac{1}{100\text{ k}\Omega} = 10\,\mu\text{mho}$$

$$\beta = \frac{10}{250} = \frac{1}{25}$$

and

$$A_{1o} = \frac{280 \times 250}{250 + 100 + 10} = 194$$

$\therefore$ from equations 6.20 and 6.21,

$$A_{1f} = \frac{194}{1 + \dfrac{194}{25}} = 22\cdot1$$

$$Y_{1nf} = 1\,000\left(1 + \frac{194}{25}\right) = 8\,780\,\mu\text{V}$$

(representing an input impedance of 114 Ω).

The output admittance with feedback can now be determined from equation 6.27.

$$A_1'' = A_1' \quad \text{and} \quad G_F' = \frac{1}{101\text{ k}\Omega} = 9\cdot9\,\mu\text{mho}$$

$$Y_{ot} = 100 + 250 + 9\cdot9(1 + 280) = 3\,130\,\mu\text{mho}$$

(representing an output impedance of 320 Ω).

Summarizing these results:

No feedback

$$A_1 = \frac{280 \times 250}{250 + 100} = 200, \; Y_{1n} \; 1\;000 \; \mu\text{mho}, \; Y_0 \; 350 \; \mu\text{mho}.$$

(a) Current derived feedback

A_1 110, Y_{1n} 1 786 μmho, Y_0 305 μmho.

(b) Voltage derived feedback

A_1 22·1, Y_{1n} 8 780 μmho, Y_0 3 130 μmho.

SUMMARY OF EFFECTS OF FEEDBACK ON AMPLIFIER IMPEDANCES

In the previous section we have seen by mathematical analysis how negative feedback modifies the gain and input and output impedances of amplifiers. We shall now review these results and see how the effects upon input and output impedance can be explained.

Whenever the feedback signal is applied in series with the input, the voltage gain is reduced, and the input impedance is increased. A series addition would be expected to increase the input impedance, and since the input voltage V' is given by iZ_{1n}, the series voltage will be $\beta A_v i Z_{1n}$ leading to the result given in equation 6.8.

If however the feedback signal is added in parallel, the current gain and input impedance are reduced. A parallel addition would similarly be expected to reduce the input impedance. In this case the input current I' is given by $v Y_{1n}$, and the parallel current by $\beta A_{1v} Y_{1n}$ leading to the increase in input admittance shown by equation 6.21.

Now moving to the output terminals; when the feedback network is connected in parallel with the output (voltage derived), the output impedance is reduced. A parallel circuit would normally reduce an impedance, but in this case the reduction is amplified by the active circuit. With negative feedback provided by a positive β and negative A_v, a reduction (say) in terminal output voltage causes a reduction in βv_0 and hence in V'. Phase inversion in the amplifier leads to an increase in $A_v' V'$ driving more current through Z_0'. This makes it appear to have a lower impedance, since a larger current change has been produced by the given reduction in terminal voltage. With a current amplifier, the change in feedback current is amplified

201

increasing the current in Y_0' in the same way with a similar result. These effects are shown in equations 6.10 and 6.27.

A feedback network connected in series with the output (current derived), naturally increases the output impedance. In this case the volt drop across the feedback network is effectively amplified in a similar way to the modification of the input impedance. Thus for both current and voltage amplifiers current derived negative feedback increases the output impedance.

Positive Feedback

All the effects discussed above have been the result of negative feedback. If however the feedback is positive all the effects are reversed. There is a limit to the amount of simple positive feedback that can be applied, and that is that βA must be less than one. The effects of $\beta A \geq 1$ with positive feedback will be discussed in the next section, and in Chapter 8.

FREQUENCY RESPONSE OF FEEDBACK AMPLIFIERS

We now come to the question of frequency response of amplifiers with feedback. Remember first that both β and A are vector quantities, and that the calculation of $A/(1 - \beta A)$ should be a vector calculation. This may be illustrated by an example.

Example 6.4. An amplifier having a voltage gain of $5 \angle \theta$ employs a feedback circuit having a β of $1/10 \angle \phi$.

At three different frequencies, the values of θ and ϕ are respectively (a) $+135°$, $0°$; (b) $-90°$, $-135°$; (c) $+45°$, $-90°$. By means of sketched vector diagrams, determine for each case whether the feedback can be said to be positive or negative.

First we must be clear about the information supplied; since the gain is $5 \angle \theta$, the output voltage is five times the terminal input voltage V', and the angle θ is expressed with V' as the reference vector. βv_0 is one tenth of v_0 and the angle ϕ is expressed with v_0 as the reference vector.

The procedure in each case is to draw respectively vectors representing V', $v_0 = A \angle \theta V'$, and the feedback signal $\beta \angle \phi v_0$. Then since

$$V' = e_s + \beta v_0$$
$$e_s = V' - \beta v_0$$

and by vector subtraction the e_s vector may be determined.

202

The gain without feedback is given by $|v_0/V'|$ and that with feedback by $|v_0/e_s|$.

$\therefore$ if $\left|\dfrac{V_0}{e_s}\right| < \left|\dfrac{V_0}{V'}\right|$ the feedback is negative and $|V'| < |e_s|$

or if $\quad \left|\dfrac{V_0}{e_s}\right| > \left|\dfrac{V_0}{V'}\right|$ the feedback is positive and $|V'| > |e_s|$

Figure 6.8 shows the constructed vector diagrams from which the results are (*a*) negative, (*b*) negative, and (*c*) positive.

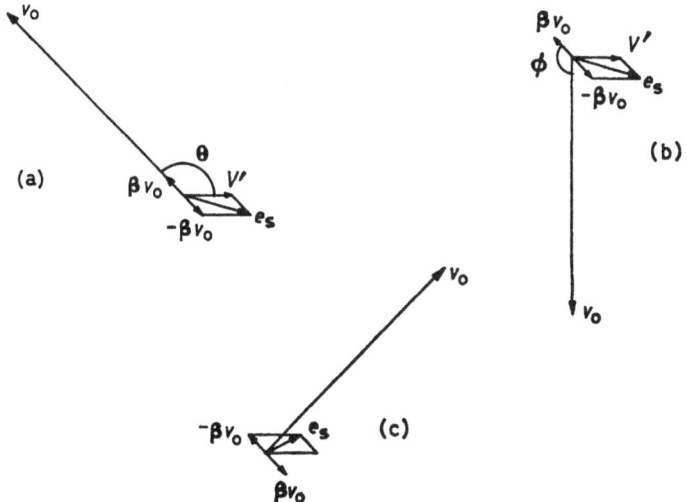

Figure 6.8. Vector diagrams for the feedback amplifiers in Example 6.4

To investigate the frequency response of a multi-stage RC coupled amplifier, the gain and phase shift at each frequency could be determined by repeated use of the general solutions found in Chapters 4 or 5. β is usually constant having an angle ϕ of 0° or 180°. Thus the procedure outlined in Example 6.4 could be repeated for a large number of frequencies, and by measurement of vectors or by calculation, the gain and phase shift with feedback determined.

An alternative procedure is to consider the locus of the gain vector as shown in *Figure 5.11* for Example 5.6. For a single stage amplifier, this is a circle with the origin on the circumference. From this we shall deduce the locus of the gain vector for amplifiers with two or more stages. But first we will use this representation to investigate the frequency response of a single stage amplifier having simple negative feedback at medium frequencies. This is shown in *Figure 6.9a*, with the vectors corresponding to three frequencies shown.

To obtain the response of the amplifier with feedback, we must find, for each frequency, the vector $(1 - \beta A)$ and divide it into the A vector for the same frequency. Assuming β to be a simple fraction having zero phase shift, the locus of the vector βA will be another circle, having its origin on the circumference. Thus *Figure 6.9a*

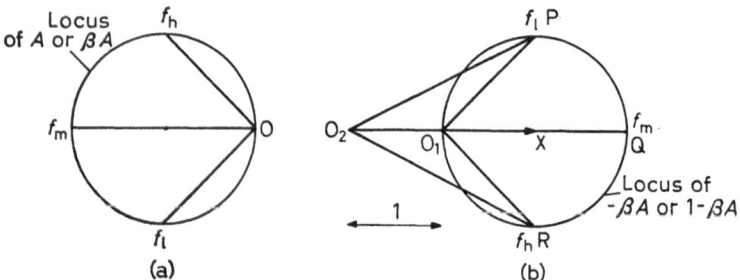

Figure 6.9. Locus of gain and feedback vectors for single stage amplifier with feedback

also represents the locus of βA provided of course that the scale is changed. Next the locus of $-\beta A$ is required. For this, each vector must be reversed leading to yet another circle as shown in *Figure 6.9b*. The origin for this locus is O_1 and note that the reversal results in f_1 appearing at the top of the diagram instead of f_h as in *Figure 6.9a*.

Since this is the locus of $(-\beta A)$, and we require $(1 - \beta A)$, we must add the vector (1) to each vector $(-\beta A)$. Assuming that the values of β and A are such that this unit vector is given by the line O_1X, then the required addition will lead to yet another circle with its origin at X. The origin of the $(1 - \beta A)$ vector would remain at O_1. The resulting diagram becomes somewhat confusing, and the same result may be achieved by shifting the origin by one unit to the left. In *Figure 6.9b*, origin O_2 is the origin for the $(1 - \beta A)$ vector. O_2P, O_2Q, and O_2R are the $(1 - \beta A)$ vectors corresponding to f_1, f_m, and f_h respectively.

Now if A_m were $100 \angle 180°$ and $\beta\frac{1}{50} \angle \theta°$, the scale of *Figure 6.9b* is correct; at f_m, $(1 - \beta A)$ is $3 \angle 0°$, and at f_1 and f_h it is $2.28 \angle +26.5°$

and 2·28 $\angle$ −26·5° respectively. A table may now be drawn up, showing the gain and phase shift with and without feedback

	f_m	f_1	f_h
A	100 $\angle$ 180°	70·7 $\angle$ 225°	70·7 $\angle$ 135°
$(1 - \beta A)$	3 $\angle$ 0°	2·28 $\angle$ 26·5°	2·28 $\angle$ −26·5°
A_f	33·3 $\angle$ 180°	31 $\angle$ 198·5°	31 $\angle$ 161·5°

Thus changes in both gain and phase shift have been reduced by the application of feedback. Examination of *Figure 6.9b* shows that $|1 - \beta A|$ can never be less than one; the feedback is therefore negative for all frequencies.

In practice if feedback is used to improve frequency response, it is applied over two or more stages, the connection of the β network ensuring simple negative feedback at medium frequencies. This will now be illustrated by an example.

Example 6.5. A two stage amplifier, having transistors and components identical to stage one of the amplifier analysed in Example 5.6, employs negative feedback to improve the frequency response. If the β factor is −8·5 × 10⁻⁴, draw the overall gain and phase response, and hence find the 3 db bandwidth of the system.

Before we can apply the methods outlined above, the locus of the gain vector for a two stage amplifier must be determined. Since at each frequency, the gain must be squared, the resultant vectors will be $|A|^2 \angle 2\theta$. This may be constructed by drawing a circle of any convenient diameter, say 2 cm and drawing the gain vectors for a single stage at 10° intervals. The length of these vectors may then be squared and the angles doubled to construct the cardioid which is the locus of the gain vector for the two stage amplifier. This is shown in *Figure 6.10a* and *b*. If *Figure 6.10b* is to apply to the problem in hand, the medium frequency gain vector must represent

$(59·3)^2$ or 3 520. The scale must therefore be 1 cm = $\dfrac{3\ 520}{4}$ = 880.

With reference to *Table 6.1* the frequencies corresponding to points on the cardioid can be inserted. The same cardioid may be used to represent $(-\beta A)$ and $(1 - \beta A)$ by changing the scale and shifting the origin. At our medium frequency (5 kHz) point

$$-\beta A = -(3\ 520 \times -8·5 \times 10^{-4}) = +3$$

Since this is represented by 4 cm, the scale must be one centimetre represents $\frac{3}{4}$ of one unit. The origin must be shifted by one unit or by $\frac{4}{3}$ cm to O_2 shown on the diagram.

205

For any frequency the complex values of A and $(1 - \beta A)$ may be found and the resulting gain and phase shift determined. For example consider the 122 kHz point:

$$|A| = 3 \cdot 22 \text{ cm} \times 880 = 2\,820$$
$$\theta = 53°$$
$$|1 - \beta A| = 4 \cdot 18 \text{ cm} \times \tfrac{3}{4} = 3 \cdot 24$$

The angle of $(1 - \beta A)$, $\phi' = 38°$.

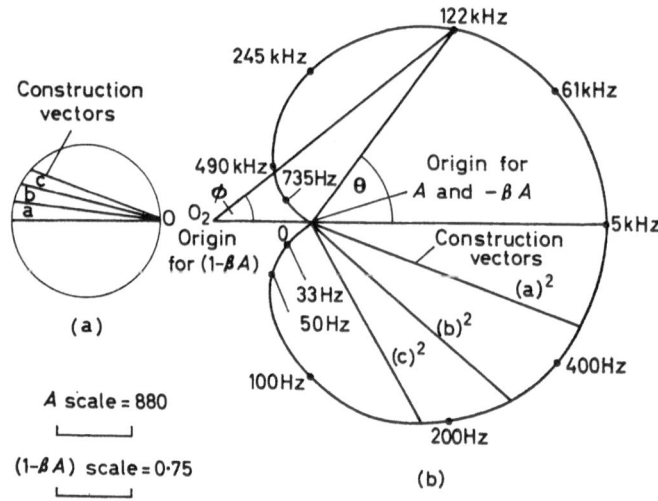

Figure 6.10. Locus of gain and feedback vectors for Example 6.5

Therefore at 122 kHz gain with feedback

$$A_f = \frac{2\,820 \angle 53°}{3 \cdot 24 \angle 38°} = 870 \angle 15°$$

Repeating this procedure for the remaining spot frequencies shown lead to the results shown in *Table 6.1*.

Table 6.1

f(kHz)	0·033	0·05	0·01	0·02	0·03	5	61	122	245	490	735
$|A_f|$	342	869	972	870	879	880	879	870	972	869	342
θ_f	−129°	−87°	−32°	−15°	−7°	0°	7°	15°	32°	87°	129°

To compare these results with the gain without feedback, it is convenient to plot graphs of gain, phase shift, and relative gain against log frequency. To obtain the relative gain in each case, $|A|$ at any particular frequency is divided by $|A|$ for the medium frequency. *Figure 6.11* shows graphs of gain, relative gain and phase

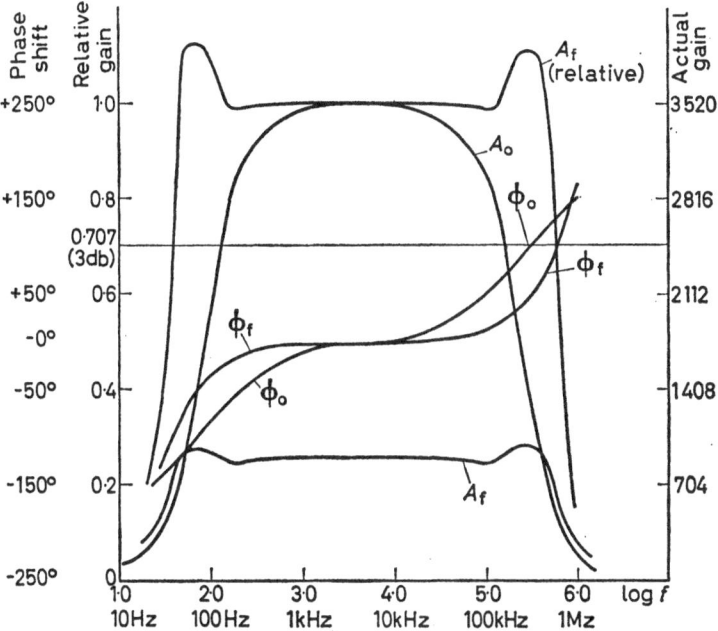

Figure 6.11. Graphs of gain, relative gain and phase shift for two stage amplifier with and without feedback

shift plotted against log frequency with and without feedback. The increase in bandwidth is most evident from the comparison of relative gains with and without feedback. From these curves the following results may be obtained.

	Lower 3 db frequency	Upper 3 db frequency	Bandwidth
No FB	130 Hz	150 kHz	$\simeq$150 kHz
FB	38 Hz	600 kHz	$\simeq$600 kHz

From these results we can see that from a practical point of view the bandwidth of an RC coupled amplifier is given by the upper 3 db frequency.

207

It is of interest to consider 'how' the application of feedback has increased the bandwidth. Examination of *Figure 6.10b* shows that a change in frequency reduces the gain A_0; the 'amount' of negative feedback $|1 - \beta A_0|$ is also reduced, allowing the gain with feedback to be maintained. Over certain sections the reduction in negative feedback is more rapid than the reduction in gain since the $(1 - \beta A)$ vector is nearly tangential to its locus. This gives rise to the two 'humps' shown in *Figure 6.11*. Remembering the definitions of negative and positive feedback we can see that in this region the feedback changes from negative to positive. This is shown on the graphs of actual gain where A_t becomes greater than A_0. Consideration of *Figure 6.10b* shows that this will occur when $|1 - \beta A|$ is less than one which obtains when the locus of the $(1 - \beta A)$ vector falls within a circle of unit radius. Such a circle should be drawn about the O_2 origin using the $(1 - \beta A)$ scale. In this region the locus only touches the real axis at O_1, the origin for $-\beta A$. This is the point corresponding to infinite and zero frequency where in any case the gain is zero. If the origin for $(1 - \beta A)$, O_2, was cut by the locus, then $|1 - \beta A|$ would be zero and the gain would become infinite. This can only occur if O_2 is moved to O_1 representing infinite feedback which is impossible. Thus at the highest and lowest frequency ranges, we can expect the gain to be greater than that without feedback, but it cannot rise to infinity. In general this is true for any two stage RC coupled amplifier with negative feedback at medium frequencies.

Instability of Feedback Amplifiers

If feedback is applied over more than two stages the above is not necessarily true, and a study of multistage feedback amplifiers will provide a useful introduction to the phenomena of oscillation. *Figure 6.12* shows the form of the locus of $-\beta A$ and $(1 - \beta A)$ for three and four stage RC coupled amplifiers employing overall feedback.

Figure 6.12a is the locus for an amplifier having three identical stages and coupling networks.

For the locus shown in *Figure 6.12b* the amplifier must have four stages, and since it is not symmetrical, the components in the coupling networks vary from stage to stage. In each case the medium frequency point occurs in the right hand plane since the feedback network ensures negative feedback at this frequency.

The origin O_1 is that applicable to the $-\beta A$ locus; O_2, O_3, O_4 etc. are possible origins for the $(1 - \beta A)$ locus depending upon the amount of feedback applied. In each case the distance to O_1 must be unity, and if this distance is smaller, $-\beta A$ and hence β must be larger, i.e. O_3 is the origin for $(1 - \beta A)$ with more feedback than O_2. Now let us consider what happens as the feedback is increased, moving the origin from O_2 to O_3 with the three stage case. With the origin at O_2 $|1 - \beta A|$ cannot equal zero, but it can be less than one, giving positive feedback. The unit circle drawn on O_2 shows the frequency range over which this applies, i.e. for those frequencies less than f_1 and those greater than f_2 the gain will be increased.

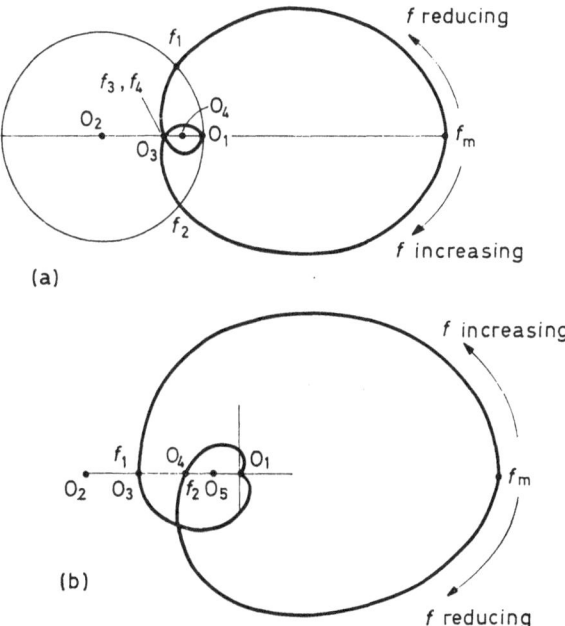

Figure 6.12. Locus of $(1 - \beta A)$ for three and four stage feedback amplifiers

For this particular case the minimum value of $|1 - \beta A|$ is about 0·75. If however β is increased so that the origin $|1 - \beta A|$ is at O_3, then the minimum value of $|1 - \beta A|$ is zero and

$$A_f = \frac{A}{1 - \beta A} = \frac{A}{O} = \infty$$

Thus at frequencies f_3 and f_4, since the gain is infinite, no input is required for an output to be present. In other words the amplifier has become a generator of alternating voltages at two frequencies simultaneously.

Since these oscillations are present at all times, the amplifier can no longer be used for amplification and is said to be unstable.

If the feedback is increased still further moving the origin to O_4, we might expect a stable condition again, but in practice the amplifier gain adjusts itself until the oscillating condition is maintained. This aspect will be considered in more detail in Chapter 8.

These remarks are all applicable to the four stage case as well and a summary of the results referring to *Figure 6.12b* is given below.

Origin Position	*Result*
O_2–O_3	Negative FB over certain ranges, positive FB at high and low frequencies, bandwidth increased.
O_3–O_4	Unstable, oscillation at f_1 only.
O_4–O_5 and beyond	Unstable, oscillation at f_1 and f_2 simultaneously.

Criterion of Stability

From the situation discussed above we can deduce the condition under which a feedback system is stable as follows: If the locus of $(1 - \beta A)$ is drawn in the complex plane, the amplifier will be unstable if the locus encloses or cuts the origin.

An alternative form known as Nyquist's criterion of stability is given as: If the locus of βA is drawn in the complex plane and it encloses or cuts the point $1 + j0$, the amplifier will be unstable.

From the above discussion we can see that feedback may be safely applied over two stages of amplification, but if it is applied over three or more stages, care must be taken to ensure the stability of the amplifier.

REDUCTION OF DISTORTION BY FEEDBACK

The last property of amplifiers to be considered is the introduction of harmonic distortion by the amplifier. This discussion will also loosely apply to the introduction of electrical noise and 'mains hum' within the amplifier.

In general the non-linearity of the device characteristics result in harmonic distortion. This means that the output voltage will contain not only a voltage at the signal frequency, but also voltages at twice, three times and higher multiples of the signal frequency. With most devices it is only the double frequency component, or second harmonic, that is troublesome, but in some instances the third harmonic must also be allowed for. Such distortion is usually expressed as a percentage; for example an amplifier having an input of 10 mV at 1 kHz might have outputs of 5 V at 1 kHz, 0·5 V at 2 kHz, and 0·1 V at 3 kHz. This would be expressed as 10 per cent second harmonic distortion and 2 per cent third harmonic distortion. The overall distortion D is then given by:

$$D = \sqrt{(D_2{}^2 + D_3{}^2)}$$
$$= \sqrt{(100 + 4)} = 10\cdot2 \text{ per cent}$$

In Chapter 1 we found that appreciable distortion only occurred with large signals and that if the amplification at the output is reduced the distortion is also reduced. With a negative feedback system the gain is reduced, and the input signal amplitude must be increased to restore the output signal to the original level. Under these conditions the distortion D can be greatly reduced by feedback. Consider the system shown in *Figure 6.13*.

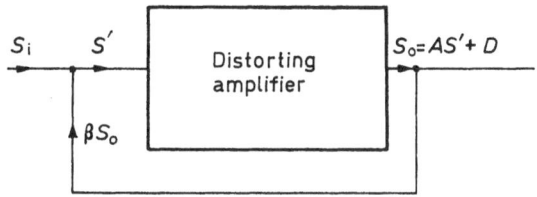

Figure 6.13. Block diagram for demonstration of the effect of feedback on non-linear distortion

Since we have an amplifier, S_0 must contain a component AS'. In addition there will be a distortion signal D whose amplitude is proportional to the amplitude of S_0.

$$\therefore \qquad\qquad S_0 = AS' + D \qquad\qquad (6.28)$$

Now writing the usual equation for the input junction:

$$S' = S_i + \beta S_0$$
$$= S_i + \beta A S' + \beta D \qquad\qquad (6.29)$$

$$\therefore \qquad\qquad S'(1 - \beta A) = S_i + \beta D$$

and
$$S' = \frac{S_i + \beta D}{1 - \beta A} \qquad\qquad (6.30)$$

Substituting for S' from equation 6.30 into equation 6.28

$$S_0 = \frac{AS'}{1 - \beta A} + \frac{A\beta D}{1 - \beta A} + D$$

211

Putting the last two terms over a common denominator

$$S_0 = \frac{AS_1}{1 - \beta A} + \frac{A\beta D + D - A\beta D}{1 - \beta A}$$

$$= \frac{AS_1}{1 - \beta A} + \frac{D}{1 - \beta A} \qquad (6.31)$$

With simple negative feedback equation 6.31 becomes:

$$S_0 = \frac{AS_1}{1 + \beta A} + \frac{D}{1 + \beta A} \qquad (6.32)$$

If S_1 is increased to restore the output to the original level then D will also be restored to the original level. But from equation 6.32 the output distortion is now only $\dfrac{D}{1 + \beta A}$, and has therefore been reduced by the use of feedback.

We shall now verify these feedback formulae by solving a problem graphically, using the methods described in Chapter 1. This will also lead to an explanation of how negative feedback reduces harmonic distortion.

Example 6.6. A triode amplifier has the anode characteristics shown in *Figure 6.14*. It is connected in series with a load R_L and a cathode bias resistor R_K having values 38 kΩ and 2 kΩ respectively.

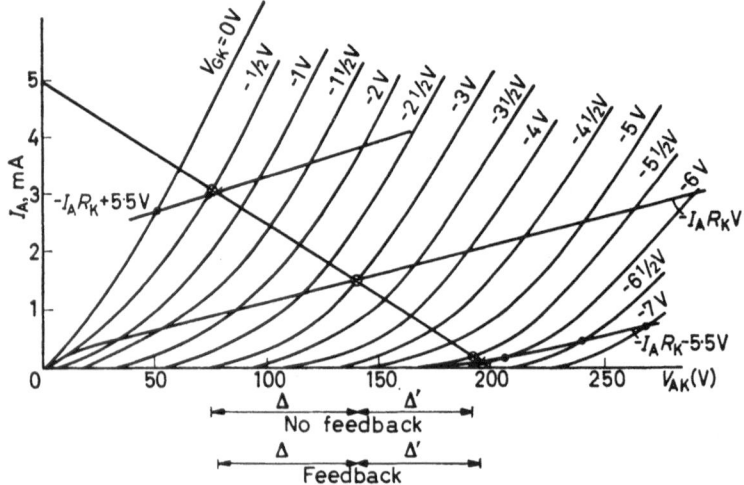

Figure 6.14. Characteristics and graphical solution for Example 6.6

The H.T. supply is 200 V and the signal e_s is sinusoidal. Calculate the voltage gain and harmonic distortion (a) when R_K is decoupled by a suitable capacitor, and e_s is 2·5 sin $\omega t V$, and (b) when negative feedback is applied by the removal of the decoupling capacitor and e_s is increased to give the same output voltage. Compare the results with those obtained by feedback theory.

Before this problem can be tackled two facts concerning second harmonic distortion must be stated. In Chapter 1 it was shown that the distortion introduced amplified one half cycle more than the

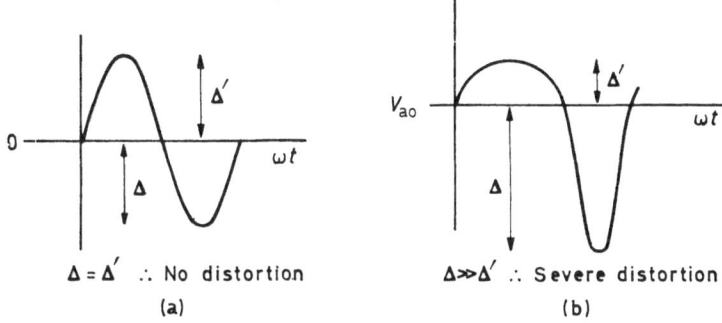

$\Delta = \Delta'$ ∴ No distortion

(a)

$\Delta \gg \Delta'$ ∴ Severe distortion

(b)

Figure 6.15. Input and output waveforms for distorting amplifier

other. It may be shown that if the two half cycles are denoted Δ and Δ' as shown in *Figure 6.15* then the fundamental or required signal output is given by

$$\frac{\Delta + \Delta'}{2}$$

and the second harmonic distortion by

$$\frac{\Delta - \Delta'}{2(\Delta + \Delta')} \times 100 \text{ per cent}$$

Further, since the value of D depends only upon the ratio of Δ to Δ', these lengths may be measured directly along the load line. *Figure 6.15a* shows the applied distortionless signal varying about zero. In *Figure 6.15b* V_{ao} is the quiescent anode voltage in the absence of any signal, and Δ and Δ' are the two peak values when the signal is present.

Now to return to the problem; on *Figure 6.14* the following steps have been taken:

213

(a) The d.c. load line for 40 kΩ, has been drawn, as has the d.c. bias line for 2 kΩ. The resulting d.c. operating point is V_{AK} 139 V, I_A 1·5 mA, V_{GK} −3 V.

(b) Without feedback, the operating point moves between V_{BK} −0·5 V to V_{GK} −5·5 V, along a 38 kΩ a.c. load line. The change from 40 kΩ to 38 kΩ is small and has been neglected. From the graph the extremes of the V_{AK} excursions are 76 V and 191 V.

$$\therefore \qquad \text{Peak } v_0 = \frac{191 - 76}{2} = 57 \cdot 5 \text{ V}$$

and therefore

$$\text{Voltage gain } A_{vo} = \frac{57 \cdot 5}{2 \cdot 5} = 23$$

Also along the load line

$$\Delta = 2 \cdot 32 \text{ cm}$$
$$\Delta' = 1 \cdot 85 \text{ cm}$$
$$\therefore \quad \text{S.H.D.} = \frac{0 \cdot 47}{2 \times 4 \cdot 17} \times 100 \text{ per cent} = 5 \cdot 7 \text{ per cent}$$

Now when the capacitor is removed, current derived negative feedback is applied. This form of circuit will be discussed in the next chapter, but β may be obtained from equation 6.11.

$$\beta = \frac{Z_F}{Z_L} = \frac{2}{38} = \frac{1}{19}$$

(c) For the graphical solution we must plot two further lines of

$$V_{GK} = -I_A R_K + \hat{e}_s \quad \text{and} \quad -I_A R_K - \hat{e}_s$$

Since the resultant v_0 must be unchanged, $\hat{e}_s$ must be increased. Trial values show that if $\hat{e}_s$ is increased to 5·5 V a satisfactory solution is obtained. The resulting construction lines are shown on the graph.

Now the V_{AK} excursion is from 78·5 V to 194 V.

$$\therefore \qquad \text{Peak } v_0 = \frac{194 - 78 \cdot 5}{2} = 57 \cdot 75 \text{ V}$$

which is approximately the same as in the first case.

But now $\qquad A_{vf} = \dfrac{57 \cdot 75}{5 \cdot 5} = 10 \cdot 4$

If feedback formulae are used

$$A_{vt} = \cfrac{23}{1 + \cfrac{19}{23}}$$

$$= 10 \cdot 4$$

Also working from the load line,

$$\Delta = 2 \cdot 24 \text{ cm}$$

$$\Delta' = 2 \cdot 0 \text{ cm}$$

$$\therefore \quad \text{S.H.D.} = \frac{0 \cdot 24}{2 \times 4 \cdot 24} \times 100 \text{ per cent} = 2 \cdot 8 \text{ per cent}$$

But from feedback formulae

$$D_t = \cfrac{5 \cdot 7\%}{1 + \cfrac{23}{19}}$$

$$= 2 \cdot 6 \text{ per cent}$$

The measured distortion is slightly larger, since the output voltage is also slightly larger in the second case.

Figure 6.16 shows the actual waveforms for e_s and v_0, with and without feedback and for βv_0 and V' in the feedback case. From this we can determine how feedback reduces distortion.

v_0 with feedback is slightly distorted; βv_0 is also slightly distorted. The terminal input signal V' given by $e_s + \beta v_0$ also contains a distortion component. When this signal is amplified and inverted it tends to cancel out the distortion produced by the amplifier. In other words, negative feedback results in an input signal distortion such that the distorting amplifier produces an undistorted output. The distortion in v_{gk} can actually be measured on the graph, Δ and Δ' being 2·4 V and 2·8 V respectively.

Notice that in this instance Δ' is the larger implying a negative distortion. In practice this merely means the second harmonic component is 180° out of phase with that introduced by the valve itself.

In this chapter we have discussed the effect of feedback upon amplifiers. In general the application of negative feedback appears to have considerable advantages provided the changes in impedance and gain are acceptable. Caution must be taken if the feedback is applied over more than two stages or instability may result. In the next chapter we shall consider how negative feedback may be applied to practical amplifier circuits, and how the properties of such amplifiers may be determined.

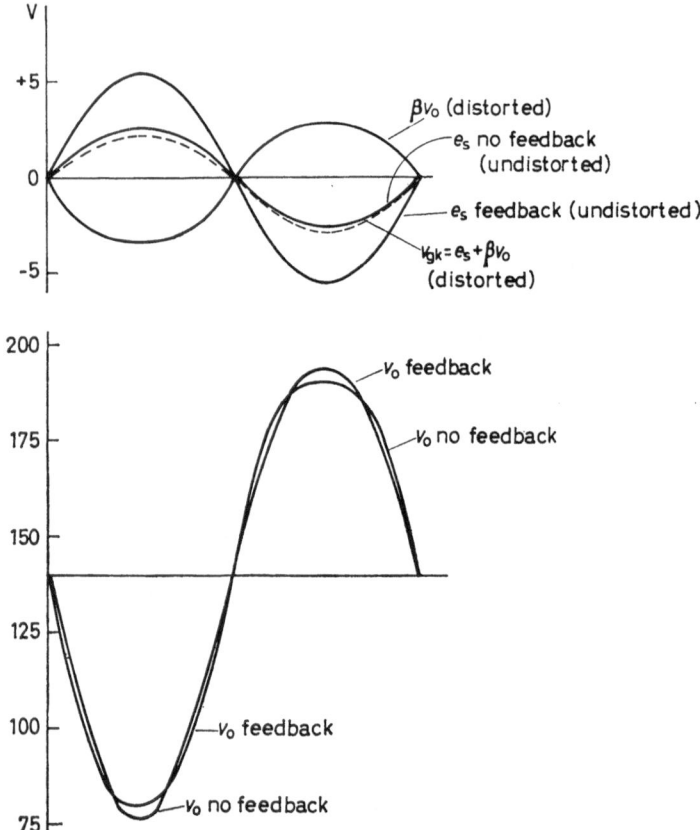

Figure 6.16. Waveforms for Example 6.6

EXAMPLES

Example 6.7. A two stage amplifier is designed to use a matched pair of transistors having h_{te} in the range 50 to 150. In a prototype, transistors having h_{te} 100 produce an overall current gain of 2 500. Feedback is to be used to minimize gain variation with spread of h_{te} such that in no amplifier will the gain be more than 10 per cent less than that with the highest gain. Calculate the required β factor and the resulting maximum gain.

Ans. 0·012, 80.

216

Example 6.8. An amplifier having input impedance $800\,\Omega$ has an open circuit voltage gain and output impedance of 5 000 and $12\,\text{k}\Omega$ respectively. Voltage derived feedback is provided by a potential divider chain of $19\cdot5\,\text{k}\Omega$ and $500\,\Omega$. This $500\,\Omega$ resistor is connected in series with the amplifier input such that simple negative feedback is applied. Determine the overall voltage gain and input impedance if the amplifier is loaded with $5\,\text{k}\Omega$. Find also the amplifier output impedance when the driving source has an internal impedance of $200\,\Omega$.

Ans. $38\cdot7$, $25\cdot8\,\text{k}\Omega$, $169\,\Omega$.

Example 6.9. The amplifier described in Example 6.8 is modified by replacing the feedback system with a current derived system. This consists of a $200\,\Omega$ resistor placed in series with the equivalent output generator and load. The voltage across this component is added to the input to provide negative feedback as before. If the load and source are unchanged, find the new voltage gain and input and output impedance.

Ans. $24\cdot5$, $47\cdot5\,\text{k}\Omega$, $4\cdot96\,\text{k}\Omega$.

Example 6.10. A current amplifier having an open circuit output admittance of 130 μmho achieves a current gain of 950 in to a load of $1\,\text{k}\Omega$. The amplifier input admittance is 2 mmho and feedback is provided by a $50\,\Omega$ resistor in series with the load. A $5\,\text{k}\Omega$ resistor is connected from this $50\,\Omega$ resistor to the input terminal such that simple negative feedback occurs. Calculate the resulting overall current gain and input impedance. Find also the output admittance when the amplifier is used with a source having internal admittance 100 μmho.

Ans. $91\cdot2$, $48\cdot8\,\Omega$, 1 013 μmho.

Example 6.11. For the circuit shown in *Figure 6.17*, determine the current ratio i_0/i_s; (a) by finding A_1 and Y_{in} with feedback, and

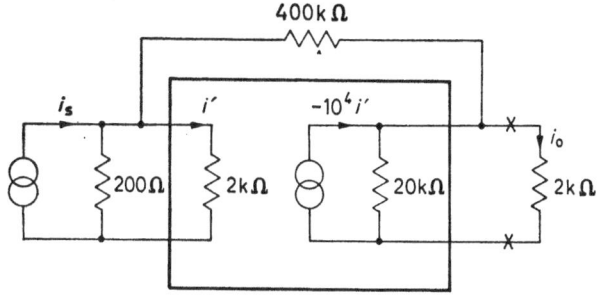

Figure 6.17. Circuit for Example 6.11

(b) by finding the equivalent Norton generator at the terminals marked XX.

Ans. 161.

Example 6.12. Amplifiers and their associated feedback systems have the following complex gain and feedback factors:

(a) $A = 35 \angle 140°, \beta = \frac{1}{7} \angle 20°$

(b) $A = 50 \angle -30°, \beta = \frac{1}{60} \angle 90°$

(c) $A = 24 \angle 200°, \beta = 0.059 \angle -245°$

By means of vector diagrams, determine for each case, whether the feedback is positive or negative. Check your answer by calculations.

Ans. (a) negative, (b) positive, (c) neither, phase shifting.

Example 6.13. An amplifier has two identical stages each having a short circuit current gain of -95 and a resistive input impedance of 1 kΩ. The total loading per stage is 600 Ω resistance in parallel with 200 pF capacitance. Assuming the coupling capacitor to have negligible reactance at all signal frequencies and using graphical methods, calculate the maximum gain and 3 db bandwidth when simple negative feedback, $1/1\,760$, is applied.

Ans. $1\,150$, 2.6 MHz.

Example 6.14. Measurements on a three stage amplifier give the following figures for gain and phase shift:

f(kHz)	0·025	0·05	0·1	0·2	0·5
A/θ	$420 \angle -45°$	$1\,100 \angle -65°$	$2\,100 \angle -90°$	$3\,300 \angle -130°$	$4\,400 \angle -155°$

f(kHz)	10·0	100	250	500	
A/θ	$5\,000 \angle 180°$	$4\,500 \angle 140°$	$3\,450 \angle 90°$	$2\,100 \angle 45°$	

f(kHz)	1\,000	2\,000
$A \angle \theta$	$1\,000 \angle 0°$	$420 \angle -45°$

If feedback is provided by a purely resistive network such that simple negative feedback occurs at medium frequencies, calculate (a) the frequency ranges over which feedback is positive if β is 5×10^{-4}, (b) the minimum value of β for oscillation to occur and the frequency of oscillation.

Ans. 0–60 Hz, 450 kHz–∞, 10^{-3}, 1 MHz.

Example 6.15. A three stage amplifier having a final load of 500 Ω introduces 7 per cent harmonic distortion in the output current. N.F.B. is to be used to reduce this figure to 4 per cent and is to be applied by connecting a resistor R_F from the final stage collector to the first stage base. If the gain and input impedance of the amplifier without feedback are $-3\,200$ and 800 Ω respectively, calculate the value required for R_F. Determine also by how much the open circuit voltage of the driving generator must be increased to restore the required output level. The source impedance is 600 Ω.

Ans. 2·12 MΩ, 32 per cent.

Example 6.16. A triode valve having the characteristics given in Example 1.8 is connected in series with 250 V H.T., R_L 13 kΩ, and R_K 1·2 kΩ. If R_K is adequately decoupled, calculate the output voltage and percentage distortion when $e_s = 1 \sin \omega t$. Use graphical methods to calculate the output voltage and percentage distortion if the decoupling capacitor is removed and e_s is increased to 3 sin ωt. Check these results using feedback theory.

Ans. 43 V peak to peak, 8·9 per cent, 47 V peak to peak, 3·4 per cent (3 per cent calculated but larger output voltage increases distortion).

7

PRACTICAL FEEDBACK AMPLIFIERS

In Chapter 6 the application of feedback to amplifiers was found to be advantageous if it was in the form of simple negative feedback at medium frequencies. In this chapter, we shall investigate the circuits and connections necessary to produce this form of feedback for various amplifier configurations. Various methods of analysis of the resulting complete circuits will also be examined.

INPUT CIRCUITS FOR FEEDBACK AMPLIFIERS

The possible forms of input circuits are determined by the original feedback definitions; if the feedback signal is added in parallel it must be a current signal, if it is added in series it must be a voltage signal. *Figure 7.1* shows possible forms of input circuit for a common emitter stage, with feedback derived from a later stage.

Figure 7.1a shows a feedback path in parallel with the input. In this situation, the amplifier must be treated as a current amplifier. The equation for terminal input current is given by:

$$i' = i_s + \beta i_0$$

and since

$$\beta i_0 = A\beta i'$$

$$i' = \frac{i_s}{1 - \beta A}$$

so for simple negative feedback, either A or β must be negative.

Figure 7.1b shows the feedback signal βv_0 in series with the input indicating a voltage amplifier. The terminal input equation in this instance is given by

$$v' = e_s + \beta v_0 \quad \text{(1:1 turns ratio)}$$

leading to

$$v' = \frac{e_s}{1 - \beta A}$$

220

and once again either β or A must be negative for simple negative feedback.

In practice the use of transformers is inconvenient, and an alternative input circuit for voltage feedback is given in *Figure 7.1c*. The arrow notation shows that the two quantities e_s and βv_o may

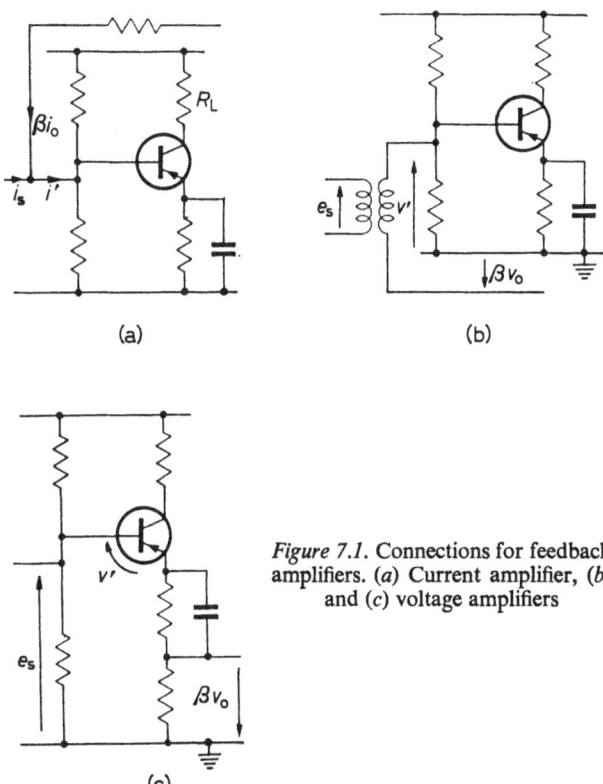

(a)

(b)

(c)

Figure 7.1. Connections for feedback amplifiers. (*a*) Current amplifier, (*b*) and (*c*) voltage amplifiers

be added in series to give v' as before. In this case however βv_o is shown as being measured at earth with respect to the emitter. If the feedback voltage is shown as being measured at the emitter with respect to earth then β would be negative.

The way in which the feedback signal is derived from the output depends upon the number of stages in the amplifier and whether it is to be considered as a current or voltage amplifier. A number of cases with examples will now be considered.

Single Stage Current Feedback Amplifier

First let us consider a single stage current amplifier. With a single stage common emitter amplifier, the current gain $\dfrac{-h_{te}\bar{Y}_L}{Y_L + h_{oe}}$ is phase inverting and therefore A is negative. For negative feedback β must be positive. The required circuit is shown in *Figure 7.2.*

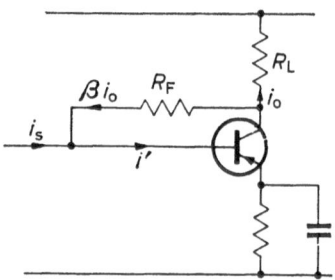

Figure 7.2. Circuit for Example 7.1

Since the feedback path is taken directly from the output terminal this is an example of voltage derived feedback. The formulae applicable to this circuit are given by equations 6.24 to 6.27 inclusive.

Example 7.1. In the circuit shown in *Figure 7.2*, the transistor parameters are h_{1e} 800 Ω, h_{te} 120, h_{oe} 90 μmho and negligible h_{re}. R_L is 2 kΩ, and the feedback resistor $R_F = 80$ kΩ. Neglecting the effect of the bias components and Y_s determine the current gain and the input and output impedances of the amplifier.

First the equivalent amplifier components are required. Since h_{re} is zero,

$$Y_{in} = \frac{1}{h_{1e}} = 1\ 250\ \mu\text{mho}$$

$$Y_o' = h_{oe} = 90\ \mu\text{mho} \quad \text{and} \quad A_1' = -h_{te} = -120$$

The feedback admittance

$$G_F = \frac{1}{80\ \text{k}\Omega} = 12{\cdot}5\ \mu\text{mho}$$

and from equation 6.25

$$\beta = \frac{G_F}{Y_L} = \frac{12{\cdot}5}{500} = 0{\cdot}025$$

Applying equation 6.26,

$$A_{10} = \frac{A_1' Y_L}{Y_0' + Y_L + G_F} = \frac{-120 \times 500}{90 + 500 + 12 \cdot 5} = -99 \cdot 6$$

The normal feedback equation now gives the current gain:

$$A_{1f} = \frac{A_{10}}{1 - \beta A_{10}} = \frac{-99 \cdot 6}{1 + 99 \cdot 6 \times 0 \cdot 025}$$

$$\therefore \qquad A_{1f} = -28 \cdot 6$$

Equation 6.21 provides the value of the input admittance,

$$Y_{1nf} = Y_{1n}(1 - \beta A_{10}) = 1\,250(1 + 99 \cdot 6 \times 0 \cdot 025)\,\mu\text{mho}$$
$$= 4\,336\,\mu\text{mho}$$

The input impedance with feedback is given by the reciprocal of Y_{1nf}, or by finding $\dfrac{Z_{1n}}{1 - \beta A_{10}}$,

$$\therefore \qquad\qquad Z_{1nf} = 231\,\Omega$$

Finally from equation 6.27:

$$Y_{of} = Y_0' + Y_L + G_F(1 - A_1')$$
$$= 90 + 500 + 12 \cdot 5(1 + 120)\,\mu\text{mho}$$
$$= 2\,100\,\mu\text{mho}$$

The resulting output impedance $\dfrac{1}{Y_{of}} = 476\,\Omega$.

(Note the value of A_1' would have to be modified for this calculation if Z_s were not much greater than Z_{1n}.)

Single Stage Voltage Feedback Amplifier

If a single stage voltage amplifier is required to have negative feedback, part or all of the emitter or cathode resistor may be left unbypassed. Examples of this were shown in Chapters 1 and 6 using graphical methods and in Chapters 4 and 5 using equivalent circuit methods. In each case the voltage gain was reduced, and with the equivalent circuit problems, the input and output impedances were increased. This suggests that the voltage feedback was current derived (equations 6.7, 6.8, 6.11, 6.12 and 6.13). *Figure 7.3* shows how the correct phase relationship arises. The feedback voltage in the sense shown is given by:

$$\beta v_0 = (i_0 + i_{1n})R_E \simeq i_0 R_E$$

But
$$v_0 = -i_0 R_L$$

$$\therefore \quad \beta = \frac{\beta v_0}{v_0} = \frac{i_0 R_E}{-i_0 R_L} = \frac{-R_E}{R_L}$$

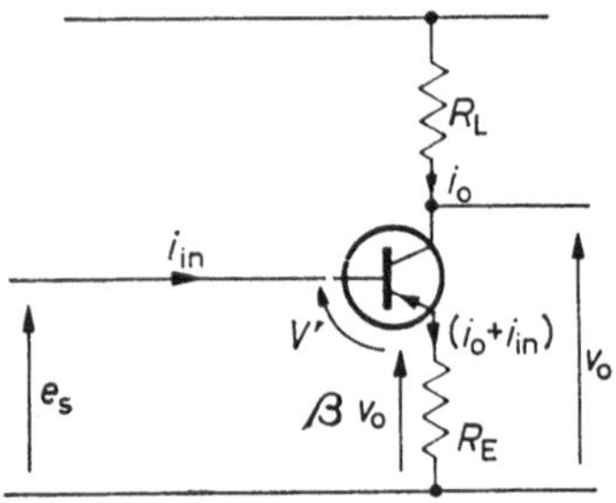

Figure 7.3. Current derived, voltage feedback on single stage amplifier

Now, using the notation shown,

$$e_s = V' - \frac{R_E}{R_L} v_0$$

$$= V' - \frac{R_E}{R_L} A_v V'$$

But A_v is negative for a single stage $\left(\dfrac{-h_{fe}}{h_{1e}(h_{oe} + Y_L) - h_{re}h_{fe}}\right)$.

$$\therefore \quad e_s = V'\left(1 + \frac{R_E}{R_L}|A_v|\right)$$

and
$$V' = \frac{e_s}{1 + |\beta A_v|}$$

Thus, since V' is less than e_s, the overall voltage gain is reduced indicating negative feedback. To verify the use of feedback methods we shall repeat Example 5.1 using the results obtained from Chapter 6.

Example 7.2. A single stage amplifier having the circuit shown in *Figure 7.4* employs a transistor with h_{1e} 900 Ω, h_{re} 5 $\times$ 10^{-4}, h_{oe} 125 μmho and h_{fe} 90. Using feedback methods, find the terminal voltage

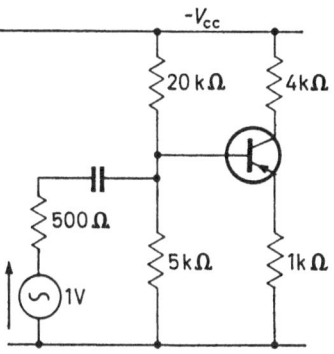

Figure 7.4. Circuit for Example 7.2

and current gain, the output voltage and current, and the output impedance.

As with the last example, we must first find the components of the appropriate equivalent amplifier.

Using the standard *h* parameter formulae:

$$Z_{1n} = 900 - \frac{90 \times 5 \times 10^{-4}}{(125 + 250)10^{-6}} = 780 \ \Omega\cdot$$

$$A_v' = \frac{-90}{900(125 + 0)10^{-6} - 90 \times 5 \times 10^{-4}} = -1\,300$$

Note, for Z_{1n}, the circuit Y_L of 250 μmho is used but A_v' is the open circuit voltage gain with Y_L zero.

$$Y_0' = 125 \times 10^{-6} - \frac{90 \times 5 \times 10^{-4}}{900} = 75 \ \mu\text{mho}$$

$$\therefore \quad Z_0' = \frac{1}{Y_0'} = 13{\cdot}33 \ \text{k}\Omega$$

Now from equations 6.12, 6.11 and 6.7

$$A_{vo} = \frac{-1\,300 \times 4}{4 + 1 + 13{\cdot}33} = -283$$

225

and
$$\beta = \frac{R_E}{R_L} = \tfrac{1}{4}$$

$\therefore$
$$A_{vt} = \frac{-283}{1 + \dfrac{283}{4}} = -3.95$$

Also from equation 6.8:

$$Z_{1nt} = 0.78 \left(1 + \frac{283}{4}\right) = 56.2 \text{ k}\Omega$$

and from equation 6.13,

$$Z_{ot} = 13.33 + 1(1 + 1\,300) = 1\,314 \text{ k}\Omega$$

But for the overall output impedance, R_L must be included in parallel

$\therefore$
$$Z_0 = \frac{1\,314 \times 4}{1\,318} \simeq 4 \text{ k}\Omega$$

Similarly, the overall input impedance should include the bias components.

Working in admittances,

$$Y_{1n} = 17.8 + 200 + 50 = 267.8 \ \mu\text{mho}$$
$\therefore$ $\qquad Z_{1n} = 3.74 \text{ k}\Omega$

Now using potential divider methods, the terminal input voltage is given by

$$\frac{1 \times 3.74}{3.74 + 0.5} = 0.883 \text{ V}$$

The resulting output voltage is therefore 0.883×3.95 V, and $v_0 = 3.48$ V.

To obtain output current:

$$i_0 = v_0 Y_L = 3.48 \times 250 \times 10^{-6} \text{ A}$$
$$= 880 \ \mu\text{A}$$

To calculate the terminal current gain, the base input current i_b is required.

$$i_b = \frac{\text{Terminal input voltage}}{Z_{1nt}} = \frac{0.883}{56.2} \cdot \text{mA}$$
$$= 15.7 \ \mu\text{A}$$

$\therefore$
$$\text{Terminal } A_1 = \frac{880}{15.7} = 56$$

Note in general terms

$$A_1 = \frac{-h_{fe}Y_L}{Y_L + h_{oe}}$$

$$= \frac{-90 \times 200}{200 + 125} = -55 \cdot 5$$

Thus we can see that the terminal current gain is not modified by voltage feedback.

The results for A_{vf} and Z_{inf} are not identical to those obtained from the equivalent circuit which were $3 \cdot 87$ and $57 \cdot 5$ kΩ. This is principally because the current in the emitter resistor is the sum of the output current i_0 and the input current i_b, which would slightly modify β. In practice the difference is negligible and if it is remembered that all components and parameters are subject to wide tolerances, an even simpler solution may be obtained as follows. Since

$$\beta A_v \gg 1 \qquad A_{vf} \simeq \frac{1}{\beta} = -4$$

$$A_{vo} \simeq \frac{-h_{fe}Z_L}{h_{ie}} = \frac{-90 \times 4\,000}{900} = -400$$

$$\therefore \qquad Z_{inf} = h_{ie}(1 + \beta A_{vo}) = 90 \text{ k}\Omega$$

This may seem a large error, but when the bias components are included, the overall input impedance becomes:

$$Z_{in} = \frac{90 \times 4}{94} = 3 \cdot 8 \text{ k}\Omega$$

$$\therefore \qquad \text{Terminal input voltage} = \frac{1 \times 3 \cdot 8}{3 \cdot 8 + 0 \cdot 5} = 0 \cdot 88 \text{ V}$$

$$\therefore \qquad v_0 = 4 \times 0 \cdot 88 = 3 \cdot 52 \text{ V}$$

which is only about 1 per cent high. Variations in components and parameters will cause a much larger margin of error than will the use of these or similar approximations.

Two Stage Current Feedback Amplifiers

Feedback is more commonly applied over two or more stages and the next problem to be considered is that of a two stage current amplifier. The current gain in this case will be positive and direct connection between final collector and first base would result in positive feedback. The required phase relationship can be obtained by connecting the feedback path to the final emitter, as shown in the simplified circuit diagram in *Figure 7.5*.

Here, A_1 and A_2 are the magnitudes of the current gains of stages 1 and 2. The interstage bias components are included in Y_{L1}. The 180° phase change per stage is shown by the $+$ and $-$ signs. Since all

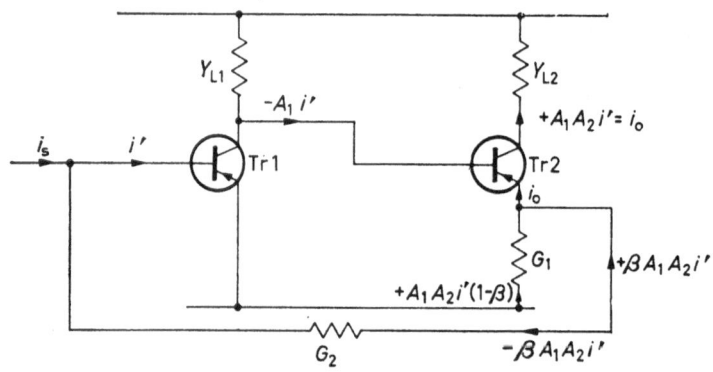

Figure 7.5. Current amplifier with current derived feedback

the output current flows into the emitter of Tr2, it must split between the two parallel paths provided by G_1 and G_2. Current splits in the direct ratio of parallel admittances; thus β is given by $G_2/(G_1 + G_2)$. The feedback current i_f is considered as flowing in the opposite sense to $\beta A_1 A_2 i'$ hence the sign change as shown. The resulting equation at the input terminal is:

$$i' = i_s + (-\beta A_1 A_2 i')$$

which upon rearrangement leads to the correct relationships for negative feedback. Since we have current derived current feedback, the solutions may be found from equations 6.15 to 6.23.

A further problem arises as a result of G_1 and G_2 in the emitter lead of Tr2. These must be allowed for in the calculation of A_1', since, although the terminal current gain will not be affected by their presence, the input impedance to Tr2 and thus the overall current gain, will be.

So, making approximations where valid, the steps in the solution will be: A_{12}, A_{v2} and Z_{1n2}. Z_{1n2f} using β for local feedback of R_E/R_L. A_{11} and Z_{1n1}. A_1' and β_1 for overall feedback of $G_2/(G_1 + G_2)$. A_{1f} and Z_{1nf}. Y_0' (approximately h_{oe} for Tr2). Y_{of}.

Example 7.3. The output current in the amplifier, shown in *Figure 7.6*, contains 20 per cent harmonic distortion at the required output signal level of 5 mA r.m.s. If the final emitter resistor is unbypassed, determine the value of the feedback resistor to be connected from the final emitter to the first base, so that the distortion may be reduced to 5 per cent. Determine also by how much the e.m.f. of a driving source, of internal impedance 600 Ω, must be changed if the

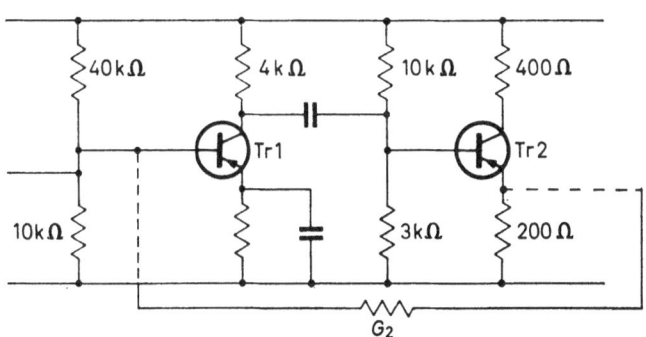

Figure 7.6. Circuit for Example 7.3

required output current is to be maintained. The transistor h parameters are h_{ie} 1·5 kΩ, h_{oe} 150 μmho, h_{fe} 140, and negligible h_{re}.

We must first determine the original current gain and hence the original source e.m.f. Since h_{re} is negligible this current gain may be written directly by current splitting techniques.

$$A_1 = \frac{Y_{\text{in1}}}{Y_{\text{in1}} + Y_{\text{B1}}} \times -h_{\text{fe}} \times \frac{Y_{\text{in2}}}{h_{\text{oe}} + Y_{\text{L1}} + Y_{\text{B2}} + Y_{\text{in2}}}$$

$$\times -h_{\text{fe}} \times \frac{Y_{\text{L2}}}{Y_{\text{L2}} + h_{\text{oe}}}$$

where Y_{B1} and Y_{B2} are the combined bias components for each stage and $Y_{\text{in1}} = Y_{\text{in2}} = 1/h_{\text{ie}}$.

$$\therefore A_1 = \frac{667}{100 + 25 + 667} \times -140$$

$$\times \frac{667}{150 + 250 + 100 + 333 + 667} \times -140 \times \frac{2\,500}{2\,500 + 150}$$

$$= 0·84 \times 140 \times 0·46 \times 140 \times 0·943 = 7\,130$$

229

The input current is given by:

$$i_{\text{in}} = \frac{5 \times 1\,000}{7\,130}\ \mu A$$

$$= 0{\cdot}7\ \mu A$$

But the overall input impedance

$$Z_{\text{in}} = \frac{1}{Y_{\text{in1}} + Y_{\text{B1}}} = 1{\cdot}26\ k\Omega$$

$$\therefore \quad \text{Source e.m.f.} = i_{\text{in}}(Z_{\text{in}} + R_s) = 0{\cdot}7(1{\cdot}26 + 0{\cdot}6)\ \text{mV}$$

$$= 1{\cdot}3\ \text{mV}$$

The unbypassed emitter resistor for Tr2 will effect the current gain without overall feedback by modifying Y_{in2}, and hence the interstage factor in the expression for A_1 above.

Since this component provides current derived voltage feedback for Tr2, we require A_{v2}.

The effective load is $600\ \Omega$ making Y_L $1\,670\ \mu\text{mho}$.

$$\therefore \qquad A_{\text{v2o}} = \frac{-140}{1\,500(150 + 1\,670)10^{-6}}$$

$$= -51$$

But

$$\beta = \frac{200}{400}$$

$$\therefore \qquad Z_{\text{in2f}} = 1{\cdot}5(1 + \tfrac{51}{2})\ k\Omega$$

$$= 40\ k\Omega$$

$$\therefore \qquad Y_{\text{in2}} = 25\ \mu\text{mho}$$

Thus the interstage factor becomes

$$\frac{25}{150 + 250 + 100 + 333 + 25} = 0{\cdot}029$$

Substituting this value in the expression above:

$$A_1 = 7\,130 \times \frac{0{\cdot}029}{0{\cdot}46} = 450$$

Now applying equation 6.32,

$$D_f = \frac{D}{1 - \beta A}$$

$$5\ \text{per cent} = \frac{20\ \text{per cent}}{1 + 450\beta}$$

$$\therefore \qquad 1 + 450\beta = 4 \quad \text{and} \quad \beta = \frac{3}{450} \quad \text{or} \quad \frac{1}{150}$$

But $\qquad \beta = \dfrac{G_2}{G_1 + G_2}$ and $G_1 = \dfrac{1}{200} = 5$ mmho

$\therefore \quad 5 + G_2 = G_2 \times 150$

and $\qquad\qquad G_2 = \dfrac{5}{149}$ mmho $= 33 \cdot 5 \ \mu$mho

This represents the overall feedback admittance or its equivalent resistance of 29·8 kΩ.

$\therefore \qquad\qquad\qquad R_F = 29 \cdot 8 \text{ k}\Omega$

In practice the nearest preferred value of 27 kΩ would be used.

With this feedback resistor in circuit, the current gain and input impedance becomes:

$$A_{1f} = \frac{450}{1 + \dfrac{450}{150}} = 112 \cdot 5$$

$$Z_{1nf} = \frac{1 \cdot 26 \text{ k}\Omega}{1 + \dfrac{450}{150}} = 315 \ \Omega$$

But the output current in the final load is to be maintained at 5 mA, requiring an input current i_{1n}, of $\dfrac{5\,000}{112 \cdot 5} \ \mu$A. The required source e.m.f. now becomes:

$$\frac{5\,000}{112 \cdot 5} (0 \cdot 6 + 0 \cdot 315) \text{ mV}$$

$\therefore \qquad\qquad\qquad e_8 = 40 \cdot 6 \text{ mV}$

Thus to reduce the distortion to 5 per cent the source e.m.f. has to be increased by a factor of approximately 30.

Three Stage Current Feedback Amplifiers

If negative feedback is to be applied over a three stage current amplifier, the phase relationships are the same as for feedback over a single stage. The procedure outline in Example 7.1 could again be applied. An alternative procedure is to regard the amplifier and the feedback network as two separate four-terminal networks, connected in parallel. In Chapter 2, we found that with this circuit arrangement, the combined network y parameters were given by the sums of the individual network y parameters. Thus if we can find the y parameters for a three stage amplifier and then for a feedback network, the general solution obtained in Chapter 2 may be applied directly.

Example 7.4. A three stage current amplifier employs identical transistors having h_{1e} 1 000 Ω, h_{fe} 90, h_{oe} 100 μmho, and negligible h_{re}. Each stage has a load of 1 kΩ and the effect of the bias components may be neglected. A 100 kΩ resistor is connected between

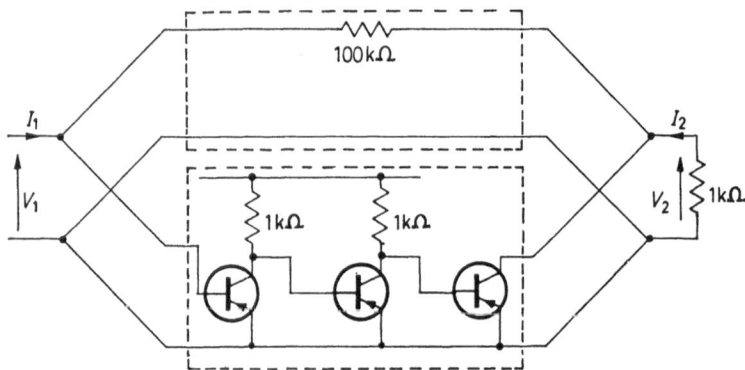

Figure 7.7. Interconnected four-terminal network circuit for Example 7.4

the final stage collector and the base of the first stage. Determine the overall current gain and the input and output admittances for the resulting circuit. Investigate the probable effect of the bias components and h_{re} on these results.

Figure 7.7 shows the circuit represented as two four-terminal networks in parallel.

It is convenient to show the load of the final transistor as the load for the combined network as shown. The y parameters for a network may be redefined.

$$y_{11} = \frac{I_1}{V_1}\bigg|_{V_2=0} \qquad y_{12} = \frac{I_1}{V_2}\bigg|_{Y_1=0}$$

$$y_{21} = \frac{I_2}{V_1}\bigg|_{V_2=0} \qquad y_{22} = \frac{I_2}{V_2}\bigg|_{V_1=0}$$

These parameters are all admittances and in each case a short circuit is applied to the opposite end of the network for the purpose of calculation.

Taking the feedback network first, and connecting a short circuit to make V_2 zero then:

$$\frac{I_1}{V_1} = y_{11} = \frac{1}{100 \text{ k}\Omega} = 10 \ \mu\text{mho}$$

Also remembering the direction or sense required for I_2,

$$I_2 = \frac{-V_1}{100 \text{ k}\Omega}$$

$$\therefore \qquad y_{21} = \frac{I_2}{V_1} = -10 \ \mu\text{mho}$$

But the network is symmetrical, so

$$y_{22} = 10 \ \mu\text{mho}, \qquad y_{12} = -10 \ \mu\text{mho}$$

Now proceeding to the amplifier, since h_{re} is zero

$$y_{11} = \frac{1}{h_{1e}} = 1 \ 000 \ \mu\text{mho}$$

where h_{1e} is a parameter of the first stage transistor, and

$$y_{22} = h_{\text{oe}} = 100 \ \mu\text{mho}$$

where h_{oe} is a parameter of the final stage transistor.

To find y_{21} we require the short circuit output current in terms of the applied input voltage V_1.

Using normal current splitting methods:

$$I_2 = \frac{V_1}{h_{1e}} \times -h_{fe} \times \frac{Y_{\text{in}2}}{h_{\text{oe}} + Y_{L1} + Y_{\text{in}2}}$$
$$\times -h_{fe} \times \frac{Y_{\text{in}3}}{h_{\text{oe}} + Y_{L2} + Y_{\text{in}3}} \times +h_{fe}$$

Note that the last h_{fe} term is positive since the normal convention requires I_s to flow into the output terminal.

$$\therefore \qquad y_{21} = \frac{I_2}{V_1} = \frac{-90}{1 \ 000} \times \frac{1 \ 000}{100 + 1 \ 000 + 1 \ 000}$$
$$\times -90 \times \frac{1 \ 000}{100 + 1 \ 000 + 1 \ 000} \times +90 \ \text{mho}$$

$$= 729 \times \frac{1}{2 \cdot 1 \times 2 \cdot 1} = 165 \cdot 5 \ \text{mho}$$

Finally since h_{re} is zero a voltage V_2 can cause no short circuit current I_1.

$$\therefore \qquad\qquad\qquad y_{12} = 0$$

The combined parameters may now be written,

$$y_{11} = 1\ 010\ \mu\text{mho} \qquad\qquad y_{12} = -10\ \mu\text{mho}$$
$$y_{21} = 165{\cdot}5 \times 10^6\ \mu\text{mho} \qquad y_{22} = 110\ \mu\text{mho}$$

The general solutions can now be applied taking Y_L as $1\ 000\ \mu\text{mho}$. From equation 2.47,

$$Y_{in} = 1\ 010 - \frac{-10 \times 165{\cdot}5 \times 10^6}{110 + 1\ 000}\ \mu\text{mho}$$
$$= 1{\cdot}01 \times 10^{-3} + 1{\cdot}49\ \text{mho}$$
$$= 1{\cdot}49\ \text{mho}$$

From equation 2.49 the current gain may be found:

$$\frac{I_2}{I_1} = \frac{V_2}{I_1}\ Y_L = \frac{y_{21} Y_L}{y_{11}(y_{22} + Y_L) - y_{21} y_{12}}$$
$$= \frac{165{\cdot}5 \times 10^{-6} \times 1\ 000}{1\ 010(110 + 1\ 000) - (-10 \times 165{\cdot}5 \times 10^6)}$$
$$= \frac{1\ 000}{\dfrac{1\ 010 \times 110}{165 \times 10^6} + 10}$$
$$= \frac{1\ 000}{10{\cdot}007} = 100$$

To determine the output admittance equation 2.50 must be used.

$$Y_0 = y_{22} - \frac{y_{21} y_{12}}{y_{11} + Y_s}$$

Y_s is not specified in the example so taking the limits of zero and infinity.

$$Y_0 = 110 + \frac{10 \times 165{\cdot}5 \times 10^6}{1\ 010} = 1{\cdot}64\ \text{mho}$$

with Y_s zero or $Y_0 = 110\ \mu\text{mho}$ with Y_s infinite. Thus the output admittance is very dependent upon the value of source admittance but with a practical value of say $1\ 000\ \mu\text{mho}$, the output admittance will be of the order of 800 mmho.

The above results suggest that if feedback formulae are to be used, the value of β is not affected by Y_s in the calculation of Y_{in} and A_1, but when calculating Y_0, the current division between Y_{in} and Y_s should be allowed for.

To complete this example, the effect of the bias components, and h_{re} would be to reduce the current gain without feedback, and under the same conditions to increase the input admittance (bias components) and reduce the output admittance (h_{re}). Since A_{1f} is approximately given by $1/\beta$ the current gain would still be 100. Y_{inf} is given by $Y_{in}(1 + |\beta A_1|)$; the first term is increased and the second reduced so there would be little change in this result. Y_{of} is, given by $Y_0' + Y_F(1 + |A_1'|)$; here both terms will be reduced. Thus we can say that the stated approximations will have negligible effect upon the input admittance and current gain but will cause the output admittance to be a little smaller than that calculated.

Multistage Current Feedback Amplifiers

If a current amplifier having more than three stages is to employ overall negative feedback, the connections obviously depend upon the number of stages. In Examples 7.1 and 7.4, the amplifier had an odd number of stages; the required phase relationship was obtained by the use of voltage derived feedback. This would equally apply to any higher odd number of stages. In Example 7.3 the amplifier had two stages with current derived current feedback. Similar circuitry would be required for four or any even number of stages. Of course the usual care would be necessary to ensure that positive feedback did not give rise to instability at the extremes of the frequency range.

Two Stage Voltage Feedback Amplifiers

Example 7.2 was concerned with a single stage voltage amplifier and the negative feedback was current derived. From the discussion above, we should expect a two stage voltage amplifier to have voltage derived feedback. This is correct, if the feedback is applied to the emitter of the first stage as in *Figure 7.1c*. If an output transformer were used, as in *Figure 7.1b*, then current derived feedback would be essential to obtain the correct phase relationship. An example of the first case will now be discussed.

Example 7.5. An ideal two stage transistor amplifier has loads of 1 kΩ per stage. Feedback is provided by a resistive potential divider chain of 9·5 kΩ and 0·5 kΩ in parallel with the output. The 0·5 kΩ

of this arrangement is connected between the emitter of the first transistor and earth. The transistor parameters may be taken as h_{ie} 1 000 Ω, h_{fe} 100, h_{oe} 100 μmho and h_{re} 0. By means of a general analysis verify a solution based on feedback theory.

The circuit arrangement for this problem is shown in *Figure 7.8*.

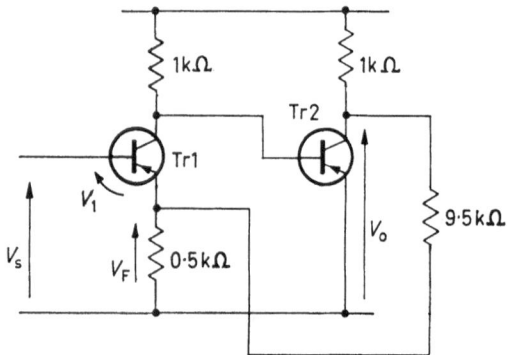

Figure 7.8. Voltage amplifier with voltage derived feedback for Example 7.5

To confirm first of all that this arrangement does provide negative feedback; the output voltage of the first stage will be A_1V_1 with 180° phase shift. V_0 will therefore be $A_1A_2V_1$ with a further 180° phase shift giving a total zero phase shift between V_0 and V_1. Initially neglecting Tr1 emitter current, V_f will be in phase with V_1 and its value will be given by:

$$V_f = \beta V_0 = \beta A_1 A_2 V_1$$

Where
$$\beta = \frac{0\cdot5}{0\cdot5 + 9\cdot5} = 0\cdot05$$

Now writing the series equation around the input circuit,

$$V_s = V_1(1 + \beta A_1 A_2)$$

or
$$V_1 = \frac{V_s}{(1 + \beta A_1 A_2)}$$

Since V_1 is less than V_s the overall gain is reduced and the feedback is negative.

In the above discussion, we have neglected the effect of Tr1 emitter current upon V_f. In practice this emitter resistor not only provides a part of the overall feedback network, it also causes 'local' negative on the first stage. The proposed technique based on feedback theory will involve the following calculations:

(1) A_{v2} and Z_{in2} using 1 kΩ in parallel with 10 kΩ as the load.

(2) $Y_{L1\ eff}$ including Z_{in2} and the 1 kΩ load for Tr1.

(3) A_{v1} and Z_{in1} without local feedback.

(4) A_{v1f} and Z_{inf} with local feedback taking β as

$$\frac{Z_F}{Z_{Leff}} = 0 \cdot 5 \text{ k}\Omega \times Y_{L1\ eff}$$

Note, in certain cases this β may be greater than one, resulting in a gain with local feedback of less than one. This does in fact occur and may be verified with measurement on practical circuits.

(5) Overall gain without overall feedback from $A_{v1f} \times A_{v2}$.

(6) Using β of 0·05, overall gain and input impedance with feedback. The output impedance can be determined in a similar manner, but strictly the overall β should be modified by the output impedance measured at the emitter of Tr1. This will reduce β for this calculation but unless the source impedance Z_s is low the effect will be negligible.

We shall now follow the procedure for the circuit shown in *Figure 7.8*.

$$Y_{L2} = (1\ 000 + 100)\ \mu mho = 1\ 100\ \mu mho$$

$$A_{v2} = \frac{-100}{1\ 000(1\ 100 + 100)10^{-6}} \quad -83 \cdot 4$$

$$Z_{in2} = 1 \text{ k}\Omega$$

$$\therefore \quad Y_{in2} = 1\ 000\ \mu mho$$

$$\therefore \quad Y_{L1\ eff} = 1\ 000 + 1\ 000 = 2\ 000\ \mu mho$$

$$A_{v1} = \frac{-100}{1\ 000(100 + 2\ 000)10^{-6}} = -47 \cdot 6$$

$$Z_{in1} = 1 \text{ k}\Omega$$

For local feedback

$$\beta = 500 \times 2\ 000 \times 10^{-6} = 1$$

$$\therefore \quad A_{v1f} = \frac{-47 \cdot 6}{1 + 47 \cdot 6} = -0 \cdot 98$$

$$Z_{in1f} = 1 \text{ k}\Omega(1 + 47 \cdot 6) = 48 \cdot 6 \text{ k}\Omega$$

Overall gain without overall feedback $= -0.98 \times -83.4 = 82$

$\therefore$ Overall gain with overall feedback $= \dfrac{82}{1 + 82 \times 0.05} = \underline{16.1}$

$$(7.1)$$

and

Overall input impedance

$$= 48.6 \text{ k}\Omega(1 + 82 \times 0.05) = \underline{248 \text{ k}\Omega} \qquad (7.2)$$

To verify this procedure we shall use the equivalent circuit shown in *Figure 7.9* and solve by nodal analysis.

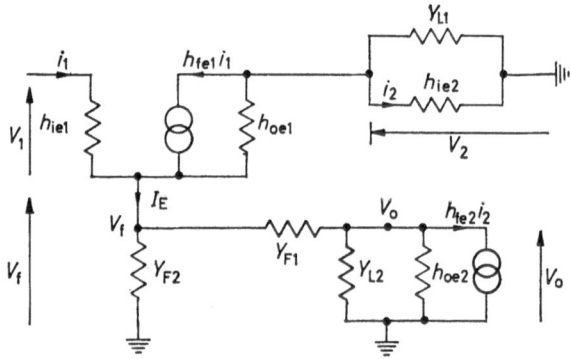

Figure 7.9. Equivalent circuit for the verification of the feedback method used in Example 7.5

First, i_1 the base current for Tr1, is given by V_1/h_{1e}. The output of Tr1 is loaded by $Y_{L1 \text{ eff}}$ in series with Y_{F2}. Strictly Y_{F1} provides additional loading on Tr1 but the effect is negligible since $Y_F \ll Y_{F2}$. This loading is given by

$$\frac{Y_{L1 \text{ eff}} Y_{F2}}{Y_{L1 \text{ eff}} + Y_{F2}} = \frac{2\,000 \times 2\,000}{2\,000 + 2\,000}\,\mu\text{mho} = 1\,000\,\mu\text{mho}$$

$$\therefore \qquad I_E = \frac{V_1}{h_{1e}}\left[1 + h_{fe}\left(\frac{1\,000}{1\,000 + h_{oe}}\right)\right]$$

Inserting values and writing I_E in mA,

$$I_E = V_1\left(1 + 100 \times \frac{1\,000}{1\,100}\right)\text{mA}$$

$$\therefore \qquad I_E = 92\,V_1\,\text{mA} \qquad (7.3)$$

$$\text{Now,} \quad V_2 = \frac{-h_{fe}V_1}{h_{1e}(h_{oe} + Y_{L1 \text{ eff}})} = \frac{-100V_1}{1\,000(100 + 2\,000)10^{-6}} \text{ V}$$

$$\text{and} \quad i_2 = \frac{V_2}{h_{1e2}} = \frac{-100V_1}{2 \cdot 1 \times 1\,000} \text{ A}$$

$$\therefore \quad i_2 = -47 \cdot 5 V_1 \text{ mA} \tag{7.4}$$

To give the correct dimensions, since currents are expressed in mA and voltage in volts, admittance must be expressed in mmho.

$$\text{Thus} \quad Y_{F2} = 2 \text{ mmho} \quad Y_{F1} = 0 \cdot 105 \text{ mmho}$$

$$Y_{L2} = 1 \text{ mmho} \quad h_{oe2} = 0 \cdot 1 \text{ mmho}$$

Now writing nodal equations for the V_f and V_o nodes and using equations 7.3 and 7.4 above:

$$92V_1 = V_f(2 + 0 \cdot 105) - 0 \cdot 105 V_o$$

$$-100(-47 \cdot 5)V_1 = -0 \cdot 105 V_f + V_o(1 + 0 \cdot 1 + 0 \cdot 105)$$

Rearranging:

$$92V_1 = 2 \cdot 105 V_f - 0 \cdot 105 V_o$$

$$4\,750V_1 = -0 \cdot 105 V_f + 1 \cdot 205 V_o$$

Solving by determinants,

$$V_o = \frac{2 \cdot 105 \times 4\,750V_1 + 0 \cdot 105 \times 92V_1}{2 \cdot 105 \times 1 \cdot 205 - 0 \cdot 105 \times 0 \cdot 105}$$

$$= \frac{10\,010}{2 \cdot 524} V_1 = 3\,960V_1 \tag{7.5}$$

$$V_f = \frac{92V_1 \times 1 \cdot 205 + 4\,750V_1 \times 0 \cdot 105}{2 \cdot 524}$$

$$= \frac{609V_1}{2 \cdot 524} = 241V_1 \tag{7.6}$$

The signal voltage $V_s = V_f + V_1$.

Substituting from equation 7.6,

$$V_s = V_1(241 + 1) = 242V_1$$

But from equation 7.5,

$$V_o = 3\,960V_1$$

$$\therefore \quad \text{Overall gain} = \frac{V_o}{V_s} = \frac{3\,960}{242} = 16 \cdot 3 \tag{7.7}$$

and

$$\text{Overall input impedance} = \frac{V_s}{i_1} = \frac{242 V_1}{V_1} \text{ k}\Omega$$

$$= 242 \text{ k}\Omega \qquad (7.8)$$

Comparing results 7.7 and 7.8 with results 7.1 and 7.2 respectively (16.1 and 248 kΩ) the difference is only of the order of 2 per cent. Since components and parameters are unlikely to be known to within 20 per cent this error is negligible and the method based upon feedback is adequately verified. The reader may have wondered why a solution based upon four-terminal network theory was not used for this example. The feedback network is connected in series with the input, and in parallel with the output of the amplifier. This suggests that if the h parameters of the network and those of the amplifier were added, the overall h parameters would be obtained. Unfortunately this is not valid, since I_1 for the amplifier is i_b, while the current flowing out of the common terminal, I_1 for the network, is $i_b + i_c$. Thus only feedback methods or circuit analysis can be used for this very common circuit. Other situations however do permit the four-terminal network approach.

The circuits for Examples 7.1 and 7.4 are the parallel input and output form permitting the summation of y parameters, and the circuit for Example 7.2 is series input and series output which is the correct configuration for summation of z parameters. The current amplifier for Example 7.3 has the feedback network in parallel with the input and in series with the output, suggesting the use of g parameters. This may be done provided the effect of the final emitter resistor is included in the calculation of input impedance for Tr2 with local feedback; the A for the calculation is the normal loaded voltage gain.

Apart from this, the calculation of g_{21} is made by neglecting the final emitter resistor and is given by $A_{v1} \times A_{v2}$ with Y_{L2} zero. The remaining procedure then follows normal four-terminal network practice, and the resulting solutions are extremely close to those obtained by feedback methods.

Three Stage Voltage Feedback Amplifiers

The one remaining simple configuration not yet considered is that of a voltage amplifier with an odd number of stages, and feedback applied to the first stage emitter (or cathode). Since voltage derivation led to negative feedback over an even number of stages, an odd number of stages will require current derived feedback. For this situation, we shall consider a valve amplifier.

240

Example 7.6. A three valve amplifier employs pentodes having g_m 3 mA/V, r_a 1 MΩ, and input capacitance 5 pF on load. If each stage is loaded with 10 kΩ and grid leak resistors of 1 MΩ, determine the overall voltage gain and the upper 3 db frequency. If negative feedback is applied as shown in the simplified circuit in *Figure 7.10* determine the new medium frequency gain and the gain at the 3 db frequency calculated above.

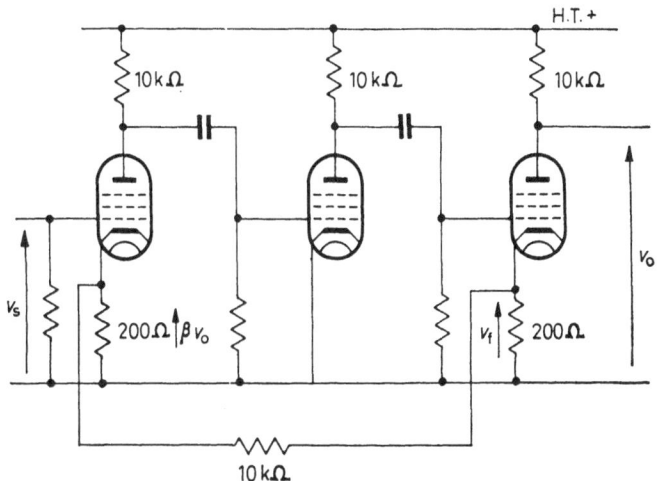

Figure 7.10. Three stage feedback amplifier for Example 7.6

When no feedback is applied, the results found in Chapter 4 may be used.

For each stage

$$r_a = R_g \gg R_L$$

$\therefore \qquad R_e \simeq R_L = 10 \text{ kΩ}$

$$A_{vm} = -g_m R_e$$
$$= -3 \times 10^{-3} \times 10^4 = -30$$

$\therefore$ Overall maximum voltage gain $= (-30)^3$
$$= -27 \times 10^3$$

To determine the 3 db frequency we need only consider the first two stages. There is no coupling network to load the final stage and so the gain may be assumed constant at -30.

At the 3 db frequency

$$|A_h| = \left| \frac{A_m}{\sqrt{2}} \right| = \frac{27 \times 10^3}{\sqrt{2}} \tag{7.9}$$

But allowing for the two interstage couplings:

$$A_h = \frac{A_m^2}{\left(1 + j\dfrac{R_e}{X_{cs}}\right)^2} \times A_m = \frac{-27 \times 10^3}{(1 + j\omega C_s R_e)^2} \qquad (7.10)$$

$\therefore$ from equation 7.9

$$\frac{27 \times 10^3}{\sqrt{2}} = \left| \frac{-27 \times 10^3}{(1 + j\omega C_s R_e)^2} \right|$$

or $\qquad |(1 + j\omega C_s R_e)^2|^2 = 2$

and $\qquad |1 + j2\omega C_s R_e - \omega^2 C_s^2 R_e^2|^2 = 2$

$\therefore \qquad (1 - \omega^2 C_s^2 R_e^2)^2 + 4\omega^2 C_s^2 R_e^2 = 2$

But $\qquad C_s R_e = 5 \times 10^{-12} \times 10^4 = 5 \times 10^{-8}$

$\therefore \; 1 - 2 \times 25 \times 10^{-16}\omega^2 + 625 \times 10^{-32}\omega^4 + 4$
$$\times 25 \times 10^{-16}\omega^2 = 2$$

or $\qquad 6{\cdot}25 \times 10^{-30}\omega^4 + 5 \times 10^{-15}\omega^2 - 1 = 0$

$\therefore \quad \omega^2 = \dfrac{-5 \times 10^{-15} \pm \sqrt{(25 \times 10^{-30} + 25 \times 10^{-30})}}{12{\cdot}5 \times 10^{-30}}$

$$= \frac{-5 \pm \sqrt{50}}{12{\cdot}5 \times 10^{-15}} = \frac{2{\cdot}07}{12{\cdot}5 \times 10^{-15}}$$

(since negative ω^2 not permissible)

$\therefore \qquad \omega = \sqrt{\left(\dfrac{2{\cdot}07}{1{\cdot}25}\right)} \times 10^7$

and $\qquad f = \dfrac{\omega}{2\pi} = 2{\cdot}04 \text{ MHz}$

Thus at 2·04 MHz the overall gain without feedback is

$$\frac{27 \times 10^3}{\sqrt{2}} = 19\ 100$$

If the feedback circuit is now considered at medium frequencies, stages one and three have local feedback due to the 200 Ω cathode resistors, and there is overall current derived feedback.

242

For the local feedback, in each case

$$\beta = \frac{200}{10^4}$$

$$\therefore \qquad A_{v1f} = A_{v3f} = \frac{-30}{1 + \dfrac{30 \times 200}{10^4}} = -18 \cdot 75 \qquad (7.11)$$

$\therefore$ Overall gain without overall feedback $= (-18 \cdot 75)^2 \times -30$

$$= -10\ 580$$

Examination of the circuit shows that

$$\beta v_0 = \frac{200}{200 + 10^4} \times v_f \quad \text{where} \quad v_f = \frac{200}{10^4} v_0$$

$$\therefore \qquad \text{Overall } \beta = \frac{4 \times 10^4}{1 \cdot 02 \times 10^8} = 3 \cdot 92 \times 10^{-4}$$

$$\therefore \qquad \text{Overall voltage gain} = \frac{-10\ 580}{1 + 10\ 580 \times 3 \cdot 92 \times 10^{-4}}$$

$$= -2\ 060$$

At the upper 3 db frequency, the overall β is unchanged, but the load on the first valve is changed modifying the local β for that stage.
At this frequency, 2·04 MHz

$$Y_{L1} = Y_{L2} = 10^{-4} + j2\pi \times 2 \cdot 04 \times 10^6 \times 5 \times 10^{-12} \text{ mho}$$
$$= (1 + j0 \cdot 64) \times 10^{-4} \text{ mho}$$
$$= 118 \cdot 5 \angle 32° \ 36' \ \mu\text{mho}$$

$$\therefore \ Z_{L1} = Z_{L2} = 8 \cdot 45 \angle -32° \ 36' \ \text{k}\Omega$$

$\therefore$ without local feedback

$$A_{v1} = A_{v2} = -g_m Z_L$$
$$= 25 \cdot 35 / 180° - 32° \ 36'$$

But local β for $A_{v1} = \dfrac{R_F}{Z_L}$

$$= \frac{0 \cdot 2}{8 \cdot 45 \angle -32° \ 36'} \qquad (7.12)$$

243

$$\therefore \quad A_{v1t} = \frac{25 \cdot 35 \ \angle \ 147° \ 24'}{1 - \dfrac{25 \cdot 35 \times 0 \cdot 2}{8 \cdot 45} \ \angle \ 147° \ 24' + 32° \ 36'}$$

$$= \frac{25 \cdot 35 \ \angle \ 147° \ 24'}{1 + 0 \cdot 6} = 15 \cdot 5 \ \angle \ 147° \ 24'$$

It is of interest to note this drop in gain due to feedback is identical to that shown in result 7.11. Feedback over a single stage does not improve the bandwidth since β changes with the load.

Now the overall gain without overall feedback at 2·04 MHz

$$A_v = 15 \cdot 5 \ \angle \ 147° \ 24' \times 25 \cdot 35 \ \angle \ 147° \ 24' \times 18 \cdot 75 \ \angle \ 180°$$

$$= 7\ 350 \ \angle \ 474° \ 48' = 7\ 350 \ \angle \ 114° \ 48'$$

The overall β is unchanged

$$\therefore \quad A_{vt} = \frac{7\ 350 \ \angle \ 114° \ 48'}{1 - 7\ 350 \times 3 \cdot 92 \times 10^{-4} \ \angle \ 114° \ 48'}$$

$$A_{vt} = \frac{7\ 350 \ \angle \ 114° \ 48'}{1 - 2 \cdot 88(\cos 114° \ 48' + j \sin 114° \ 48')}$$

$$= \frac{7\ 350 \ \angle \ 114° \ 48'}{1 + 1 \cdot 21 - j2 \cdot 62}$$

$$= \frac{7\ 350 \ \angle \ 114° \ 48'}{3 \cdot 12 \ \angle \ -49° \ 51'}$$

$$= 2\ 357 \ \angle \ 164° \ 39'$$

This is greater than the medium frequency gain with negative feedback, but is considerably less than the gain at this frequency without feedback. This result is as would be expected from the general case discussed in Chapter 6. The 3 db frequency with feedback may be found by following a similar technique. The local feedback effect on the first stage does not vary with frequency so this can be ignored.

Thus for the 3 db frequency

$$\frac{1}{\sqrt{2}} \left| \frac{A_m}{1 - \beta A_m} \right| = \left| \frac{A_h}{1 - \beta A_h} \right|$$

$$= \left| \frac{\dfrac{A_m}{(1 + j\omega C_s R_e)^2}}{1 - \dfrac{\beta A_m}{(1 + j\omega C_s R_e)^2}} \right|$$

$$= \left| \frac{A_m}{(1 + j\omega C_s R_e)^2 - \beta A_m} \right|$$

$$\therefore \quad \sqrt{2}|1 - \beta A_m| = |(1 + j\omega C_s R_e)^2 - \beta A_m|$$

But
$$\beta A_m = 3.92 \times 10^{-4} \times 10\,580 = -4.2$$

$$\therefore \quad \sqrt{2} \times 5.2 = |(1 + j\omega C_s R_e)^2 + 4.2|$$
$$= |1 + j2\omega C_s R_e - \omega^2 C_s^2 R_e^2 + 4.2|$$

Taking the modulus and squaring:

$$2 \times 27 = (5.2 - \omega^2 C_s^2 R_e^2)^2 + 4\omega^2 C_s^2 R_e^2$$
$$= 27 - 10.4\omega^2 C_s^2 R_e^2 + \omega^4 C_s^4 R_e^4 + 4\omega^2 C_s^2 R_e^2$$

$$\therefore \quad \omega^4 C_s^4 R_e^4 - 6.4\omega^2 C_s^2 R_e^2 - 27 = 0$$

But $C_s R_e = 5 \times 10^{-8}$,

$$\therefore \quad 625 \times 10^{-32}\omega^4 - 1.6 \times 10^{-14}\omega^2 - 27 = 0$$

$$\omega^2 = \frac{1.6 \times 10^{14} \pm \sqrt{(2.6 \times 10^{-28} + 1.69 \times 10^{-28})}}{12.5 \times 10^{-30}}$$

$$\omega^2 = \frac{3.66}{12.5} \times 10^{16}$$

$$\therefore \quad \omega = 5.31 \times 10^7 \text{ rad/sec}$$

and
$$f = \frac{\omega}{2\pi} \simeq 10 \text{ MHz}$$

Thus since the low frequency 3 db frequency may be ignored, negative feedback has increased the bandwidth from 2 MHz to 10 MHz while reducing the medium frequency gain from 27 000 to 2 060 which is in approximately the same ratio.

Complex Feedback Factors

So far in this chapter we have only considered β factors with real values. In the last example, at high frequency, β for the local

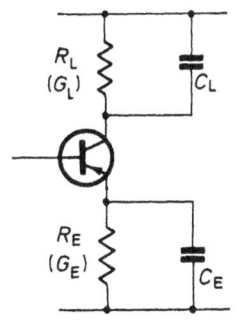

Figure 7.11. Single stage amplifier with frequency compensation by negative feedback

245

feedback on the first stage was complex (7.12). This resulted in no improvement in frequency response for that stage. The frequency response for a single stage may be improved if the cathode or emitter resistor is shunted by a small capacitance.

Consider the circuit shown in *Figure 7.11*. The voltage gain for a single stage grounded emitter amplifier is approximately $-h_{te}Z_L/h_{ie}$ and β for the current derived feedback shown is Z_E/Z_L.

$$\therefore \qquad \text{Overall gain} = \frac{\dfrac{-h_{te}Z_L}{h_{ie}}}{1 + \dfrac{h_{te}Z_L}{h_{ie}}\dfrac{Z_E}{Z_L}}$$

The Z_Ls in the denominator cancel, and if $h_{te}Z_E/h_{ie} \gg 1$ the overall gain $\simeq Z_L/Z_E = Y_E/Y_L$.

Taking account of the components shown,

$$A_v = \frac{G_E + j\omega C_E}{G_L + j\omega C_L}$$

This ratio will be constant if

$$\frac{\omega C_E}{G_E} = \frac{\omega C_L}{G_L}$$

or if

$$R_E C_E = R_L C_L$$

This implies a flat frequency response so long as

$$\frac{h_{te}Z_E}{h_{ie}} \gg 1$$

Let h_{te} be 100, h_{ie} 1 kΩ, R_L 4 kΩ and R_E 1 kΩ. If the shunt capacitance C_L is 0·01 μF then C_E must be 0·0025 μF. To determine the approximate useful frequency range, let

$$\frac{h_{te}Z_E}{h_{ie}} = 5 \quad \text{or} \quad \frac{h_{te}}{h_{ie}Y_E} = 5$$

$$\therefore \qquad 5 = \left| \frac{100}{1\,000(10^{-3} + j\omega 25 \times 10^{-10})} \right|$$

$$\therefore \qquad |10^{-2} + j\omega 2\cdot5 \times 10^{-8}| = 0\cdot2$$

Since the real term can have little effect,

$$\omega = \frac{0\cdot2}{2\cdot5} \times 10^8$$

and

$$f = \frac{20}{2\pi \times 2\cdot5} \simeq 1\cdot3 \text{ MHz}$$

Another way in which a capacitor may be used in a feedback path, is to prevent instability. If overall feedback is applied to, say, a four stage amplifier, and oscillation results at a particular high frequency, a shunt capacitor may be used to reduce β at the high frequencies in this range. This will reduce the amount of *positive* feedback at these frequencies eliminating the instability and flattening the gain frequency response. Similar modifications for low frequencies may be achieved with suitable capacitors or inductors. Many such combinations are possible and may be found in practical circuits.

COMPOSITE FEEDBACK

Other forms of feedback circuit may also be found when two types of feedback may be included on the same amplifier. This is sometimes referred to as composite feedback. *Figure 7.12* shows some examples of composite feedback circuits.

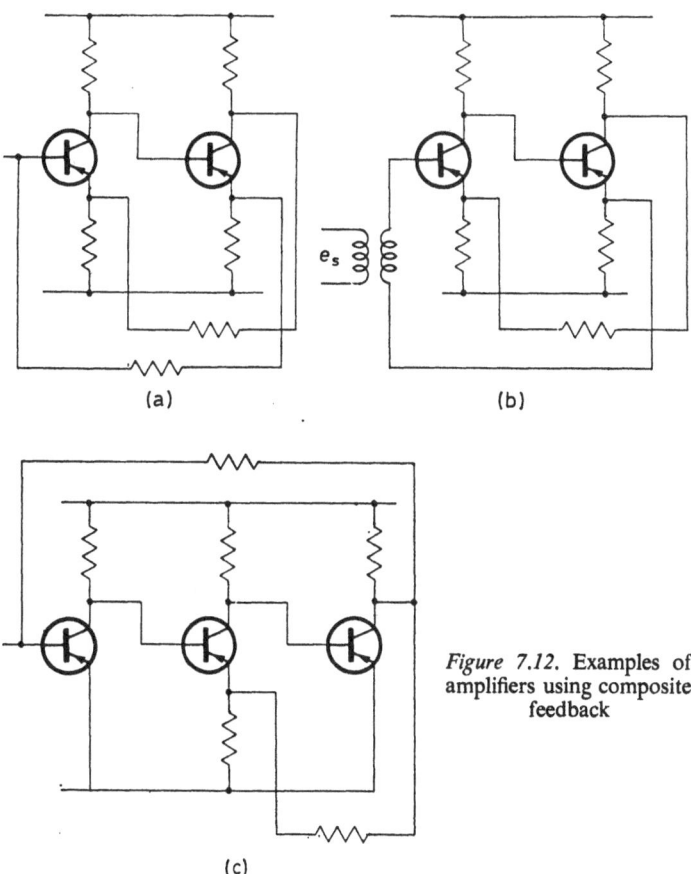

Figure 7.12. Examples of amplifiers using composite feedback

In each case only negative feedback at medium frequencies is employed. In *Figure 7.12a* both current derived current feedback and voltage derived voltage feedback are present. Since the voltage feedback is applied to the first emitter, the current feedback loop is 'outside' the voltage feedback loop. The procedure would be to determine first, A_{v2} and A_{12} accounting for the local feedback present; second to find A_{v1} and A_{11} including local feedback; third to calculate the overall gain and input impedance with voltage feedback, finally to account for the effect of current feedback.

The circuit shown in *Figure 7.12b* employs both voltage derived and current derived voltage feedback and a similar procedure should be used taking the voltage derived loop first.

The three stage amplifier in *Figure 7.12c* includes voltage derived voltage and current feedback. The voltage feedback is applied only over the last two stages. A final example, based on this circuit, will now be considered.

Example 7.7. The amplifier shown in *Figure 7.13* employs overall feedback through R_1 to reduce the input impedance to 10 Ω and to

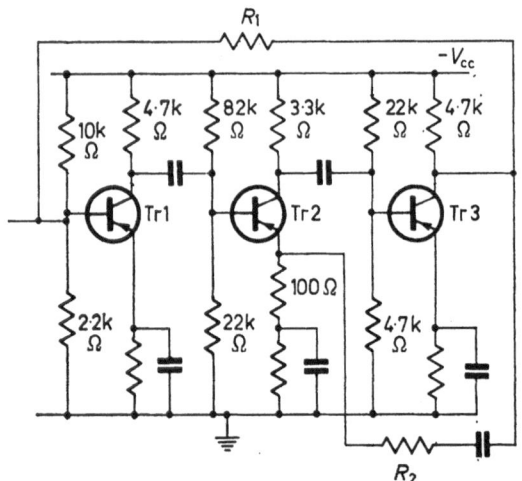

Figure 7.13. Circuit for Example 7.7

stabilize the circuit against changes in the d.c. supply voltage V_{cc}. Feedback through R_2 is provided to reduce the second harmonic distortion in the output voltage by a factor of 5. Determine suitable values for R_1 and R_2, and the output voltage if the amplifier is driven from a source which may be represented by a current generator of 1 μA in parallel with an impedance of 50 Ω. All capacitors may be assumed to have negligible reactance at the signal frequency. The transistor parameters are h_{1e} 1 200 Ω, h_{fe} 140, h_{oe} 120 μmho and h_{re} negligible.

Assuming that R_1 and R_2 will be much greater than 4·7 kΩ (212 μmho).

$$A_{v3} = \frac{-140}{1\,200(120 + 212)10^{-6}} = -352$$

and

$$A_{13} = \frac{-140 \times 212}{212 + 120} = -89$$

and

$$Y_{1n3} = \frac{1}{1\,200}\text{mho} = 834\,\mu\text{mho}$$

$\therefore$

$$Y_{L2} = 834 + 212 + 46 + 303 = 1\,395\,\mu\text{mho}$$

$\therefore$

$$A_{v2} = \frac{-140}{1\,200(120 + 1\,395)10^{-6}} = -77$$

$Z_{1n2} = 1\,200\,\Omega$

and

$$A_{12} = \frac{-140 \times 834}{1\,395 + 120} = -77\cdot5$$

For local feedback

$$\beta = Z_F Y_L = 100 \times 1\,395 \times 10^{-6} = 0\cdot1395$$

$$A_{v2f} = \frac{-77}{1 + 77 \times 0\cdot1395} = -6\cdot55$$

$$Z_{1n2f} = 1\,200(1 + 77 \times 0\cdot1395)\,\Omega = 14\cdot1\,\text{k}\Omega$$

Now for the voltage feedback loop

$$A = A_{v3} \times A_{v2f}$$
$$= 2\,300$$

But distortion must be reduced by a factor of 5

$\therefore$

$$1 + \beta A = 5 \quad \text{and} \quad \beta A = 4$$

$$\beta = \frac{4}{2\,300}$$

But

$$\beta = \frac{100}{R_2 + 100} = \frac{4}{2\,300}$$

$\therefore$

$$R_2 + 100 = \frac{2\,300 \times 100}{4} = 57\,500\,\Omega$$

and

$$\underline{R_2 = 57\cdot4\,\text{k}\Omega}$$

This feedback path further modifies the input impedance to

$$Z_{1n2f}(1 + \beta A) = 14 \cdot 1 \times 5 \text{ k}\Omega$$
$$= 70 \cdot 5 \text{ k}\Omega$$

$$\therefore \qquad Y_{1n2} = 14 \cdot 2 \text{ } \mu\text{mho}$$

By current division

$$A_{11} = \frac{-140 \times 14 \cdot 2}{120 + 14 \cdot 2 + 212 + 12 \cdot 2 + 45 \cdot 5} = -4 \cdot 9$$

Also $Y_{1n1} = 834 + 100 + 455 \text{ } \mu\text{mho}$
$$= 1 \text{ } 389 \text{ } \mu\text{mho}$$

Now allowing for current splitting at the input to Tr1,

$$\text{Overall } A_1 = \frac{834}{1 \text{ } 389} \times -4 \cdot 9 \times -77 \cdot 5 \times -89$$
$$= -20 \text{ } 300$$

But if the input impedance is to be 10 Ω, then

$$Y_{1n} = 10^5 \text{ } \mu\text{mho} = 1 \text{ } 389(1 + \beta A_1)$$

or

$$1 + \beta A_1 = \frac{10^2}{1 \cdot 389} = 72$$

and

$$\beta = \frac{71}{20 \text{ } 300} = 3 \cdot 5 \times 10^{-3}$$

But for voltage derived current feedback

$$\beta = \frac{Y_F}{Y_L}$$

$$\therefore \qquad Y_F = 3 \cdot 5 \times 10^{-3} \times 212 \text{ } \mu\text{mho} = 0 \cdot 74 \text{ } \mu\text{mho}$$

The feedback resistor

$$R_1 = \frac{1}{Y_F} = 1 \cdot 35 \text{ M}\Omega$$

Now all that remains is to find the load current for the complete amplifier. Since the source impedance is 50 Ω and the input impedance of the amplifier is 10 Ω, then the input current

$$i_{1n} = \frac{50}{50 + 10} \mu\text{A}$$

But

$$\text{Current gain with feedback} = \frac{-20\ 300}{1 + 20\ 300 \times 3\cdot5 \times 10^{-3}}$$

$$= \frac{-20\ 300}{72} = -282$$

∴ Load current $= -282 \times \tfrac{50}{60}\ \mu A = 235\ \mu A$

∴ Output voltage $= i_o Z_L = 0\cdot235\ \text{mA} \times 4\cdot7\ \text{k}\Omega$

$$= 1\cdot1\ \text{V}$$

For comparison, if the problem had been attempted using the complete equivalent circuit, the solution would have involved not less than five simultaneous equations.

In this chapter we have considered the practical circuits for providing negative feedback on a number of amplifier configurations.

Many other forms are used, including those with common base or common collector stages, but the methods applied here may be simply extended in most cases. For convenience, in any particular situation, feedback formulae carefully used probably gives the best approach. For more general analysis, the interconnected four-terminal network is probably better.

EXAMPLES

Example 7.8. Repeat Example 4.11 using feedback methods.
Ans. −10·5, 17·5 kΩ.

Example 7.9. Repeat Example 4.19 using feedback methods.
Ans. −9·62, 4·88 kΩ.

Example 7.10. Repeat Example 5.10 using feedback methods.
Ans. 6·8 kΩ, −1·97, −6·72.

Example 7.11. Repeat Example 5.13 using feedback methods.
Ans. −11·4, 264 Ω, 656 Ω.

Example 7.12. A two stage common emitter amplifier is constructed with the following components; R_{L1} 4 kΩ, R_{L2} 1 kΩ, combined shunt bias resistors 10 kΩ per stage. Feedback is provided by connecting a 100 Ω resistor in the emitter lead of Tr2 and a 100 kΩ resistor between Tr2 emitter and Tr1 base. The transistors are identical and have h_{ie} 1 250 Ω, h_{oe} 125 μmho, h_{fe} 90 and negligible h_{re}. Calculate the amplifier input impedance and current gain.
Ans. 490 Ω, 525.

Example 7.13. A two stage common emitter amplifier is constructed with the following components; R_{L1} 2 kΩ, R_{L2} 4 kΩ, combined shunt bias resistors 8 kΩ per stage. Feedback is provided by a 400 Ω resistor in the emitter lead of Tr1 and a 20 kΩ resistor between Tr2 collector and Tr1 emitter. Calculate the overall voltage gain and input impedance. The transistor parameters are h_{ie} 900 Ω, h_{oe} 100 μmho, h_{fe} 75, and h_{re} 0.

Ans. 43·5, 200 kΩ in parallel with 8 kΩ.

Example 7.14. A three stage common emitter amplifier employs identical transistors having h_{ie} 1 000 Ω, h_{oe} 125 μmho, h_{fe} 60 and h_{re} 0. Each stage has a collector load of 4 kΩ and the shunt bias components total 12·5 kΩ per stage. In the emitter leads of Tr1 and Tr3 are 50 Ω and 100 Ω resistors respectively and the two emitters have a 10 kΩ resistor connected between them. If the amplifier is driven by a source of e.m.f. 0·1 mV and internal impedance 2 kΩ, calculate the output voltage and the amplifier output impedance.

Ans. 0·475 V, 3·98 kΩ.

Example 7.15. The three stage amplifier described in Example 7.14 is converted into a current amplifier by removing the emitter resistors on Tr1 and Tr3 and the 10 kΩ feedback resistor. Feedback is now provided by connecting a 1 MΩ resistor between Tr3 collector and Tr1 base. Calculate the terminal current gain and the input impedance. Find also the output impedance if the source impedance is 50 Ω.

Ans. −250, 0·004 Ω, 0·224 Ω.

Example 7.16. Repeat Example 7.15 using interconnected four-terminal network theory.

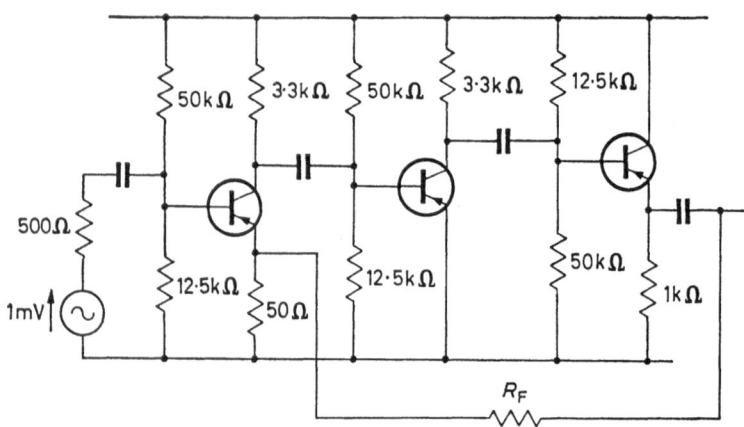

Figure 7.14. Circuit for Example 7.17

252

EXAMPLES

Example 7.17. The amplifier shown in *Figure 7.14* is required to have an output impedance of 1 Ω. If the transistor parameters are h_{ie} 1·5 kΩ, h_{fe} 140, h_{oe} 80 μmho, and negligible h_{re}, calculate the required value of R_{F}. Find also the resulting output voltage.

Ans. 4·5 kΩ, 83 mV.

Example 7.18. Each stage of a multistage amplifier has Z_{in} 1 kΩ, Z_{out} 2 kΩ, and an open circuit voltage gain of −150. Calculate A_{v}, A_{i}, Z_{in} and Z_{out} for the configurations shown in *Figure 7.15*.

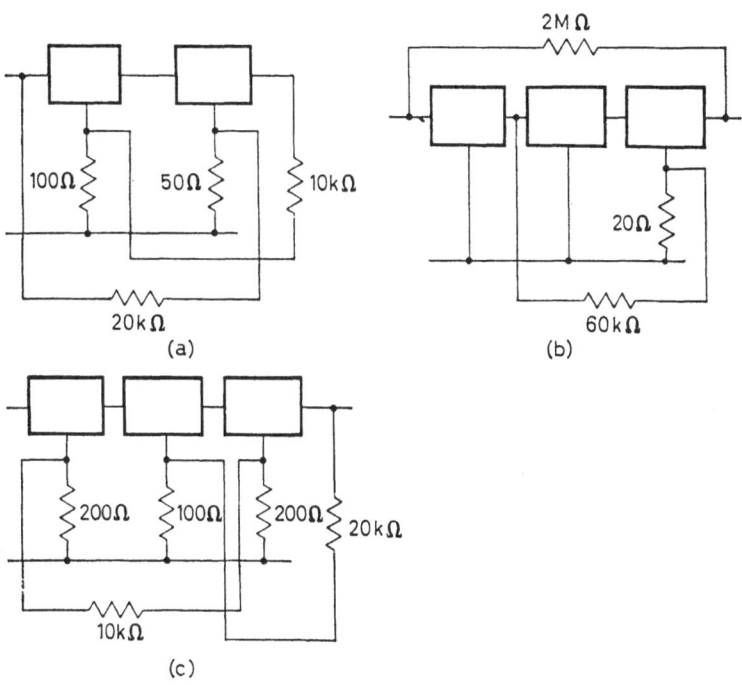

Figure 7.15. Circuits for Example 7.18

Assume that the collector load for each stage is 1 kΩ and that the driving source impedance is also 1 kΩ.

Ans. (a) 46·1, 287, 6·23 kΩ, 1·65 kΩ in parallel with 1 kΩ.

(b) −18 400, −1 820, 88·5 Ω, 95 Ω.

(c) −77·3, −1 310, 17 kΩ, 71·5 kΩ in parallel with 1 kΩ.

253

Example 7.19. The amplifier shown in *Figure 7.16* employs transistors with h_{ie} 1 000 Ω, h_{fe} 100, h_{oe} 125 μmho and h_{re} 0. The effect of the bias components may be neglected and the effective capacitance

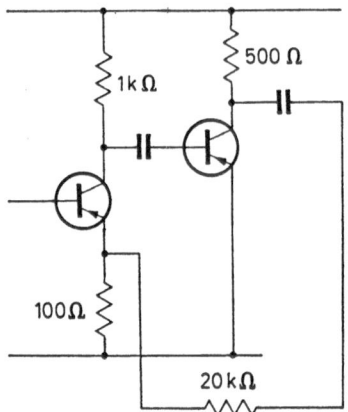

Figure 7.16. Circuit for Example 7.19

in parallel with each load is 500 pF. Determine the maximum gain and the 3 db bandwidth assuming that the coupling capacitors have negligible reactance at all signal frequencies.

Ans. 212, 0·99 MHz.

254

8

OSCILLATORS

In Chapter 6, we found that under certain conditions positive feedback could result in an amplifier having infinite gain at certain frequencies. This resulted in self oscillation, or an output when no input was present. For many electronic systems a source of alternating e.m.f. or current is required and oscillators based on the positive feedback principle are suitable for this purpose. Electronic oscillators can be constructed to work at frequencies as low as one cycle in 10 minutes or longer and as high as 200,000 MHz. For frequencies higher than a few 100 MHz special forms of valve and circuitry are required, but the basic principles for all frequencies are the same.

Consider once again the expression for the gain of an amplifier with feedback:

$$A_f = \frac{|A| \angle \theta}{1 - |\beta A| \angle \theta + \phi} \tag{8.1}$$

If the gain, A_f, is to be infinite, two conditions must be fulfilled; $|\beta A|$ must equal 1 and $(\theta + \phi)$ must be zero. In general θ will be either $0°$ or $180°$ depending upon the number of amplifier stages. For oscillation at a particular frequency, ϕ must be $180°$ or $0°$ at that frequency only. Then if $A \geq (1/\beta)$ at that frequency, the system will oscillate.

RC OSCILLATORS

A common class of oscillators employs feedback networks consisting of resistors and capacitors only. We shall consider four cases; voltage phase shift networks giving $0°$ or $180°$ and current phase shift networks giving $0°$ or $180°$.

These feedback circuits may then be employed with one or two stage voltage or current amplifiers respectively.

255

Voltage Wien Bridge Oscillator

Consider the voltage phase shift network shown in *Figure 8.1*. Using normal potential division methods, we can write

$$V_2 = \frac{\dfrac{-jX_{C2}R_2}{R_2 - jX_{C2}}}{R_1 - jX_{C1} - \dfrac{jX_{C2}R_2}{R_2 - jX_{C2}}} \times V_1$$

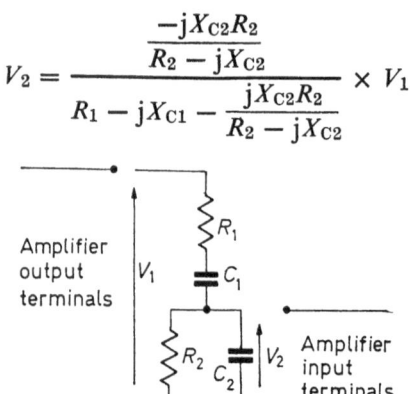

Figure 8.1. Wien bridge voltage phase shift network

Multiplying by $(R_2 - jX_{C2})$,

$$\frac{V_2}{V_1} = \frac{-jX_{C2}R_2}{R_1R_2 - X_{C1}X_{C2} - jR_2X_{C1} - jR_1X_{C2} - jX_{C2}R_2} \quad (8.2)$$

and if
$$\beta = \frac{V_2}{V_1}$$

The angle of this β factor must be either $0°$ or $180°$. Thus the whole expression must be real, having no imaginary component. In equation 8.2, the angle of the numerator is $-90°$; if the angle of the denominator can be $\pm 90°$, β will have the required angle. This can only be true if the real part of the denominator equals zero, i.e.

$$R_1R_2 - X_{C1}X_{C2} = 0$$

$$\therefore \qquad R_1R_2 = \frac{1}{\omega C_1} \times \frac{1}{\omega C_2}$$

The frequency at which this is true is given by

$$\omega^2 = \frac{1}{R_1R_2C_1C_2} \text{ (rad/sec)}^2 \quad (8.3)$$

and
$$f = \frac{1}{2\pi\sqrt{(R_1R_2C_1C_2)}} \text{ Hz} \quad (8.4)$$

Note if $R_1 = R_2$ and $C_1 = C_2$,

$$f = \frac{1}{2\pi CR} \text{ Hz} \tag{8.5}$$

Now at this frequency, the feedback factor β is given by:

$$\beta = \frac{V_2}{V_1} = \frac{-jX_{C2}R_2}{-jR_2X_{C1} - jR_1X_{C2} - jR_2X_{C2}}$$

The minimum amplifier gain A_v must be equal to or greater than $1/\beta$. So, cancelling the $-j$ throughout and inverting

$$A_{v\ min} = \frac{R_2X_{C1} + R_1X_{C2} + R_2X_{C2}}{R_2X_{C2}}$$

$$= \frac{X_{C1}}{X_{C2}} + \frac{R_1}{R_2} + 1$$

putting $X_C = (1/\omega C)$,

$$A_{v\ min} = \frac{C_2}{C_1} + \frac{R_1}{R_2} + 1 \tag{8.6}$$

If, as before, identical capacitors and resistors are used:

$$A_{v\ min} = 1 + 1 + 1 = +3$$

Since this result is positive, the amplifier required must have no phase shift and a voltage gain of at least 3. This suggests either a single stage common base transistor amplifier or a grounded grid valve amplifier. Unfortunately these circuits are not suitable since the very low input impedance in each case must effectively become R_2. Suitable values of R_1, C_1 and C_2 for the required frequency then make $A_{v\ min}$ more the amplifier can provide when loaded with the feedback network.

The practical solution is to use a two stage common emitter or grounded cathode amplifier, and to ensure that the β network has a negligible loading effect upon the final stage. To find the order of the loading effect we will consider the case when $R_1 = R_2 = R$ and $C_1 = C_2 = C$.

The impedance Z presented by the feedback network to the amplifier output terminals is given by:

$$Z = R - \frac{j}{\omega C} - \frac{\dfrac{jR}{\omega C}}{R - \dfrac{j}{\omega C}}$$

Substituting for ω from equation 8.5,

$$Z = R - \frac{jCR}{C} - \frac{\dfrac{jR^2C}{C}}{R - \dfrac{jCR}{C}}$$

$$= R\left[(1 - j) - \frac{j}{(1 - j)}\right] = \frac{3}{2}R(1 - j) \qquad (8.7)$$

But in practice R will be the value of the amplifier input impedance. Thus a suitable amplifier will be one having an output impedance much less than its input impedance with a voltage gain greater than 3. *Figure 8.2* shows valve and transistor circuits based upon these principles.

Figure 8.2a shows a valve oscillator designed to operate at 1 kHz. The feedback network has equal capacitors, and resistors making

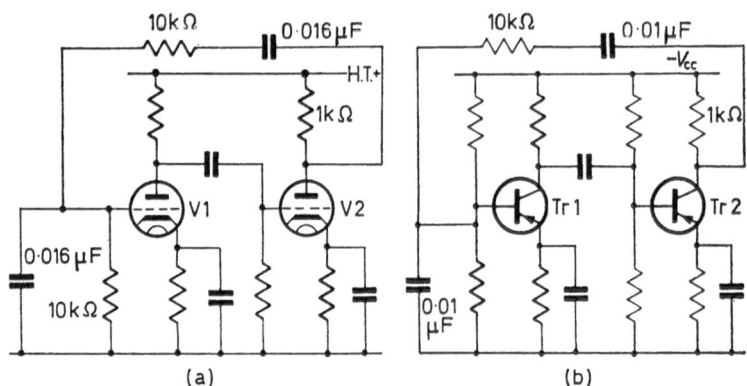

Figure 8.2. Wien bridge oscillators

the network impedance $15\sqrt{2} \angle 45°$ kΩ, and the minimum amplifier voltage gain 3. The final load of 1 kΩ makes the amplifier output impedance much less than the network impedance, and the amplifier input impedance excluding the R_g of 10 kΩ which forms part of the β network is infinite. Taking typical μ and r_a values of 30 and 10 kΩ respectively, the second stage gain

$$|A_{v2}| = \frac{30 \times 1}{10 + 1} = 2\cdot7$$

258

So provided the gain of the first stage is greater than 1·1 the circuit will oscillate.

Figure 8.2b shows a transistor circuit designed to oscillate at 5 kHz. R_2 in this case is the amplifier input impedance which will be approximately h_{1e}, say 1 kΩ. Making R_1 10 kΩ ensures that the network impedance Z will be much greater than the final load of 1 kΩ.

Applying equation 8.4 gives the common value of the equal C_1 and C_2 from

$$C = \frac{1}{2\pi f \sqrt{(R_1 R_2)}} \simeq 0\cdot01 \ \mu F$$

Now applying equation 8.6 to find the minimum voltage gain:

$$A_{v \ min} = \frac{10}{1} + 1 + 1 = 12$$

If the transistors have h_{1e} 50, and

$$A_v \simeq \frac{h_{1e}}{h_{1e}} Z_L$$

Then $A_{v2} \simeq 50$ giving more than sufficient voltage gain.

Amplitude Stability

This raises the question of the behaviour of these circuits if the gain is more than the minimum required. First consider an amplifier circuit with the d.c. supplies switched off. The gain will be zero or very much less than one. When the supplies are switched on, a finite time will elapse before the direct currents build up to their steady value. During this time the gain will rise from zero towards the final steady value calculated for the circuit. It must pass through the value which makes $\beta A \angle \theta + \phi$ equal to $1 \angle 0°$ where the gain with feedback becomes infinite. The circuit begins to oscillate, and the amplitude of the oscillating sinusoidal signal tends to rise to infinity. But as this signal amplitude rises the gain will fall, either because of change in parameters with large signals, or because the devices run into cut off and bottoming. A stable condition will be reached when $\beta A \angle \theta + \phi$ is exactly equal to $1 \angle 0°$. This process is shown in *Figure 8.3*. Suppose the overall signal gain of the amplifier is 6 without feedback and equal resistors and capacitors are used as in *Figure 8.2a*.

259

Assume our amplifier is a two stage transistor amplifier working from an eight volt d.c. supply and that the d.c. operating point for the second transistor is given by $V_{CE} = -4$ V. *Figure 8.3a* shows the input signal shortly after oscillation has commenced. This is amplified by 6 to give the output of 2·4 volts peak to peak shown in

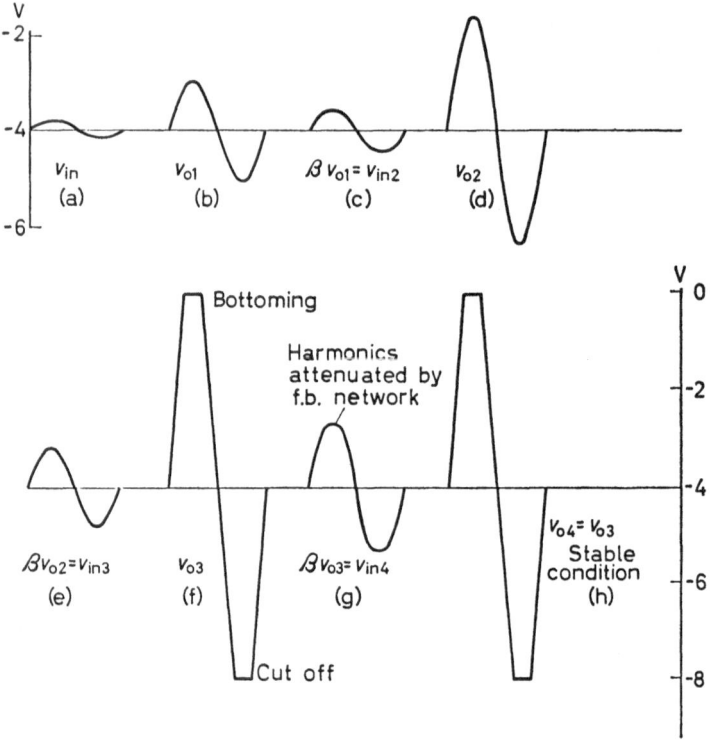

Figure 8.3. Limitation of amplitude by d.c. conditions

Figure 8.3b. Since β is one third, the new signal at the input is 0·8 V peak to peak resulting in an output of 4·8 V (*Figure 8.3d*). In *Figure 8.3e* the resulting input is 1·6 V peak to peak but if it was multiplied by 6 the output would be 9·6 V which is greater than the d.c. supply voltage. The final transistor therefore cuts off and bottoms as shown in *Figure 8.3f*. The fundamental of the waveform is attenuated by one third but the harmonics are more severely attenuated thus

V_{in4} in *Figure 8.3g* is nearly sinusoidal. This will again cause bottoming and cut off resulting in the same value of βv_0 as before. The gain is now

$$\frac{\text{Fundamental o/p}}{\text{i/p}} = \frac{8}{2 \cdot 67} = 3$$

The resulting distortion in the output is undesirable and additional techniques must be introduced to eliminate it. Firstly negative feedback can be included in the circuit to reduce the small signal gain to just greater than $1/\beta$. Then as the signal amplitude increases, change of parameters can be sufficient to reduce the gain to exactly

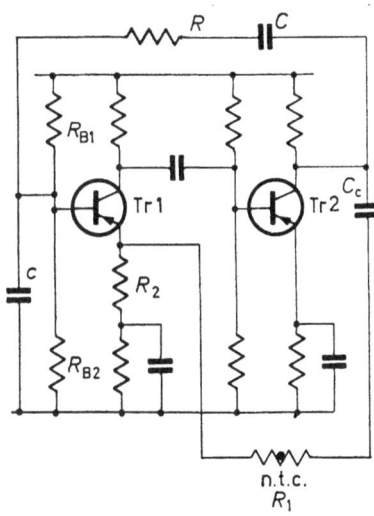

Figure 8.4. Wien bridge oscillator employing negative feedback for amplitude stabilization

$1/\beta$ before excessive distortion occurs. Unfortunately the amplitude of the resulting output signal will be extremely sensitive to any change in temperature or supply voltage. This fault may in turn be eliminated by making the amount of negative feedback proportional to the signal amplitude. *Figure 8.4* shows a circuit employing this technique.

Since negative feedback is employed, the input impedance for Tr1 will be large. The shunt R of the positive feedback network will therefore be given by R_{B1} and R_{B2} in parallel. Voltage negative feedback is provided by R_1 and R_2 with C_c to eliminate any d.c.

261

path. R_1 is a thermistor which is a resistor having a very high temperature coefficient of resistance, in this case negative (n.t.c.). The negative feedback β is given by $R_2/(R_1 + R_2)$ making the gain greater than 3 for small signals. When the output voltage approaches the desired maximum value the dissipation V^2/R_1 in R_1 rises and with it the temperature of R_1. The resistance of R_1 falls increasing β and thus reducing the gain. A stable condition is now reached without distortion of the output signal.

Current Wien Bridge Oscillator

The oscillator circuits discussed so far are known as Wien bridge oscillators. An alternative form of Wien bridge oscillator is based on a current amplifier. In this case, the phase shift network is required to give zero phase shift between input and output currents, at the required frequency. A suitable circuit is shown in *Figure 8.5*.

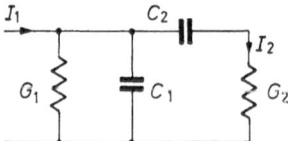

Figure 8.5. Wien bridge current phase shift network

Since we are concerned with current division, it is more convenient to work with admittances as shown.

$$I_2 = \frac{I_1 Y_2}{Y_1 + Y_2} = \frac{I_1 \dfrac{j\omega C_2 G_2}{G_2 + j\omega C_2}}{G_1 + j\omega C_1 + \dfrac{j\omega C_2 G_2}{G_2 + j\omega C_2}}$$

Multiplying numerator and denominator by $G_2 + j\omega C_2$

$$\frac{I_2}{I_1} = \frac{j\omega C_2 G_2}{G_1 G_2 - \omega^2 C_1 C_2 + j\omega C_1 G_2 + j\omega C_2 G_1 + j\omega C_2 G_2} \quad (8.8)$$

This expression has exactly the same form as equation 8.2. Following the same reasoning, we may deduce that for zero phase shift

$$\omega = \sqrt{\left(\frac{G_1 G_2}{C_1 C_2}\right)} = \frac{1}{\sqrt{(R_1 R_2 C_1 C_2)}} \quad (8.9)$$

If $R_1 = R_2 = R$ and $C_1 = C_2 = C$,

$$f = \frac{1}{2\pi CR} \text{ as before} \quad (8.10)$$

At this frequency, the minimum current gain A_{1min} is given by

$$A_{1min} = \frac{1}{\beta} = \frac{I_1}{I_2} = 1 + \frac{G_1}{G_2} + \frac{C_1}{C_2}$$

$$= 1 + \frac{R_2}{R_1} + \frac{C_1}{C_2} \tag{8.11}$$

which, for equal components $= 3$.

The input admittance may also be determined for equal components and referring to equation 8.7 the reader can show this to be

$$Y_{in} = \tfrac{3}{2}G(1 + j) \tag{8.12}$$

The amplifier requirements can now be considered; it must have zero phase shift, and the input impedance must either be less than R_1 or become R_1. Since the load is reactive (8.12), if the short circuit current gain has zero phase shift, the same phase shift will be obtained on load when the amplifier has a low output admittance. This may be verified from an expression for current gain:

$$A_1 = \frac{A_{1,s/c} Y_L}{Y_0 + Y_L} \tag{8.13}$$

If the amplifier is a two stage transistor amplifier the shortcircuit current gain $A_{1s/c}$ will have zero phase shift. If $Y_L \gg Y_0$, Y_0 may be neglected, and the phase shift will be zero even if Y_L is complex. Such an amplifier will have a current gain far greater than that required to sustain oscillation, therefore negative feedback may be used to reduce the input impedance, and to reduce the output admittance as required. The correct modifications will be achieved by using current derived, current feedback. A thermistor can be used to limit the amplitude in the same way as for the voltage amplifier.

Example 8.1. The amplifier shown in *Figure 8.6* is to be converted into an oscillator by (*a*) using a voltage phase shift network, and

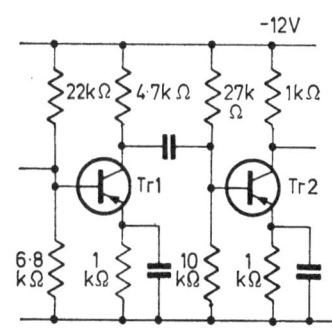

Figure 8.6. Amplifier circuit for
Example 8.1

263

(b) a current phase shift network. All the resistors shown are required to provide the correct d.c. conditions and stability. Design suitable positive and negative feedback circuits for each case if the required oscillator frequency is 5 kHz. Assume transistor h parameters of h_{ie} 5 kΩ, h_{oe} 90 μmho, h_{fe} 100 and h_{re} 0.

Consider first the voltage phase shift circuit: voltage derived, series applied feedback will be employed to reduce the gain to the required value of 3 for equal capacitors and resistors in the phase shift network. This will make the input impedance at the base of Tr1 very large and R for the network can be taken as the 6·8 kΩ and 22 kΩ bias resistors in parallel. The negative feedback will also make the output impedance very much less than the 1 kΩ final load, removing any possibility of loading by the network.

Working on the assumption that the final gain will be approximately $1/\beta$, the emitter resistor of Tr1 may be unbypassed and used for the R_2 of the feedback network. But β will be approximately $1/3$ so R_1 will be of the order of 2 kΩ. Thus for the calculation of A_{vo}, the load can be taken as 1 kΩ in parallel with $(2 + 1)$ kΩ, i.e. 750 Ω.

Now to determine A_{vo}:

$$A_{v2} = \frac{-100}{1\,500 \left(90 + \frac{1\,000}{0·75}\right) 10^{-6}} = -47·6$$

$$Y_{L1} = 667 + 37 + 100 + 213 \ \mu\text{mho}$$
$$= 1\,017 \ \mu\text{mho}$$

$$\therefore \qquad A_{v1o} = \frac{-100}{1\,500(1\,100)10^{-6}} = -60·5$$

But local $\qquad \beta = R_E Y_L \simeq 1$

$$\therefore \qquad A_{v1f} = \frac{-60·5}{1 + 60·5} \simeq -1$$

but $\qquad Z_{inf} = 1·5(1 + 60·5) \text{ kΩ} = 92 \text{ kΩ}$

Overall gain without feedback = 47·6.

But feedback must reduce this to 3,

$$\therefore \qquad 3 = \frac{47·6}{1 + 47·6\beta}$$

and $\qquad \beta = \frac{(47·6/3) - 1}{47·6} = 0·313$

But $\qquad \beta = \dfrac{R_2}{R_1 + R_2} = \dfrac{1}{1 + R_1}$ (since $R_2 = 1\ \text{k}\Omega$)

$\therefore \qquad R_1 = \dfrac{1}{0\cdot313} - 1 = 2\cdot2\ \text{k}\Omega$

which compares favourably with the estimated $2\ \text{k}\Omega$, and the resulting change in A_{v2} will have a negligible effect upon the calculation. Now,

$$Z_{\text{inf}} = 92(1 + 0\cdot313 \times 47\cdot6)\ \text{k}\Omega$$
$$= 1\cdot5\ \text{M}\Omega$$

Thus R for the phase shift network is given by the bias components alone.

$\therefore \qquad R = \dfrac{6\cdot8 \times 22}{22 + 6\cdot8}\ \text{k}\Omega = 5\cdot2\ \text{k}\Omega$

The required frequency,

$$f = 5\ \text{kHz} = \dfrac{1}{2\pi CR} \text{ (equation 8.5)}$$

$\therefore \qquad C = \dfrac{10^6}{2\pi \times 5\,000 \times 5\,200}\ \mu\text{F}$

$\qquad\qquad = 0\cdot006\ \mu\text{F}$

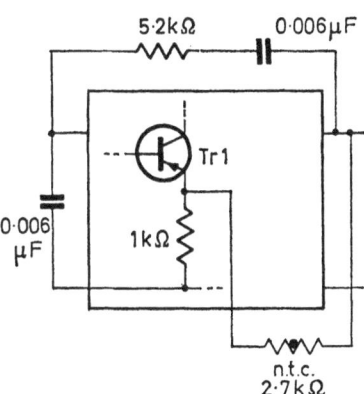

Figure 8.7. Solution for Example 8.1(a)

In practice β would be adjusted so that the small signal gain was greater than 3, by making R_1 a negative temperature coefficient thermistor of say $2\cdot7\ \text{k}\Omega$ cold. The completed arrangement is shown diagrammatically in *Figure 8.7*.

Finally, to confirm that the required output'impedance condition is satisfied; taking Z_0 without feedback to be approximately $750 \ \Omega$

$$Z_{ot} = \frac{750}{1 + 0\cdot313 \times 47\cdot6} \Omega = 47 \ \Omega$$

But the load provided by the phase shift network is

$$\tfrac{3}{2}R(1 - j) = 7\cdot8(1 - j) \ k\Omega$$

which will cause a negligible change to the gain and phase shift of the feedback amplifier.

Turning now to the alternative solution with a current phase shift network, the R will be the load of Tr2, 1 kΩ. Current derived current feedback can be obtained by unbypassing the emitter resistor of Tr2 and connecting an additional resistor from Tr2 emitter to Tr1 base. The load admittance is given by $\tfrac{3}{2}G(1 + j)$ (equation 8.12) and putting G as 1 mmho.

$$\therefore \qquad\qquad Y_L = 1\cdot5 + j1\cdot5 \ \text{mmho}$$

$$\simeq 2 \ \angle 45° \ \text{mmho}$$

This is very much greater than h_{oe} (0·09 mmho); the current gain of Tr2 is therefore approximately h_{te}. Tr$_2$ input impedance will be increased by the local voltage feedback provided by Tr$_2$ emitter resistor. For this calculation, the voltage gain is required. Neglecting h_{oe},

$$A_{vo} = \frac{-h_{te}}{h_{le} Y_L}$$

and
$$\beta_v = Z_e Y_L$$

$$\therefore \qquad (1 - \beta_v A_{vo}) = 1 + \frac{h_{te} Z_e Y_L}{h_{le} Y_L}$$

$$= 1 + 100 \times \frac{1 \ 000}{1 \ 500} \simeq 68$$

The input impedance and admittance are therefore given by,

$$Z_{in} = 1\cdot5(1 + 67) \ k\Omega \quad \text{and} \quad Y_{in} = 9\cdot7 \ \mu\text{mho}$$

The current gain, including current splitting for both sets of bias components is given by:

$$A_1 = \underbrace{\frac{667}{667 + 45\cdot5 + 14\cdot7}}_{\text{i/p bias}} \times \underbrace{100}_{h_{te}} \times \underbrace{\frac{9\cdot7}{9\cdot7 + 90 + 212 + 37 + 100}}_{\text{interstage bias}} \times \underbrace{100}_{h_{te}}$$

$$= 0\cdot918 \times 100 \times 2\cdot05 = 188$$

But the required gain of 3 must be given by

$$A_{1f} = \frac{A_1}{1 - \beta A_1}$$

$$\therefore \qquad 3 = \frac{188}{1 + 188\beta}$$

and

$$\beta = \frac{\dfrac{188}{3} - 1}{188} = 0\cdot33$$

But β is given by $Y_2/(Y_1 + Y_2)$ when Y_1 is the 1 kΩ emitter resistor.

$$\therefore \qquad 0\cdot33\,Y_1 = Y_2(1 - 0\cdot33)$$

$$Y_2 = \frac{0\cdot33}{1 - 0\cdot33} \text{ mmho} \backsimeq 0\cdot5 \text{ mmho}$$

This would be provided by a 2 kΩ resistor but the input impedance without feedback must be included in this value. Now

$$Y_{in} = 667 + 45\cdot5 + 147 = 859\cdot5 \,\mu\text{mho}$$

and $\qquad Y_{in} = 1\cdot16 \text{ k}\Omega$

The required feedback resistor is therefore just less than 1 kΩ. In practice a thermistor having a cold resistance of 1 kΩ would be satisfactory.

The overall input impedance will now be given by

$$Z_{inf} = \frac{Z_{in}}{1 - \beta A_1} = \frac{1\cdot16 \text{ k}\Omega}{1 + 188 \times 0\cdot33} \backsimeq 18 \,\Omega$$

This is very much less than the series R for the phase shift network which must therefore be included externally. The output admittance with feedback is given by

$$\frac{Y_0}{1 - \beta A} = \frac{h_{oe}}{1 + 188 \times 0.33} \simeq 1.5 \ \mu\text{mho}$$

satisfying the required condition that $Y_0 \ll Y_L$.

All that remains is to determine the value of C for the network. From equation 8.10,

$$f = \frac{1}{2\pi CR}$$

$$\therefore \qquad C = \frac{10^6}{2\pi \times 5\,000 \times 1\,000} \ \mu\text{F}$$

$$= 0.03 \ \mu\text{F}$$

The complete arrangement is shown diagrammatically in *Figure 8.8.*

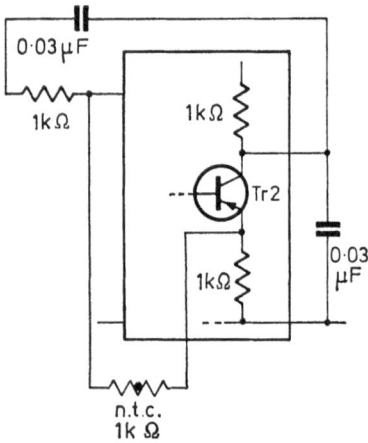

Figure 8.8. Solution for Example 8.1(*b*)

Single Stage RC Oscillators

An alternative RC network may be used, to give 180° phase shift, at a particular frequency. The attenuation is greater than that found for the Wien bridge networks, but a single stage amplifier can be constructed to give sufficient gain for the combined circuit to oscillate.

As with the Wien bridge circuits, such networks may be either voltage, or current, phase shifting. The four basic configurations are shown in *Figure 8.9.*

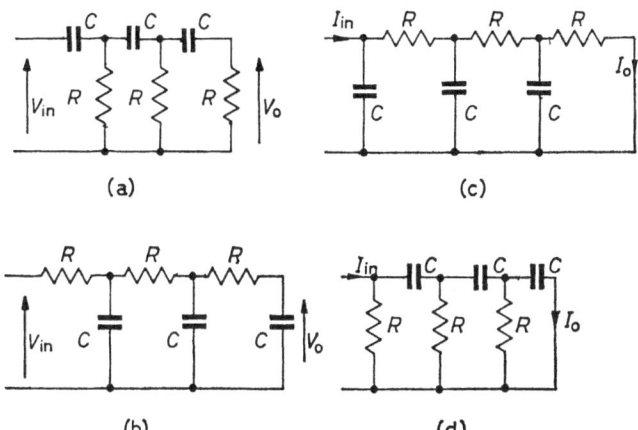

Figure 8.9. Voltage and current phase shifting networks

Figures 8.9a and *b* are both voltage shifting and the frequency for 180° phase shift is given by

$$\frac{1}{2\pi CR\sqrt{6}} \text{ Hz} \quad \text{and} \quad \frac{\sqrt{6}}{2\pi CR} \text{ Hz}$$

respectively for equal capacitors and resistors. For the same conditions the voltage attentuation, β, is 1/29th. With suitable ratios between the C and R values, the attenuation may be as low as 1/8th but the frequency is of course modified. With the circuit in *Figure 8.9a*, the final R will be the amplifier input impedance, but for that in *Figure 8.9b*, the input impedance should be very much greater than X_c at the oscillatory frequency. In each case the network input impedance should not load the amplifier. *Figure 8.9c* and *d* show current phase shifting networks having 180° phase shift at

$$\frac{\sqrt{6}}{2\pi CR} \text{ Hz} \quad \text{and} \quad \frac{1}{2\pi CR\sqrt{6}} \text{ Hz}$$

respectively. The current attenuation will once again be 1/29th for equal capacitors and resistors. With these circuits, the amplifier input impedance should be very much lower than that of the final

269

network component, and the network input impedance should be much less than the amplifier output impedance.

The analysis of the voltage networks is by mesh analysis, while that of the current networks is more conveniently achieved by nodal analysis. Only one example will be considered here, but the other forms may be analysed by similar methods.

Example 8.2. From first principles, determine the frequency of oscillation and minimum value for h_{te} for the circuit shown in *Figure 8.10.* h_{1e} and h_{oe} may be taken as 1 kΩ and 100 μmho respectively and h_{re} can be neglected.

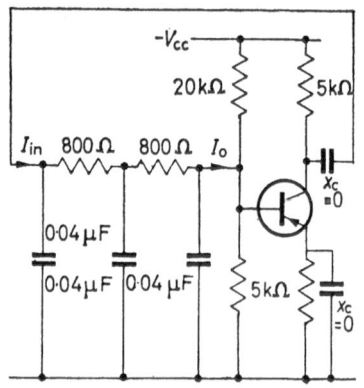

Figure 8.10. Oscillator circuit for Example 8.2

The amplifier input impedance is given by h_{1e} in parallel with the bias components.

$$\therefore \qquad Z_{1n} = \frac{\dfrac{20 \times 5}{25} \times 1}{\dfrac{20 \times 5}{25} + 1} \; k\Omega = 800 \; \Omega$$

Thus the phase shift network has identical components and may be analysed by nodal analysis as follows:

$$\left.\begin{array}{lll}
\text{Node 1:} & I_{1n} = V_1(G + j\omega C) - GV_2 \\
\text{Node 2:} & 0 = -GV_1 + V_2(2G + j\omega C) - GV_3 \\
\text{Node 3:} & 0 = \qquad\qquad -GV_2 + V_3(2G + j\omega C)
\end{array}\right\} \qquad (8.14)$$

270

Now using determinants, equation 8.14 may be solved for V_3 and hence I_0 may be determined.

$$I_0 = V_3 G = G \times \dfrac{\begin{vmatrix} G + j\omega C & -G & I_{1n} \\ -G & 2G + j\omega C & 0 \\ 0 & -G & 0 \end{vmatrix}}{\begin{vmatrix} G + j\omega C & -G & 0 \\ -G & 2G + j\omega C & -G \\ 0 & -G & 2G + j\omega C \end{vmatrix}}$$

$$I_0 = \dfrac{0 + 0 + GI_{1n}(G^2 - 0)}{(G + j\omega C)[(2G + j\omega C)^2 - G^2] + G[-G(2G + j\omega C) - 0] + 0}$$

and

$$\beta = \frac{I_0}{I_{1n}} = \frac{G^3}{(G + j\omega C)(3G^2 + j4\omega CG - \omega^2 C^2) - 2G^3 - j\omega CG^2} \tag{8.15}$$

$$= \frac{G^3}{3G^3 + j4\omega CG^2 - \omega^2 C^2 G + j3\omega CG^2 - 4\omega^2 C^2 G - j\omega^3 C^3 - 2G^3 - j\omega CG^2}$$

$$= \frac{G^3}{G^3 - 5\omega^2 C^2 G + j(6\omega CG^2 - \omega^3 C^3)} \tag{8.16}$$

If β is to be real, the imaginary terms in the denominator must be zero

$$\therefore \qquad\qquad 6\omega CG^2 = \omega^3 C^3$$

and

$$\omega = \frac{G\sqrt{6}}{C} \text{ rad/sec} \tag{8.17}$$

or

$$f = \frac{\sqrt{6}}{2\pi CR} = 12 \cdot 2 \text{ kHz} \tag{8.18}$$

At this frequency,

$$\beta = \frac{G^3}{G^3 - 5\omega^2 C^2 G}$$

Dividing through by G and substituting from equation 8.18

$$\beta = \frac{G^2}{G^2 - 30G^2} = \frac{-1}{29} \tag{8.19}$$

271

To determine the approximate minimum h_{fe}, first note that the reactance of one shunt capacitor is $1/\omega C$,

∴ from equation 8.17

$$X_c = \frac{1}{\dfrac{GC\sqrt{6}}{C}} = \frac{R}{\sqrt{6}} = 327\ \Omega$$

Since the total network impedance must be less than this, and since R_L is parallel with h_{oe} is much greater than this, the current gain may be taken as

$$A_1 = -h_{fe} \times \frac{Y_{in}}{Y_{in} + Y_{bias}}$$

$$= -h_{fe} \times \frac{1\ 000}{1\ 000 + 250}$$

For oscillation A_1 must be -29.

∴ $\qquad$ Minimum $h_{fe} = 29 \times \dfrac{1\ 250}{1\ 000} = 36$

In practice both the frequency and the minimum h_{fe} would be modified by the load and h_{oe} and this could be allowed for in the first term in equation 8.14 by replacing $(G + j\omega C)$ by

$$(G + j\omega C + Y_L + h_{oe})$$

Putting $G' = G + Y_L + h_{oe}$ and re-writing equation 8.15

$$\frac{I_o}{I_{in}} = \frac{G^3}{(G' + j\omega C)(3G^2 + j4\omega CG - \omega^2 C^2) - 2G^3 - j\omega CG^2}$$

$$= \frac{G^3}{\begin{aligned}3G^2G' &+ j4\omega CGG' - \omega^2 C^2 G' + j3\omega CG^2 \\ &- 4\omega^2 C^2 G - j\omega^3 C^3 - 2G^3 - j\omega CG^2\end{aligned}}$$

Equating the imaginary term to zero as before:

$$4\omega CGG' + 2\omega CG^2 = \omega^3 C^3$$

∴ $\qquad$ $4GG^2 + 2G^2 = \omega^2 C^2$

$$\omega = \frac{\sqrt{(4GG' + 2G^2)}}{C}$$

But $G = 1.25$ mmho and $G' = 1.55$ mmho.

$$\therefore \ f = \frac{\sqrt{[(4 \times 1.938 \times 10^{-6}) + (2 \times 1.56 \times 10^{-6})]}}{2\pi \times 4 \times 10^{-8}} \text{ Hz}$$

$= 13.1$ kHz

and

$$\frac{I_0}{I_{\text{in}}} = \frac{G^3}{3G^2G' - \omega^2C^2G' - 4\omega^2C^2G - 2G^3}$$

$$= \frac{G}{3G' - \dfrac{\omega^2C^2G'}{G^2} - \dfrac{4\omega^2C^2}{G} - 2G}$$

$$= \frac{1.25}{4.65 - 10.7 - 34.5 - 2.5} = \frac{1}{-34.5}$$

$$\therefore \qquad \text{Minimum } h_{\text{te}} = 34.5 \times \frac{1\,250}{1\,000} = 43$$

It is interesting to note that although the load admittance and h_{oe} are much less than the network admittance, the combined effect is to change both the frequency and minimum h_{te} by about 10 per cent.

Single stage RC oscillators are not often used in practice for two reasons; first, frequency adjustment requires the simultaneous switching of three components; second, the single stage amplifier is not as suitable for stabilization by means of negative feedback.

LC OSCILLATORS

Wien bridge oscillators are widely used for audio frequency signal generators. The useful range of up to 1 MHz is limited by stray capacitance and amplifier input impedance. Radio frequency oscillators are usually of the LC type. With these, the oscillating frequency is approximately the resonant frequency of the LC circuit involved, and the feedback circuit can be either capacitive or through mutual inductance. The simplest form employs a parallel tuned circuit as the load of a single stage amplifier and inductive coupling between input and output for the feedback network. Consider first the tuned anode oscillator shown in *Figure 8.11*. At the resonant frequency of the anode circuit, the anode load impedance is purely resistive and is given by $(L_1/Cr) \ \Omega$. Thus, if an a.c. signal at this frequency is applied as v_{gk}, the anode voltage $(-\mu Z)/(r_a + Z)$ will be 180° out of phase with this signal. Neglecting the effect of r, the current i_{L}

through L_1 will lag the anode voltage by a further 90°. If two coils have mutual inductance between them, a current i in one induces a voltage of $\pm j\omega M i$ V in the other. The sign here depends only upon the sense of winding of the two coils. If this is such that v_{gk} is given by $-j\omega M i_L$ then the total phase shift is zero. Now if M and the valve gain are sufficiently large, the system will oscillate. This current may be analysed using normal equivalent circuit

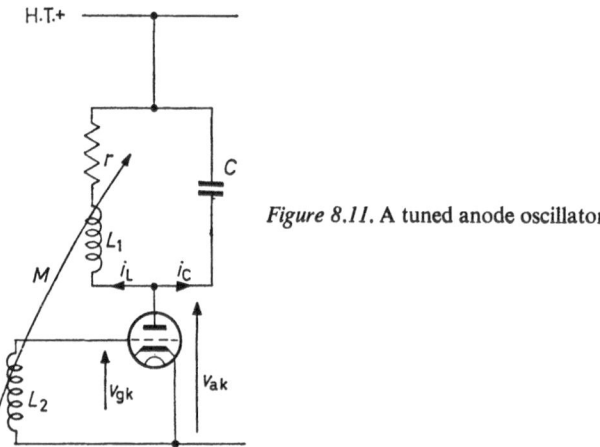

Figure 8.11. A tuned anode oscillator

methods, taking either the constant current or the constant voltage form.

Figure 8.3. A tuned anode oscillator is to be constructed using a triode valve having μ 40 and r_a 20 kΩ. The available coil assembly has two windings of 5 mH and 0·2 mH with a maximum coupling factor K of 0·1. At the required frequency of 159 kHz the Q factor of each coil is 20. Determine from first principles which of the two coils should be used in the anode circuit and the correct value of tuning capacitor C.

This problem is best solved by finding a general solution for the frequency of oscillation and the maintenance condition using symbols. The equivalent circuit is shown in *Figure 8.12*.

As a result of the mutual inductance M, v_{gk} is given by:

$$v_{gk} = \pm j\omega M i$$

nd

$$i = \frac{V_a}{r + j\omega L_a}$$

274

Writing a nodal equation:

$$\pm j\omega M g_m \frac{V_a}{r + j\omega L_a} = \frac{V_a}{r_a} + jV_a\omega C + \frac{V_a}{r + j\omega L_a}$$

One solution of this equation would be given by $V_a = 0$, but if the circuit is oscillating $V_a \neq 0$, therefore V_a may be cancelled.

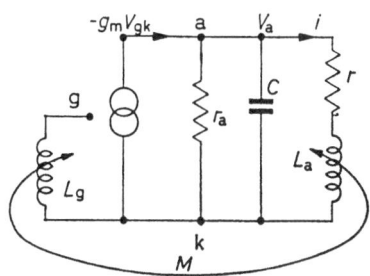

Figure 8.12. The equivalent circuit for a tuned anode oscillator

If both sides of the equation are now multiplied by $r_a(r + j\omega L_a)$ we obtain

$$\pm j\omega M g_m r_a = (r + j\omega L_a)(1 + j\omega C r_a) + r_a$$
$$= r + r_a - \omega^2 L_a C r_a + j\omega L_a + j\omega C r r_a$$

Now, by equating the real and imaginary parts of this equation, the required results can be obtained.

Real parts,

$$0 = r + r_a - \omega^2 L_a C r_a$$

$$\therefore \qquad \omega^2 = \frac{r + r_a}{L_a C r_a} = \frac{1}{L_a C}\left(1 + \frac{r}{r_a}\right) \qquad (8.20)$$

∴ The frequency of oscillation

$$f = \frac{1}{2\pi\sqrt{(L_a C)}}\sqrt{\left(1 + \frac{r}{r_a}\right)} \text{ Hz} \qquad (8.21)$$

Imaginary parts,

$$\pm \omega M g_m r_a = \omega L_a + \omega C r r_a$$

Putting $g_m r_a = \mu$, dividing by ω, and taking the negative M we obtain:

$$M\mu = L_a + C r r_a \qquad (8.22)$$

which is known as the maintenance condition.

275

In the problem M and μ are known, and for each coil C and r can be found from equation 8.20. Thus for each set of values, we can see that if μ is sufficiently large, equation 8.22 can be rearranged to,

$$\mu = \frac{L_a}{M} + \frac{Crr_a}{M} \qquad (8.23)$$

First, for each coil we can find r. Let the 5 mH coil be L_1 and the associated resistance be r_1. Similarly let L_2 and r_2 be the inductance and resistance of the 0·2 mH coil. Now from

$$r = \frac{\omega L}{Q}$$

$$r_1 = \frac{10^6 \times 5 \times 10^{-3}}{20} = 250 \; \Omega$$

and

$$r_2 = \frac{10^6 \times 2 \times 10^{-4}}{20} = 10 \; \Omega$$

Referring to equation 8.21, $(r/r_a) \ll 1$ for both r_1 and r_2.

$\therefore$

$$f \simeq \frac{1}{2\pi\sqrt{(L_aC)}} \; \text{Hz}$$

and rearranging,

$$C = \frac{1}{4\pi^2 f^2 L_a}$$

Substituting values, and putting $4\pi^2 f^2 = 10^{12}$

$$C_1 = \frac{10^{12}}{10^{12} \times 5 \times 10^{-3}} \, \text{pF} = 200 \; \text{pF}$$

and

$$C_2 = \frac{10^{12}}{10^{12} \times 2 \times 10^{-4}} = 5\,000 \; \text{pF}$$

Also

$$M = K\sqrt{(L_1 L_2)}$$

$\therefore$

$$M = 0 \cdot 1\sqrt{(1 \cdot 0)} = 0 \cdot 1 \; \text{mH}$$

If L_1 is used in the anode circuit, from equation 8.23:

$$\mu = \frac{5 \times 10^{-3}}{10^{-4}} + 200 \times \frac{10^{-12} \times 250 \times 2 \times 10^4}{10^{-4}} = 60$$

But the available valve has a μ of only 40, so this arrangement would not oscillate.

276

If L_2 is used in the anode circuit, the same equation yields:

$$\mu = \frac{0 \cdot 2 \times 10^{-3}}{10^{-4}} + \frac{5\,000 \times 10^{-12} \times 10^4 \times 10 \times 2}{10^{-4}} = 12$$

Thus if the smaller coil is used in the anode circuit, the μ of 40 will be more than sufficient to ensure oscillation.

The Tuned Collector Oscillator

The transistor equivalent to the tuned anode oscillator is the tuned collector oscillator. The mechanism of the operation is identical to that of the valve circuit, but the analysis is a little more involved as alternating currents flow in both coils. The circuit and the h parameter equivalent are shown in *Figure 8.13*.

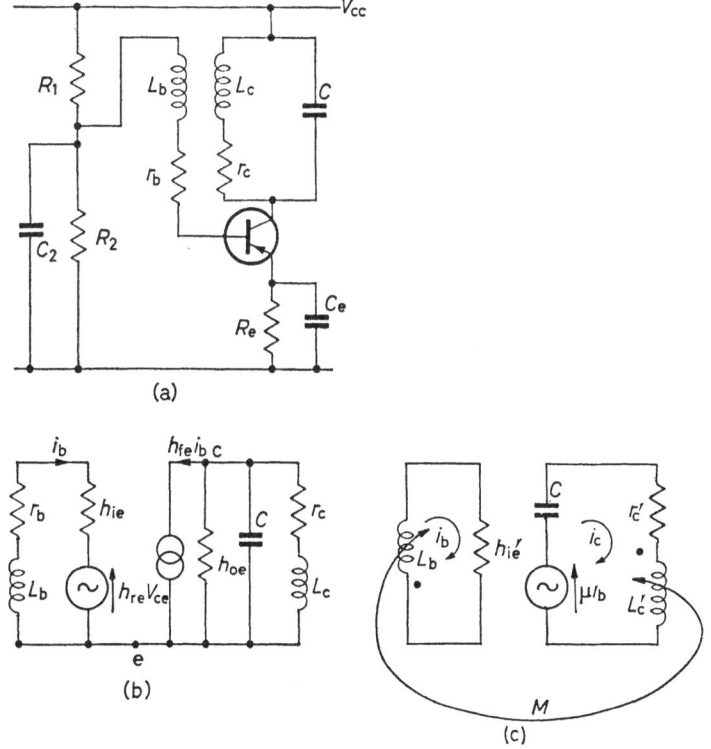

Figure 8.13. A tuned collector oscillator. (*a*) Full circuit.
(*b*) h parameter equivalent circuit. (*c*) Simplified equivalent circuit

277

Figure 8.13a shows the complete circuit; R_1, R_2, R_e and C_e provide the correct d.c. operating point. C_2 has a very low reactance at the oscillatory frequency, effectively earthing one end of coil L_b. The full equivalent circuit is shown in *Figure 8.13b*. The simplified equivalent circuit shown in *Figure 8.13c* is obtained as follows: h_{re} is neglected and r_b is added to h_{ie} to make the new component h_{ie}'. h_{oe} is combined with r_c and L_c modifying them to r_c' and L_c'. This operation will be shown in an example. Thèvenin's theorem is applied to the remainder of the collector circuit resulting in the voltage generator of:

$$\mu_{1b} = -h_{fe}i_b \times \frac{1}{j\omega C}$$

$$\therefore \qquad \mu = \frac{jh_{fe}}{\omega C}$$

The impedance in series with the generator is given by $1/j\omega C$ and is represented by C. The dot notation on the coils L_c' and L_b indicates the sense of the mutually induced e.m.f.: since both currents shown enter the coil at the end indicated, both $j\omega Mi$ terms will be positive.

By mesh analysis:

$$0 = (h_{ie}' + j\omega L_b)i_b + j\omega Mi_c \qquad (8.24)$$

and

$$\frac{jh_{fe}i_b}{\omega C} = i_c\left(r_c' + j\omega L_c' + \frac{1}{j\omega C}\right)i_b + j\omega Mi_b \qquad (8.25)$$

Rearranging equation 8.24

$$i_c = \frac{-(h_{ie}' + j\omega L_b)i_b}{j\omega M} = +j\frac{(h_{ie}' + j\omega L_b)i_b}{\omega M} \qquad (8.26)$$

Substituting for i_c in 8.25 from 8.26,

$$\frac{jh_{fe}i_b}{\omega C} = j\frac{(h_{ie}' + j\omega L_b)i_b}{\omega M}\left(r_c' + j\omega L_c' + \frac{1}{j\omega C}\right) + j\omega Mi_b \quad (8.27)$$

If the circuit is to oscillate $i_b \neq 0$ and can be cancelled. Thus multiplying equation 8.27 by $-j\omega^2 MC$ leads to:

$$\omega Mh_{fe} = (h_{ie}' + j\omega L_b)(\omega Cr_c' + j\omega^2 L_c'C - j) + \omega^3 M^2C \quad (8.28)$$

Without expansion, the real and imaginary components may be selected and equated.

Imaginary terms:

$$0 = \omega^2 L_b C r_c' + \omega^2 L_c' C h_{1e}' - h_{1e}'$$

$$\therefore \qquad \omega^2 = \frac{h_{1e}'}{L_b C r_c' + L_c' C h_{1e}'}$$

Dividing by h_{1e}' and rearranging,

$$\omega^2 = \frac{1}{L_c' C \left(1 + \dfrac{L_b r_c'}{L_c' h_{1e}}\right)} \qquad (8.29)$$

In practice $L_b r_c'/L_c' h_{1e}'$ will usually be much less than one.

$$\therefore \qquad \omega^2 \simeq \frac{1}{L_c' C} \qquad (8.30)$$

and the frequency of oscillation

$$f = \frac{1}{2\pi \sqrt{(L_c' C)}} \text{ Hz} \qquad (8.31)$$

Real terms:

$$\omega M h_{fe} = \omega C h_{1e}' r_c' + \omega^3 M^2 C + \omega L_b - \omega^3 L_c' L_b C$$

Dividing through by ω and substituting for ω^2 from equation 8.30 leads to

$$M h_{fe} = h_{1e}' r_c' C + \frac{M^2 C}{L_c' C} - \frac{L_c' L_b C}{L_c' C} + L_b$$

Therefore the minimum h_{fe} necessary to maintain oscillation is given by

$$h_{fe} = \frac{h_{1e}' r_c' C}{M} + \frac{M}{L_c'} - \frac{L_b}{M} + \frac{L_b}{M}$$

$$= \frac{h_{1e}' r_c' C}{M} + \frac{M}{L_c'} \qquad (8.32)$$

Example 8.4. A tuned collector oscillator employs a collector coil of inductance 1·6 mH and resistance 100 Ω tuned to a nominal frequency of $10^6/2\pi$ Hz by a shunt capacitor. The base coil has inductance 0·1 mH and 10 Ω resistance with coupling factor K of 0·01. If the transistor h_{1e} is 1 000 Ω and the effect of h_{oe} is neglected, determine the actual frequency of oscillation and the minimum value of h_{fe}. Find also how these values are modified if the h_{oe} of 125 μmho, and an external shunt load of 2 kΩ are included.

First the values of C and M are required. Since

$$\omega = 10^6, \ C = \frac{1}{\omega^2 L_c} = \frac{10^{12}}{10^{12} \times 1 \cdot 6 \times 10^{-3}} = 625 \text{ pF}$$

and $\quad M = K\sqrt{(L_b L_c)} = 0\cdot01\sqrt{(1\cdot6 \times 0\cdot1)} \text{ mH} = 0\cdot004 \text{ mH}$

From 8.29

$$\omega^2 = \frac{1}{L_c C\left(1 + \dfrac{0\cdot1 \times 100}{1\cdot6 \times 1\,000}\right)} = \frac{10^{12}}{1 + 6\cdot25 \times 10^{-3}} \simeq 10^{12}$$

$$\therefore \qquad\qquad\qquad f = \frac{10^6}{2\pi} \text{ Hz}$$

Also from 8.32,

$$h_{fe} = \frac{1\,010 \times 100 \times 625 \times 10^{-12}}{4 \times 10^{-6}} + \frac{4 \times 10^{-6}}{1\cdot6 \times 10^{-3}}$$

$$= \frac{1\cdot01 \times 62\cdot5}{4} + 2\cdot5 \times 10^{-3}$$

$$\therefore \qquad\qquad \text{Minimum } h_{fe} = 15\cdot8$$

If h_{oe} and Y_L are included we must find the modified values of L_c' and r_c'. Consider the circuit shown in *Figure 8.14*.

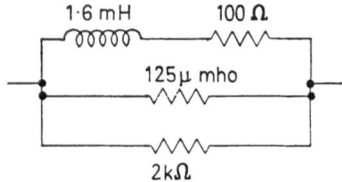

Figure 8.14. Circuit for the solution of Example 8.4

At the nominal frequency of $\omega = 10^6$ rad/sec the circuit will have an impedance which may be represented by r_c' and L_c' in series. First Y_L and h_{oe} may be lumped together:

$$Y_{eff} = (125 + 500) \ \mu\text{mho} = 625 \ \mu\text{mho}$$

$$\therefore \qquad R_{eff} = 1\cdot6 \text{ k}\Omega$$

280

Also at $\omega = 10^6$ rad/sec

$$X_{\mathrm{L}} = 10^6 \times 1 \cdot 6 \times 10^{-3} = 1\,600\,\Omega$$

$$\therefore \text{ Overall } Z = \frac{(100 + \mathrm{j}1\,600)1\,600}{1\,700 + \mathrm{j}1\,600}\,\Omega$$

$$= \frac{1 \cdot 6(100 + \mathrm{j}1\,600)(1 \cdot 7 - \mathrm{j}1 \cdot 6)}{1 \cdot 7^2 + 1 \cdot 6^2}\,\Omega$$

$$= 0 \cdot 294(170 + 2\,560 + \mathrm{j}2\,720 - \mathrm{j}160)\,\Omega$$

$$= 800 + \mathrm{j}753\,\Omega$$

$$\therefore \qquad r_{\mathrm{c}}' = 800\,\Omega$$

$$L_{\mathrm{c}}' = 0 \cdot 753\,\text{mH}$$

Now
$$\omega^2 = \frac{1}{0 \cdot 753 \times 10^{-3} \times 625 \times 10^{-12}\left(1 + \dfrac{0 \cdot 1}{0 \cdot 753} \times \dfrac{800}{1\,000}\right)}$$

$$= \frac{10^{12}}{0 \cdot 47(1 + 0 \cdot 106)}$$

$$\therefore \qquad \omega = \frac{10^6}{\sqrt{5}} \quad \text{and} \quad f = \frac{10^3}{2\pi\sqrt{5}}\,\text{kHz} = 71\,\text{kHz}$$

This is an approximation, since this value of ω should have been used in the calculation of L_{c}' and r_{c}'.

The new value of

$$M = 0 \cdot 01\sqrt{(0 \cdot 753 \times 0 \cdot 1)} = 2 \cdot 74 \times 10^{-6}\,\text{H}$$

$$\therefore \text{ Minimum } h_{\mathrm{fe}} = \frac{1\,010 \times 800 \times 625 \times 10^{-12}}{2 \cdot 74 \times 10^{-1}} + \frac{2 \cdot 74 \times 10^{-6}}{7 \cdot 53 \times 10^{-4}}$$

$$= \frac{808 \times 62 \cdot 5}{2 \cdot 74} + 0 \cdot 36 \times 10^{-2}$$

$$= 184$$

These results show that if the design frequency is to be maintained, the tuning capacitor will have to be increased by a factor of approximately 2. Also the available h_{fe} is unlikely to be as high as 184 suggesting a maximum shunt loading of say 5 kΩ.

Hartley and Colpitts Oscillators

There are many other forms of LC oscillator two of which are shown in *Figure 8.15*.

Analysis in each case may be accomplished using normal equivalent circuit methods. For both circuits the transistor employs the normal bias circuit with the emitter capacitor having negligible reactance at the oscillatory frequency. The feedback capacitor C' will also

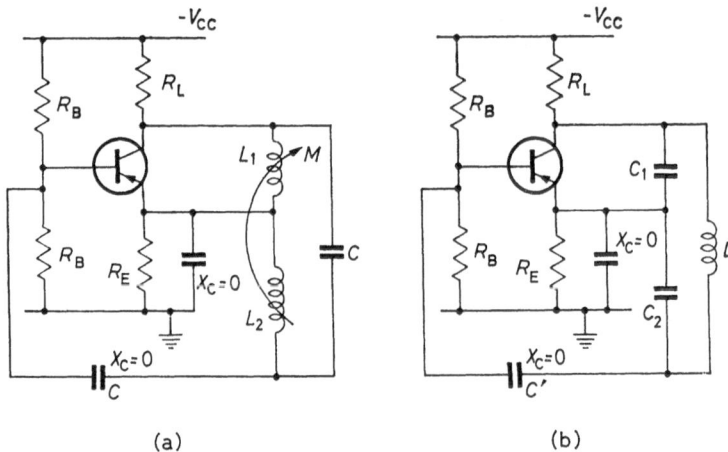

(a)   (b)

Figure 8.15. Hartley and Colpitts oscillators

have negligible reactance, and the bias resistors may be neglected or included in the transistor h_{ie} and h_{fe} parameters. One further example will be considered.

Example 8.5. The Colpitts oscillator shown in *Figure 8.16a* employs a transistor having h_{fe} 40, h_{ie} 1 300 Ω, h_{oe} 125 μmho and negligible h_{re}. Determine the unloaded frequency of oscillation, and the maximum loading for which the frequency will be changed by less than 5 per cent. Find also the maximum loading beyond which oscillation will cease.

Figure 8.16b shows the complete equivalent circuit for the given information. The simplified version in *Figure 8.16c* is obtained by combining h_{oe}, Y_{L1} and Y_{L2}, by expressing the current generator in terms of i_1 instead of i_b, and by combining R_{B1}, and R_{B2}, and h_{ie} together as h_{ie}'.

282

The values of these components are obtained as follows:

$$Y_L' = 125 + 200 + Y_{L2} = (325 + Y_{L2})\, \mu\text{mho} \qquad (8.33)$$

and

$$\frac{1}{h_{1e}'} = \frac{1}{2\,700} + \frac{1}{10\,000} + \frac{1}{1\,300}\, \text{mho} \qquad (8.34)$$

$$\therefore \qquad h_{1e}' = 800\ \Omega$$

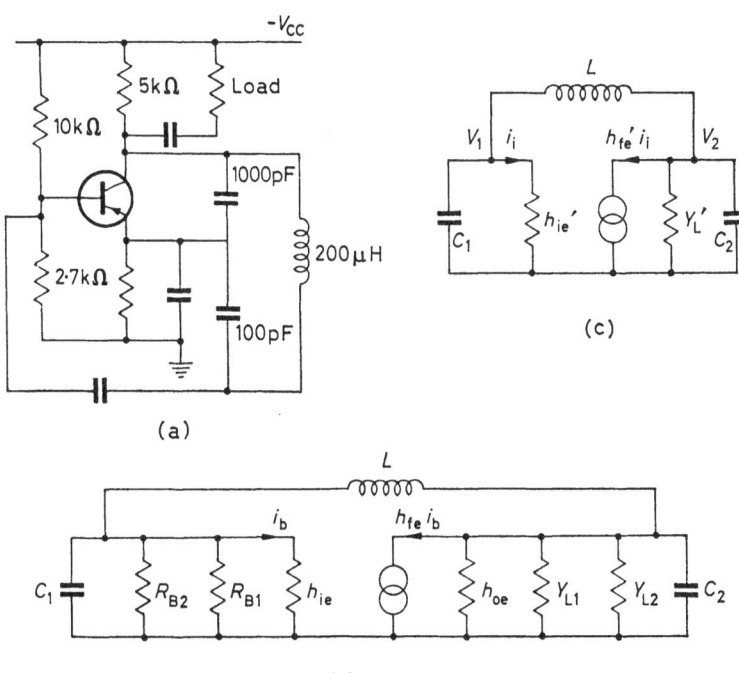

Figure 8.16. (a) Colpitts oscillator for Example 8.5. (b) Equivalent circuit. (c) Simplified equivalent circuit

Also by expressing i_b in terms of i_1,

$$h_{fe}' = 40 \times \frac{\dfrac{1}{1\,300}}{\dfrac{1}{2\,700} + \dfrac{1}{10\,100} + \dfrac{1}{1\,300}} \simeq 25 \qquad (8.34)$$

Our equivalent circuit is now in the correct form for nodal analysis.

Node 1:
$$0 = V_1\left(\frac{1}{h_{1e}'} + j\omega C_1 + \frac{1}{j\omega L}\right) - \frac{V_2}{j\omega L} \tag{8.36}$$

Node 2:
$$-h_{fe}'i_1 = \frac{-V_1}{j\omega L} + V_2\left(Y_L' + j\omega C_2 + \frac{1}{j\omega L}\right) \tag{8.37}$$

But
$$i_1 = \frac{V_1}{h_{1e}'} \tag{8.38}$$

and from equation 8.36

$$V_2 = \left(\frac{j\omega L}{h_{1e}'} - \omega^2 L C_1 + 1\right) V_1 \tag{8.39}$$

Substituting for V_2 from 8.39 and i_1 from 8.38 into 8.37,

$$\frac{-h_{fe}'V_1}{h_{1e}'} = \frac{jV_1}{\omega L} + \left(\frac{j\omega L}{h_{1e}'} - \omega^2 L C_1 + 1\right)\left(Y_L' + j\omega C_2 + \frac{1}{j\omega L}\right) V_1 \tag{8.40}$$

Following the usual technique, V_1 cannot be zero and may therefore be cancelled. Now equating the imaginary components:

$$0 = \frac{1}{\omega L} + \frac{\omega L Y_L'}{h_{1e}'} + (1 - \omega^2 L C_1)\left(\omega C_2 - \frac{1}{\omega L}\right)$$

$$\therefore \quad 0 = \frac{1}{\omega L} + \frac{\omega L Y_L'}{h_{1e}'} + \omega C_2 - \frac{1}{\omega L} - \omega^3 L C_1 C_2 + \omega C_1$$

Rearranging and dividing by ω,

$$\omega^2 L C_1 C_2 = \frac{L Y_L'}{h_{1e}'} + C_1 + C_2$$

$$\therefore \quad \omega^2 = \frac{1}{L\left(\dfrac{C_1 C_2}{C_1 + C_2}\right)} + \frac{Y_L'}{C_1 C_2 h_{1e}'} \tag{8.41}$$

Putting
$$\frac{C_1 C_2}{C_1 + C_2} = C_{eff}$$

$$\omega^2 = \frac{1}{L C_{eff}}\left(1 + \frac{L Y_L'}{(C_1 + C_2)h_{1e}'}\right) \tag{8.42}$$

284

From the information supplied,

$$C_{eff} = 91 \text{ pF}$$

and taking Y_L' when Y_{L2} is zero.

$$\omega^2 = \frac{10^{12}}{200 \times 10^{-6} \times 91} \left(1 + \frac{200 \times 10^{-6} \times 325 \times 10^{-6}}{1\,100 \times 10^{-12} \times 800}\right)$$

$$= \frac{10^{14}}{1 \cdot 82}(1 + 0.074)$$

$$\therefore \qquad \omega = \frac{10^7}{1 \cdot 35} \times 1 \cdot 036 \text{ rad/sec}$$

and the frequency

$$f = \frac{10 \times 1 \cdot 036}{2\pi \times 1 \cdot 35} \text{ MHz} = 1 \cdot 22 \text{ MHz}$$

If this frequency changes by 5 per cent, the term $1 \cdot 036$ must change to $1 \cdot 036 \times 1 \cdot 05 = 1 \cdot 09$, which in equation 8.42 becomes $1 \cdot 09^2 = 1 \cdot 18$.

$$\therefore \qquad \frac{200 \times 10^{-6} Y_L'}{1\,100 \times 10^{-12} \times 800} = 0 \cdot 18$$

and

$$Y_L' = 0 \cdot 18 \times 4 \times 1\,100 \text{ } \mu\text{mho}$$

$$= 793 \text{ } \mu\text{mho}$$

$$\therefore \qquad Y_{L2} = 793 - 325 = 468 \text{ } \mu\text{mho}$$

which represents a shunt load of $2 \cdot 1 \text{ k}\Omega$.

For the second part of the problem we must equate the real parts of equation 8.40.

$$\frac{-h_{te}'}{h_{1e}'} = (1 - \omega^2 L C_1) Y_L' + \frac{j\omega L}{h_{1e}'}\left(j\omega C_2 - \frac{j}{\omega L}\right)$$

Minimum $h_{te}' = h_{1e}' Y_L'(\omega^2 L C_1) + \omega^2 L C_2 - 1$ \qquad (8.43)

But from equation 8.41

$$\omega^2 \simeq \frac{1}{L\left(\dfrac{C_1 C_2}{C_1 + C_2}\right)}$$

$$\therefore \qquad h_{te}' = h_{1e}' Y_L'\left(\frac{C_1 + C_2}{C_2} - 1\right) + \left(\frac{C_1 + C_2}{C_1}\right) - 1$$

$$= h_{1e}' Y_L' \frac{C_1}{C_2} + \frac{C_2}{C_1} \qquad (8.44)$$

Once again inserting values:

$$25 = \frac{800\,Y_{\mathrm{L}}'}{10} + 10$$

$$\therefore \qquad \frac{150}{800} = Y_{\mathrm{L}}' = 187 \text{ mmho}$$

This represents a shunt load of approximately 5 Ω. In practice a load of this magnitude would certainly stop oscillation. The resistance of the coil has been ignored in the analysis since the loading effects of Y_{L}' and h_{1e}' would make this negligible compared with loads greater than 1 kΩ. In the extreme case shown above, this is no longer true, and an accurate analysis should include these components.

The choice of oscillatory circuit for any particular application is beyond the scope of this book, but some of the factors involved are as follows:

Frequency stability with change of load.
Frequency stability with change of d.c. supplies.
Tuning range for available variable capacitor.
Effects of valve or transistor capacitances.

This last effect may be utilized in the design of other forms of oscillator circuits, and an example of this will be discussed in Chapter 9.

EXAMPLES

Example 8.6. A two stage valve amplifier has loads of 20 kΩ and the first stage has an undecoupled cathode resistor of 1 kΩ. The valves have μ 80 and r_a 20 kΩ and the effects of R_G, C_c and C_s may be neglected. A Wien bridge feedback network is connected to convert the circuit into an oscillator. It has a series arm of 5 kΩ and 0·001 μF and the parallel arms are 20 kΩ and 0·02 μF. Determine the minimum gain required from the amplifier and hence design a suitable negative feedback network so that the oscillatory output will be undistorted. Find also the frequency of oscillation.

Ans. 21·25, P.D. network 22·2 kΩ, 1 kΩ, 3·5 kHz.

Example 8.7. A two stage transistor amplifier has a first stage voltage gain of $1 \angle 180°$. The final stage collector load is 500 Ω and the transistors have h_{ie} 1 000 Ω and h_{oe} 100 μmho. If the Wien bridge network shown in *Figure 8.17* is connected between output and input, determine the minimum h_{fe} for the second transistor

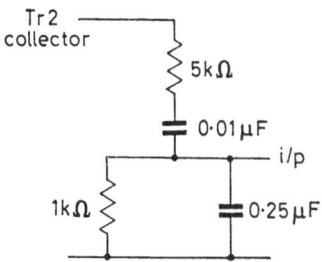

Figure 8.17. Circuit for Example 8.7

and the frequency of oscillation. Assume that the input impedance of the first stage is much greater than 1 kΩ.

Ans. 65, 1·43 kHz.

Example 8.8. The transistor oscillator shown in *Figure 8.18* employs transistors with h_{fe} 90, h_{ie} 1·3 kΩ, h_{oe} 125 μmho and h_{re} 0. If it is required to oscillate at 2 kHz, calculate the values of C_1 and

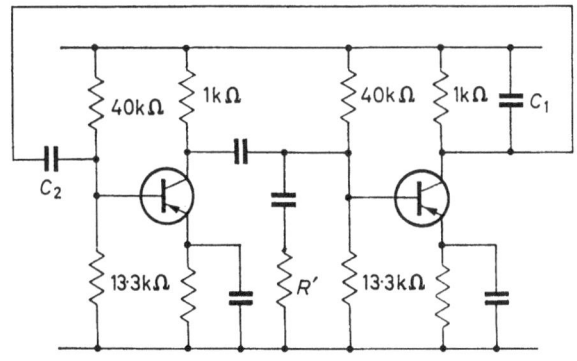

Figure 8.18. Circuit for Example 8.8

C_2. In the absence of a negative feedback network, R' is included to reduce the current gain to the required level. Calculate the value of R'. Assume all other capacitors to have negligible reactance at 2 kHz.

Ans. 0·074 μF, 0·56 Ω.

Example 8.9. Figure 8.19 shows a phase shift oscillator employing a pentode having g_m 9 mA/V and r_a 380 kΩ at the d.c. bias voltage of −4 V. Calculate the minimum value of R_L and the frequency of oscillation. What effect will the capacitive loading have on this

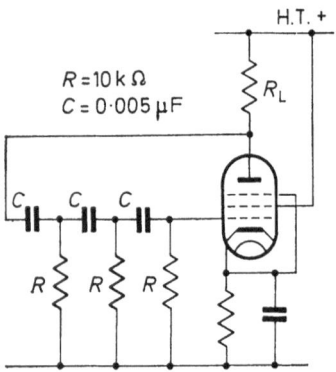

R = 10 k Ω
C = 0·005 μF

H.T. +

Figure 8.19. Circuit for Example 8.9

frequency? Determine the r.m.s. output voltage if R_L is exactly 3·5 kΩ and the pentode mutual characteristic is given by:

I_A (mA)	65	52·3	40	27·5	17·5	9·5	5·0	2·5	1·0	0·3	0
V_{GK} (V)	0	1	2	3	4	5	6	7	8	9	10

(Construct a graph of g_m against peak to peak v_{gk}.)

Ans. 1·3 kHz, 3·23 kΩ, lower it, 61·5 V.

Example 8.10. A single stage common emitter amplifier is to be used as a low frequency oscillator. The feedback network is to be a three stage voltage phase shift network having equal resistors and capacitors. The transistor h parameters are h_{ie} 1·5 kΩ, h_{oe} 150 μmho, h_{fe} 120, h_{re} 0 and the shunt bias resistors are 47 kΩ and 12 kΩ. Determine the value of the capacitors and the minimum value for R_L if the frequency is to be 175 Hz.

Ans. 0·286 μF, 383 Ω.

Example 8.11. The oscillator shown in *Figure 8.20* employs a transistor with h_{ie} 900 Ω and h_{oe} 100 μmho. If the oscillator frequency is to be 3 kHz, calculate the value of C and the minimum value of h_{fe}.

Ans. 0·0066 μF, 35.

288

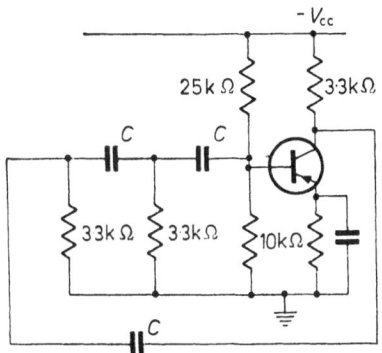

Figure 8.20. Circuit for Example 8.11

Example 8.12. A current amplifier, together with the phase shift network shown in *Figure 8.21*, is to be used as an oscillator. Determine the required current gain and the frequency of oscillation.

Ans. −16, 8·2 kHz.

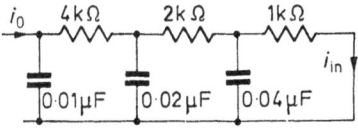

Figure 8.21. Circuit for Example 8.12

Example 8.13. A tuned anode oscillator employs anode and grid coils of Q factor 10 and inductance 0·3 mH and 0·7 mH respectively. If the frequency of oscillation is to be 250 kHz and the valve parameters are μ 60, r_a 5 kΩ. Determine (*a*) the tuning capacitor, (*b*) the minimum coupling between the coils.

Ans. 1 380 pF, 0·02.

Example 8.14. If the valve in Example 8.13 was replaced by a transistor having h_{ie} 500 Ω, find the values of C and h_{fe} for the same frequency of oscillation. Neglect the effect of h_{oe}, h_{re} and the bias components. How would these results be modified if a load of 1 000 Ω was capacitively coupled to the collector and the coupling factor K between the coils was reduced to 0·005?

Ans. 1 160 pF, 0·785, 970 pF, 46.

Example 8.15. Figure 8.22 shows a Colpitts oscillator employing a transistor with h_{1e} 1 500 Ω and h_{oe} 125 μmho. Calculate the value of C_1 which will result in an oscillatory frequency of 600 kHz in the absence of the load. Estimate the minimum value of the load

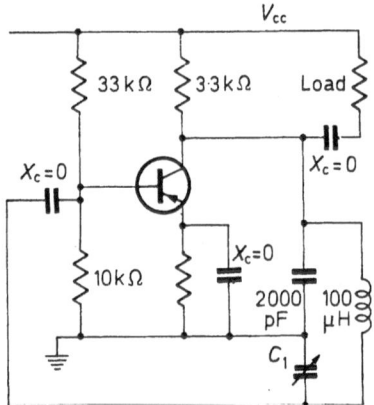

Figure 8.22. Circuit for Example 8.15

if the transistor h_{fe} of 30 is to maintain oscillations. Find also the frequency when this minimum value is connected.

Ans. 1 080 pF, 28 Ω, 840 kHz.

Example 8.16. Show that the approximate frequency and maintenance conditions for a Hartley transistor oscillator are given by:

$$\omega = \frac{1}{\sqrt{[C(L_b + L_c \pm 2M)]}} \text{ rad/sec}$$

$$h_{fe} = \frac{L_b}{L_c} \pm \frac{M}{L_c}$$

Neglect h_{oe}, h_{re}, bias components and coil resistance.

9

MODIFICATIONS TO EQUIVALENT CIRCUITS FOR HIGH FREQUENCY OPERATION

The use of the equivalent circuits discussed in the preceding chapters becomes inaccurate at higher frequencies for a number of reasons. These are the effects of the various reactances associated with the construction of valves and transistors, and the transit time for which the electrons or holes are crossing the active region of the device. The reactances are due to capacitance between the various electrodes and to lead inductance. If the transit time is of the same order as a single period of the signal, μ for a valve and α for a transistor will be considerably reduced. In this chapter, we shall consider how the small signal equivalent circuits must be modified for use at high frequencies and we shall examine the necessary techniques for the solution of various circuit configurations with such modified circuits.

VALVE EQUIVALENT CIRCUIT AT HIGH FREQUENCIES

We shall first consider in detail the effect of the interelectrode capacitance of a triode valve. The electrodes of a valve are conductors separated by an insulating medium. The resulting capacitances are denoted C_{ag}, C_{gk} and C_{ak} and they can be shown diagrammatically as in *Figure 9.1a*.

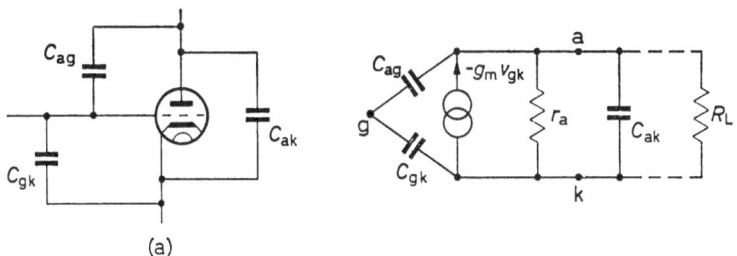

(a)

Figure 9.1. Triode valve and equivalent circuit for high frequencies

291

Figure 9.1b shows the constant current equivalent circuit with the interelectrode capacitances included. Typical values are C_{ag} 1·5 pF, C_{gk} 2·5 pF and C_{ak} 1 pF. C_{ak} forms part of C_s and may be allowed for in the normal calculation of high frequency gain. The reactance of C_{ag} is much greater than R_L at normal operating frequencies, but together with C_{gk} it produces a shunt capacitance across the input. This capacitance provides the principal component of C_s for the previous stage. In addition it may result in a shunt conduc-

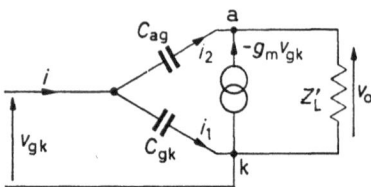

Figure 9.2. Equivalent circuit for the derivation of valve input admittance

tance further modifying the gain of the preceding stages. To analyse this situation, we shall assume initially that C_{ag} provides a negligible load on the output, and with reference to *Figure 9.2*, calculate the input admittance of the amplifier shown.

By inspection,

$$Y_{in} = \frac{i}{v_{gk}} \tag{9.1}$$

and

$$i = i_1 + i_2 \tag{9.2}$$

and

$$i_1 = j\omega C_{gk}v_{gk} \tag{9.3}$$

Applying the sense of i_2 shown,

$$i_2 = j\omega C_{ag}(v_{gk} - v_0) \tag{9.4}$$

But

$$v_0 = A \cdot v_{gk} = -g_m Z_L' v_{gk} \tag{9.5}$$

where Z_L' is the total effective load, including r_a, C_{ak} and any external load present.

From equations 9.2, 9.3, 9.4 and 9.5,

$$i = j\omega C_{gk}v_{gk} + j\omega C_{ag}v_{gk}(1 + g_m Z_L') \tag{9.6}$$

292

Now applying equation 9.1,

$$Y_{\text{in}} = \frac{i}{v_{\text{gk}}} = j\omega C_{\text{gk}} + j\omega C_{\text{ag}}(1 + g_{\text{m}}Z_{\text{L}}') \qquad (9.10)$$

If Z_{L}' is approximately resistive, $g_{\text{m}}Z_{\text{L}}'$ will be a real number and the input circuit appears as the parallel combination of two capacitors, C_{gk} and $C_{\text{ag}}(1 + g_{\text{m}}Z_{\text{L}}')$. Thus since $g_{\text{m}}Z_{\text{L}}'$ may be quite large the input capacitance may be as high as 100 pF. In general $Z_{\text{L}}' = R' + jX'$ where X may be positive or negative.

Equation 9.10 becomes

$$Y_{\text{in}} = j\omega C_{\text{gk}} + j\omega C_{\text{ag}}(1 + g_{\text{m}}R' + jg_{\text{m}}X')$$
$$= j\omega C_{\text{gk}} + j\omega C_{\text{ag}}(1 + g_{\text{m}}R') - \omega C_{\text{ag}}g_{\text{m}}X' \qquad (9.11)$$

If Z_{L}' is capacitive, X' will be negative and the input admittance includes a positive conductance. If Z_{L}' is inductive, X' is positive and the resulting input conductance is negative. The input circuit obtained is shown in *Figure 9.3*.

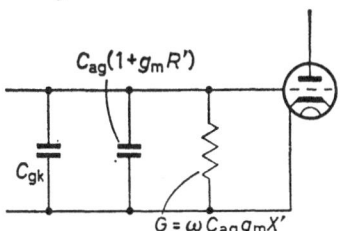

Figure 9.3. Effective input circuit for valve at high frequencies

Example 9.1. A two stage amplifier employs triodes having g_{m} 10 mA/V, r_{a} 10 kΩ, C_{ag} 1·5 pF, C_{ak} 1 pF, C_{gk} 2·5 pF. Each stage has a resistive load of 20 kΩ and a grid resistor of 100 kΩ. Interstage wiring capacitance totals 10 pF. Determine the overall voltage gain and input impedance at an angular frequency ω of 10^8 rad/sec.

Since the gain of a stage is given by $-g_{\text{m}}Z_{\text{L}}'$ we must first find Z_{L}' for the second valve. This will be composed of the valve r_{a} and C_{ak}, and the 20 kΩ load.

$$\therefore \qquad Y_{\text{L}}' = 100 + 50 + j10^8 \times 10^{-12} \times 10^6 \ \mu\text{mho}$$
$$= 150 + j100 = 100\sqrt{(3\cdot25)} \ \angle \ 33°\ 42' \ \mu\text{mho}$$
$$Z_{\text{L}}' = \frac{1}{Y_{\text{L}}'} = 5\cdot55 \ \angle \ -33°\ 42' = 4\cdot6 - j3\cdot08 \ \text{k}\Omega$$
$$\therefore \qquad A_{\text{v2}} = 55\cdot5 \ \angle \ 180 - 33°\ 42'$$

293

Applying equation 9.11,

$$Y_{1n} = [(j10^8 \times 2\cdot5 \times 10^{-12}) + (j10^8 \times 1\cdot5 \times 10^{-12} \times 47)$$
$$+ (10^8 \times 1\cdot5 \times 10^{-12} \times 30\cdot8)] \times 10^6 \,\mu\text{mho}$$
$$= 4\cdot62 + j7\cdot3 \text{ mmho}$$

But this forms a part of Y_{L1}', the remainder being provided by R_L, r_a, R_g, C_s and C_{ak1}:

$$\therefore \ Y_{L1}' = (4\cdot62 + 0\cdot1 + 0\cdot05 + 0\cdot01)$$
$$+ (j10^8 \times 11 \times 10^{-12} \times 10^3) + j7\cdot15 \text{ mmho}$$
$$= 4\cdot78 + j8\cdot25 \text{ mmho}$$

But

$$Z_{L1}' = \frac{1}{Y_{L1}'} = \frac{4\cdot78 - j8\cdot25}{4\cdot78^2 + 8\cdot25^2} \text{ k}\Omega = 0\cdot0525 - j0\cdot091 \text{ k}\Omega$$

$$\therefore \ A_{v1} = -10Z_{L1}' = -(0\cdot525 - j0\cdot91)$$
$$= 1\cdot1 \ \angle \ 120°$$

Thus the overall gain

$$A_{v1} \times A_{v2} = 55\cdot5 \times 1\cdot1 \ \angle \ -93° \ 42'$$

For Y_{1n1} we again refer to equation 9.11:

$$Y_{1n1} = [(j10^8 \times 2\cdot5 \times 10^{-12}) + (j10^8 \times 1\cdot5 \times 10^{-12} \times 0\cdot525)$$
$$+ (10^8 \times 1\cdot5 \times 10^{-12} \times 0\cdot91)] \times 10^3 \text{ mmho}$$
$$= j0\cdot25 + j0\cdot079 + 0\cdot137 \text{ mmho}$$

Including the R_g for the first valve, this becomes,

$$Y_{1n} = 0\cdot147 + j0\cdot33 \text{ mmho}$$
$$= 0\cdot361 \ \angle \ 66° \text{ mmho}$$
$$\therefore \qquad Z_{1n} = 2\cdot77 \ \angle \ -66° \text{ k}\Omega$$

Thus at the upper figure limits of a triode, in the RC coupled common cathode configuration, the addition of extra stages produces little increase in gain and a very low capacitive input impedance. The changes in input admittance discussed above are the result of internal feedback, and this may be utilized to design an oscillator circuit. If the anode load is inductive, the resulting negative input conductance may be used to neutralize the losses in a parallel tuned grid circuit. A lossless tuned circuit can maintain a non-decaying sinusoidal signal at the resonant frequency. A suitable circuit is shown in *Figure 9.4a*. In practice, the parallel tuned circuit is provided by a quartz crystal and the inductive load is a parallel LCR circuit as shown in *Figure 9.4b*. This anode load is tuned to a frequency above that of the oscillator and is thus inductive at the required oscillator frequency. This method is preferable since it eliminates the effect of coil self capacitance.

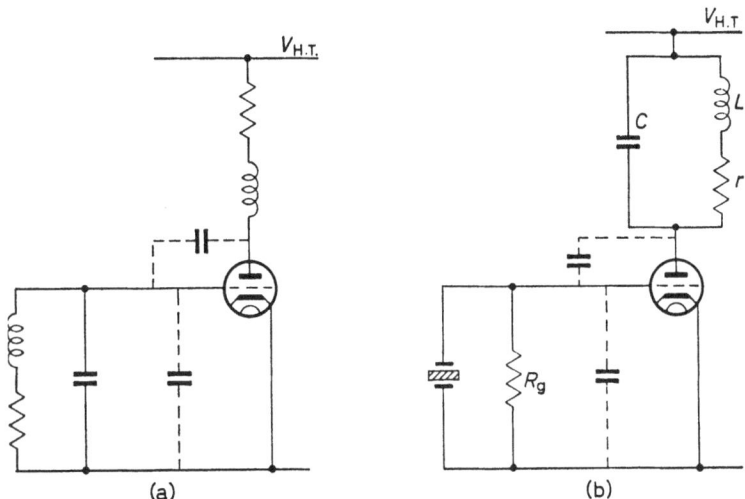

Figure 9.4. Oscillator incorporating Miller feedback

Example 9.2. Taking the simple circuit of *Figure 9.4a* and the valve used in Example 9.1, determine the values of X' and R' for an oscillatory frequency of 10^7 rad/sec. The parallel tuned circuit consists of a coil of inductance 0.1 mH, Q 50 in parallel with a capacitor of 50 pF.

First we must find the value of the total shunt capacitance to give the required frequency.

$$\omega^2 \simeq \frac{1}{LC}$$

$$\therefore \qquad C = \frac{10^{12}}{10^{14} \times 10^{-4}} \, \text{pF} = 100 \, \text{pF}$$

$\therefore$ Valve input capacitance $= 100 - 50 = 50$ pF. But valve input capacitance $= C_{gk} + C_{ag}(1 + g_m R')$.

Inserting values, $50 = 2 + 1 \cdot 5(1 + 10R') \, \text{pF}$

from·which $\qquad R' = 3 \cdot 1 \, \text{k}\Omega$

At resonance, the conductance of a parallel tuned circuit is given by

$$G_0 = \frac{CR}{L}$$

But $\qquad\qquad R = \frac{\omega_0 L}{Q} = \frac{10^7 \times 10^{-4}}{50} = 20 \, \Omega$

$\therefore \qquad\qquad G_0 = \frac{20 \times 10^{-10}}{10^{-4}} \, \text{mho}$

For oscillation, the negative input conductance to the valve, G_1, must equal G_0.

$\therefore \quad \dfrac{20 \times 10^{-10}}{10^{-4}} = \omega C_{ag} \times 1 \cdot 5 \times 10^{-12} \times 10 X' \qquad (X' \text{ in } \text{k}\Omega)$

Inserting values, $\qquad X' = 0 \cdot 133 \, \text{k}\Omega$

from which $\qquad\qquad L' = \dfrac{133}{10^7} = 13 \cdot 3 \, \mu\text{H}$

Pentode Valves

The effect of C_{ag} is considerably reduced by the use of pentode valves where this parameter may be of the order of 0·005pF. Pentodes also have the advantage of very high μ and r_a, values being typically 2 000 and 500 kΩ respectively. At very high frequencies of the order of 50 kHz, two further effects must be considered. Detailed treatment is beyond the scope of this book, but these are the cathode lead inductance (0·005 μH) and the electron transit time. These properties both have the effect of introducing a shunt conductance in the input circuit. The resulting parallel resistance may be of the order of a few 1 000 Ω.

HIGH FREQUENCY TRANSISTOR EQUIVALENT CIRCUITS

When working with transistors at high frequencies, similar modifications must be made to the equivalent circuits. An exact model suitable for all frequencies is exceedingly complex, and its form depends upon the way in which the transistor was manu-factured. In general there will be capacitances associated with the emitter base, and the collector base junctions, and there is the resistance of the base material between the active region and the external connection. In addition the current gain α or α', depending on configuration, is reduced with increase in frequency.

For any particular transistor type, manufacturers quote the upper frequency limit in one of three ways: f_α is the frequency at which α has fallen by 3 db, i.e. to approximately 0·7.

f_1 is the frequency at which α' or h_{fe} has fallen to unity.

f_T is the calculated frequency at which α' will fall to unity if the rate of fall at the upper frequency end of the pass band is maintained at 6 db per octave.

In general either the hybrid π equivalent circuit or a y parameter circuit is used for high frequency work. The hybrid π is accurate

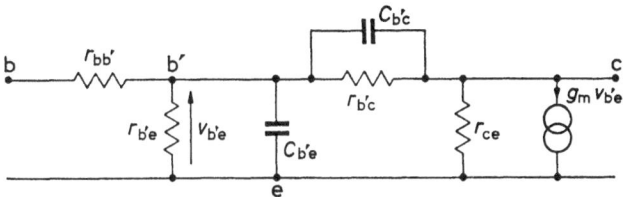

Figure 9.5. Hybrid π equivalent circuit for transistor in the common emitter configuration

from low frequencies upwards, but becomes increasingly less accurate as f_α is approached. The y parameters are frequently quoted in manufacturers' published data, but they are usually quoted for a specified frequency only. In certain cases variation of y parameters with frequency are shown graphically permitting calculation at any desired frequency.

In this section we shall consider the analysis of circuits using both hybrid π and y parameter equivalent circuits. *Figure 9.5* shows the hybrid π equivalent circuit for common emitter connection.

Typical values for the components shown are:

$r_{bb'}\ 50 - 250\ \Omega,\quad r_{b'e}\ 300 - 3\ 000\ \Omega,\quad C_{b'e}\ 250 - 5\ 000\ \mathrm{pF}$

$r_{b'c}\ 2 - 5\ \mathrm{M}\Omega,\quad C_{b'c}\ 5 - 30\ \mathrm{pF},\quad r_{ce}\ 20 - 40\ \mathrm{k}\Omega$

$g_m\ 20 - 40\ \mathrm{mA/V}$

g_m is related to α' by the equation $g_m = \alpha'/r_{b'e}$. The upper frequency to which this circuit is valid would lie in the range 10 kHz to 100 MHz depending on transistor type. We shall first examine a complete analysis at a particular frequency and then see if this suggests any approximation to simplify calculation.

Example 9.3. A common emitter amplifier is supplied from a high impedance source having short circuit current of 1 μA at a frequency of 1 MHz. If the collector load is a parallel tuned circuit of dynamic resistance 50 kΩ at the resonant frequency of 1 MHz determine the output voltage.

The transistor hybrid π parameters are: $r_{bb'}$ 100 Ω, $r_{b'e}$ 790 Ω, $C_{b'e}$ 150 pF, $r_{b'c}$ 2·6 MΩ, $C_{b'c}$ 15 pF, r_{ce} 122 kΩ and g_m 38 mA/V. Repeat with suitable approximations where the load is only 2 kΩ.

For a frequency of 1 MHz the reactances of the various capacitors may be determined.

For $C_{b'e}$, $\quad X_c = \dfrac{10^{12}}{1\,500 \times 2\pi \times 10^6} = 106\ \Omega = X_{be}$

and $\hspace{5cm} B_{be} = 9\cdot4$ mmho

For $C_{b'c}$, $\quad X_c = \dfrac{10^{12}}{15 \times 2\pi \times 10^6} = 10\cdot6\ k\Omega = X_{bc}$

and $\hspace{5cm} B_{bc} = 0\cdot094$ mmho

Since the input impedance is less than

$$r_{bb'} - \frac{jX_{be}r_{b'e}}{r_{b'e} - jX_{be}}$$

the input current may be taken as 1 μA.

We can therefore solve by nodal analysis using nodes $V_{b'e}$ and V_{ce} only.

Working in μA, volts, and μmho:

$$1 = V_{b'e}(1\,265 + 0\cdot39 + j9\,400 + j94) - V_{ce}(0\cdot39 + j94)$$
$$-38\,000V_{b'e} = -V_{b'e}(0\cdot39 + j94) + V_{ce}(8\cdot2 + 20 + 0\cdot39 + j94)$$

where $g_{b'e} = 1\,265\ \mu$mho, $g_{b'c} = 0\cdot39\ \mu$mho, $g_{ce} = 8\cdot2\ \mu$mho, and $g_L = 20\ \mu$mho.

Collecting terms:

$$1 = V_{b'e}(1\,265 + j9\,494) - V_{ce}(0\cdot39 + j94) \hspace{2cm} (9.12)$$
$$0 = V_{b'e}(38\,000 - j94) + V_{ce}(28\cdot6 + j94) \hspace{2cm} (9.13)$$

$\therefore \quad V_{ce} =$

$$\frac{(38\,000 - j94)}{(1\,265 + j9\,494)(28\cdot6 + j94) + (38\,000 - j94)(0\cdot39 + j94)}$$

$$\frac{1}{V_{ce}} = \frac{10^4 \times 9\cdot55\ \angle\ 82°\,24' \times 9\cdot83\ \angle\ 73°\,6'}{3\,800\ \angle\ -9'} + 0\cdot39 + j94$$

$$= 24\cdot8\ \angle\ 155°\,39' + 0\cdot39 + j94$$

$$= -22\cdot6 + j10\cdot2 + 0\cdot39 + j94$$

$$= -22\cdot2 + j104 = 106\ \angle\ 102°$$

∴ The a.c. output voltage

$$V_{ce} = \frac{1}{106} \angle -102° = 9.4 \angle -102° \text{ mV}$$

This calculation could have been simplified if $r_{b'c}$ had been neglected and with it the $-j94$ in the first term of the right hand side of equation 9.13.

If the collector load is sufficiently small, the additional loading due to $C_{b'c}$ may be neglected and the circuit may be treated in a similar manner to that for the triode input admittance calculation. This approach will be used for the second part of the question.

Neglecting $r_{b'c}$, the input admittance at $V_{b'e}$ is given by

$$\frac{i}{V_{b'e}} = g_{b'e} + j\omega C_{b'e} + j\omega C_{b'c}(1 + g_m Z_{L'} g_m Z_{L'})$$

where $Z_{L'}$ is the effective collector load.

In this case $Z_{L'}$ is given by r_{ce} in parallel with the R_L of 2 kΩ.

$$\therefore \qquad g_m Z_{L'} = 38 \times \frac{2 \times 122}{2 + 122} \simeq 76$$

$$\therefore \qquad Y_{inb'} = 1.265 + j9.4 + j0.094(1 + 76) \text{ mmho}$$
$$= 1.265 + j16.6 \simeq 16.6 \angle 85° 42'$$

Now $\qquad V_{b'e} = \dfrac{i}{Y_{in}} = \dfrac{10^{-6} \times 10^3}{16.6 \times 10^{-3} \angle 85° 42'} \text{ mV}$

$$= 0.06 \angle -85° 42' \text{ mV}$$

The output voltage is given by

$$V_{ce} = -g_m Z_{L'} V_{b'e} \qquad (9.14)$$
$$\therefore \qquad V_{ce} = -38 \times 2 \times 0.06 \angle -85° 42' \text{ mV}$$
$$= 4.5 \angle +94° \text{ mV}$$

If the overall voltage gain is required, the effect of $r_{bb'}$ must be included.

$$V_{b'e} = \frac{V_{in} Z_{inb'}}{r_{bb'} + Z_{inb'}}$$

where $\qquad Z_{inb'} = \dfrac{1}{Y_{inb'}} = 60 \angle -85° 42' \ \Omega$

$$\therefore \quad \frac{V_{\text{in}}}{V_{\text{b'e}}} = 1 + \frac{r_{\text{bb'}}}{Z_{\text{inb'}}}$$

$$= 1 + \frac{100}{60 \angle -85° 42'} = 1 + 1\cdot66 \angle +85° 42'$$

$$= 1\cdot25 + j1\cdot66$$

$$= 2\cdot075 \angle +48° 24' \tag{9.15}$$

$\therefore$ Overall voltage gain,

$$\frac{V_{\text{ce}}}{V_{\text{in}}} = \frac{V_{\text{ce}}}{V_{\text{b'e}}} \times \frac{V_{\text{b'e}}}{V_{\text{in}}}$$

applying equations 9.14 and 9.15

$$A_{\text{v}} = \frac{76 \angle 180°}{2\cdot075 \angle +48° 24} = 36\cdot7 \angle +131° 36'$$

The input impedance may be found from

$$Z_{\text{in}} = \frac{V_{\text{in}}}{i} = \frac{10^{-3} \times 4\cdot5 \angle +94°}{10^{-6} \times 36\cdot7 \angle +131° 36'} \, \Omega$$

$$= 123 \angle -37° 36'$$

This example shows how the hybrid π equivalent circuit may be used for a high frequency calculation. At low and medium frequencies the effects of the capacitance become negligible, and if $r_{\text{b'c}}$ is neglected calculation is extremely simple. Neglecting $r_{\text{b'c}}$ is making a similar approximation to that made by neglecting h_{re} when using the h parameter equivalent circuit.

Use of y Parameter Equivalent Circuit

Although the hybrid π circuit discussed above may be used for calculation up to the high frequency limit of any particular transistor, most manufacturers do not quote the hybrid π parameters. In the published data, for transistors intended for high frequency applications, the y parameters are usually quoted for a specific frequency and d.c. conditions. In some cases, graphs showing how the y parameters vary with frequency and operating point, are also supplied. At these high frequencies, the four parameters are all complex, and the resulting common emitter equivalent circuit is given in *Figure 9.6.*

Amplifier calculations in terms of known y parameters may be made by use of the general solution obtained in Chapter 2. If feedback is included in the circuit, either feedback theory, or interconnected four-terminal network theory, will provide the correct solution. Two examples will now be considered; first a single stage amplifier

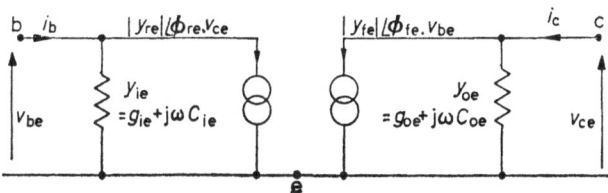

Figure 9.6. y parameter equivalent circuit for transistor in the common emitter configuration

with a resistive load, and secondly a multi-stage amplifier with feedback.

Example 9.4. An r.f. transistor is to be used as a single stage amplifier with a resistive load of 2 kΩ. At the signal frequency of 200 MHz and at the d.c. conditions applicable to the circuit, the y parameters are specified as:

$$y_{ie}(0.028 + j13\omega \times 10^{-12}) \text{ mho}$$
$$y_{re}\ 500 \times 10^{-6} \angle 250° \text{ mho}$$
$$y_{fe}\ 0.034 \angle 292° \text{ mho}$$
$$y_{oe}(220 \times 10^{-6} + j2\omega \times 10^{-12}) \text{ mho}$$

If the amplifier is driven by a source of short circuit current 10 μA and internal admittance $(0.01 + j0.005)$ mho, determine the output voltage.

Rewriting the general solutions obtained in Chapter 2,

$$y_{in} = y_{11} - \frac{y_{12}y_{21}}{y_{22} + Y_L} \tag{9.16}$$

and
$$\frac{V_2}{I_1} = \frac{-y_{21}}{y_{11}(y_{22} + Y_L) - y_{21}y_{12}} \tag{9.17}$$

We can first find the input admittance, and hence the input current. Then, using 9.17 we can find the output voltage directly.

At 200 MHz,

$$y_{1e} = 0{\cdot}028 + j200 \times 10^6 \times 2\pi \times 13 \times 10^{-12} \text{ mho}$$
$$= 0{\cdot}028 + j0{\cdot}016 \text{ mho}$$

and

$$y_{oe} = 220 \times 10^{-6} + j200 \times 10^6 \times 2\pi \times 2 \times 10^{-12} \text{ mho}$$
$$= (220 + j2\,500) \times 10^{-6} \text{ mho}$$

Now working in millimhos:

$$Y_{1n} = 28 + j16 - \frac{0{\cdot}5 \angle 250° \times 34 \angle 292°}{0{\cdot}22 + j2{\cdot}5 + 0{\cdot}5}$$

$$= 28 + j16 - \frac{17 \angle 182°}{2{\cdot}6 \angle 74°}$$

$$= 28 + j16 - 6{\cdot}5 \cos 108° - j6{\cdot}5 \sin 108°$$

$$= 28 + j16 + 2{\cdot}02 - j6{\cdot}2$$

$$= 30 + j10 \text{ mmho}$$

Note that the term due to y_{re} is by no means negligible.

The short circuit source current of 10 μA must divide between the source admittance and Y_{1n}.

$$\therefore \qquad I_{1n} = \frac{10(30 + j10)}{(30 + j10) + (10 + j5)} \mu\text{A}$$

$$= \frac{10(3 + j)(4 - j1{\cdot}5)}{16 + 2{\cdot}25} \mu\text{A}$$

$$= 0{\cdot}548(12 + 1{\cdot}5 + j4 - j4{\cdot}5) \mu\text{A}$$

$$= 7{\cdot}4 - j0{\cdot}274 \mu\text{A}$$

But for the output voltage calculation, only the magnitude of I_{1n} is required.

$$|I_{1n}| = \sqrt{(7{\cdot}4^2 + 0{\cdot}274^2)} \simeq 7{\cdot}4 \mu\text{A}$$

Applying equation 9.17 and working in mA, mmho and volts.

$$V_2 = \frac{-7{\cdot}4 \times 10^{-3} \times 34 \angle 292°}{(28 + j16)(0{\cdot}72 + j2{\cdot}5) - 34 \angle 292° \times 0{\cdot}5 \angle 250°} \text{ V}$$

$$= \frac{-0{\cdot}252 \angle 292°}{20{\cdot}2 - 40 + j11{\cdot}5 + j70 - 17(\cos 182° + j \sin 182°)} \text{ V}$$

$$= \frac{-0{\cdot}252 \angle 292°}{-19{\cdot}8 + j81{\cdot}5 + 17 + j0{\cdot}6} \text{ V}$$

$$|V_2| = \frac{252}{\sqrt{(36{\cdot}8^2 + 81{\cdot}5^2)}} \text{ mV}$$

$$= \underline{2{\cdot}8 \text{ mV}}$$

If required, the voltage gain may be obtained from

$$|V_{\text{in}}| = \frac{|I_{\text{in}}|}{|Y_{\text{in}}|} = \frac{7 \cdot 4 \times 10^{-6}}{\sqrt{(3^2 + 1^2)} \times 10^{-2}} \text{ V}$$
$$= 0 \cdot 232 \text{ mV}$$

$$\therefore \qquad |A_v| = \frac{2 \cdot 8}{0 \cdot 232} = 12$$

For our final example we shall consider a two stage amplifier with overall current feedback.

Example 9.5. An amplifier is required for signals at 450 kHz. The available transistor has a suitable d.c. operating point of $V_{\text{CE}} - 6$ V, I_c 1 mA. The average h_{fe} is quoted as 150, but to allow for spread and temperature variations a d.c. stability factor of $K = 0 \cdot 05$ is to be used. The required current gain of 100, into a resistive load of 1 kΩ is to be obtained using two stages with overall feedback to improve stability. Using the published data quoted below, design a suitable amplifier using a 10 V supply.

$$I_{\text{CO}} \; 2 \; \mu\text{A}, \quad V_{\text{BE}} - 300 \text{ mV}, \quad h_{\text{fe}} \; 150$$
$$g_{\text{ie}} \; 0 \cdot 25 \text{ mmho}, \quad C_{\text{ie}} \; 70 \text{ pF}, \quad |y_{\text{fe}}| \; 37 \text{ mA/V}, \quad \phi_{\text{fe}} \; 0°$$
$$g_{\text{oe}} \; 1 \; \mu\text{mho}, \quad C_{\text{oe}} \; 4 \text{ pF}, \quad |y_{\text{re}}| \; 4 \; \mu\text{mho}, \quad \phi_{\text{re}} \; 270°$$

The proposed circuit is shown in *Figure 9.7.*

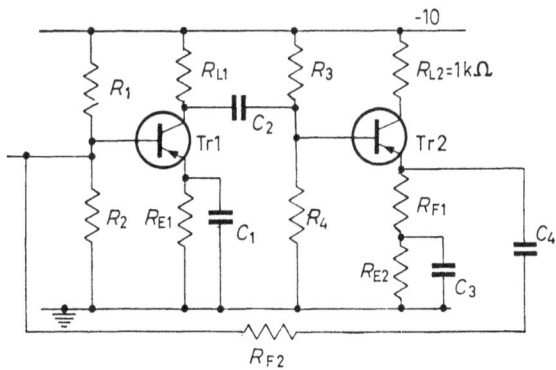

Figure 9.7. Circuit for Example 9.5

The first consideration must be to obtain the correct d.c. operating conditions. This was discussed in Chapter 1, and the results obtained will be used here.

In each case the required operating point is $V_{CE} - 6$ V and I_C 1 mA. Consider first Tr2:

$$V_{CE} = -6 = -10 + I_C R_{L2} + I_E(R_{E2} + R_{F1})$$

Taking $I_E \simeq I_C$

$$R_{E2} + R_{F1} + R_{L2} = \tfrac{4}{1} \text{ k}\Omega$$

$$\therefore \qquad\qquad R_{E2} + R_{F1} = 3 \text{ k}\Omega \qquad (9.18)$$

The stability factor

$$K = \frac{R_E + R_B}{R_E(1 + \alpha') + R_B} = 0.05$$

where

$$R_B = \frac{R_3 R_4}{R_3 + R_4} \quad \text{and} \quad R_E = (R_{E2} + R_{F1}) = 3 \text{ k}\Omega \quad (9.19)$$

Inserting values $0.05 = \dfrac{3 + R_B}{3(1 + 150) + R_B}$

$$\therefore \qquad 453 + R_B = 60 + 20R_B$$

and

$$R_B = \frac{397}{19} = 20.9 \text{ k}\Omega \qquad (9.20)$$

Also $V' = -I_B R_B - I_E R_E + V_{BE}$ (using conventional polarities)

where

$$V' = \frac{V_{CC} R_4}{R_3 + R_4} \qquad (9.21)$$

But

$$I_B = \frac{I_c}{\alpha'} - \frac{I_{co}'}{\alpha'}$$

and

$$\frac{I_{co}'}{\alpha'} \simeq I_{co}$$

$$\therefore \qquad\qquad I_B = \frac{1\,000}{150} - 2 \ \mu\text{A}$$

$$= 4.7 \ \mu\text{A} \qquad (9.22)$$

$$\therefore \qquad V' = -20.9 \times 0.0047 - 3 - 0.3$$

$$= -3.4 \text{ V}$$

Applying equations 9.19 and 9.21

$$-3.4 = \frac{-10R_4}{R_3 + R_4} \qquad (9.23)$$

and

$$20.9 = \frac{R_3 R_4}{R_3 + R_4} \qquad (9.24)$$

From equation 9.23

$$R_3 + R_4 = \frac{10}{3\cdot4}\,R_4 = 2\cdot9R_4 \qquad (9.25)$$

Substituting in equation 9.24,

$$20\cdot9 = \frac{R_3 R_4}{2\cdot94 R_4}$$

$$R_3 = 61\cdot5 \text{ k}\Omega$$

Substituting in equation 9.25

$$R_4 = \frac{R_3}{1\cdot94} = 31\cdot7 \text{ k}\Omega$$

In practice the nearest preferred values would be used, i.e.

$$R_3 = 68 \text{ k}\Omega \quad \text{and} \quad R_4 = 33 \text{ k}\Omega$$

C_3 must have a reactance that is negligible compared with R_{E2} at the signal frequency.

Let $$X_c = 30 \ \Omega = \frac{1}{2\pi 450 \times 10^3 C}$$

From which a 0·01 μF capacitor will be found to be satisfactory.

A similar procedure may now be conducted for Tr1. In this case R_L is not given, but since a high current gain is required R_{L1} should be large.

By comparison with equation 9.18

$$R_{L1} + R_{E1} = 4 \text{ k}\Omega$$

Let $$R_{L1} = 3 \text{ k}\Omega \quad \text{and} \quad R_{E1} = 1 \text{ k}\Omega$$

From stability considerations:

$$0\cdot05 = \frac{1 + R_B}{151 + R_B}$$

$$\therefore \qquad 151 + R_B = 20 + 20R_B$$

and $$R_B = 6\cdot9 \text{ k}\Omega$$

where $$R_B = \frac{R_1 R_2}{R_1 + R_2} \qquad (9.26)$$

Now applying equations 9.21 and 9.22

$$V' = -0\cdot0047 \times 6\cdot9 - 1 - 0\cdot3 = -1\cdot33 \text{ V}$$

where $$V' = \frac{-10R_2}{R_1 + R_2} \qquad (9.27)$$

From 9.27
$$R_1 + R_2 = R_2 \times \frac{10}{1 \cdot 33} = 7 \cdot 5 R_2 \qquad (9.28)$$

Substituting in 9.26,
$$6 \cdot 9 = \frac{R_1 R_2}{7 \cdot 5 R_2}$$

and
$$R_1 = 51 \cdot 7 \text{ k}\Omega$$

So, from 9.28,
$$R_2 = \frac{51 \cdot 7}{6 \cdot 5} = 8 \text{ k}\Omega$$

In this case the preferred values would be R_1 47 kΩ and R_2 8·2 kΩ. By comparison with C_3, C_1 should be 0·025 μF.

The first step in determining the current gain is to decide upon a suitable value for R_{F1} so that the input impedance to Tr2 with local voltage feedback may be found. Since $R_{E2} + R_{F1}$ must be 3 kΩ. suitable preferred values would be R_{E2} 2·7 kΩ and R_{F1} 330 Ω. This will give a β_v of 330/1 000 $\simeq \frac{1}{3}$. Next the applicable y parameters must be found.

$$y_{ie} = 0 \cdot 25 + j 2\pi 450 \times 10^3 \times 70 \times 10^{-12} \times 10^3 \text{ mmho}$$
$$= 0 \cdot 25 + j 0 \cdot 05 \text{ mmho}$$
$$y_{te} = 37 \angle 0 \text{ mmho}$$
$$y_{re} = 0 \cdot 004 \angle 270° \text{ mmho}$$
$$y_{oe} = 0 \cdot 001 + j 2\pi 450 \times 10^3 \times 4 \times 10^{-12} \times 10^3 \text{ mmho}$$
$$\simeq 0 \cdot 001 + j 0 \cdot 001 \text{ mmho}$$

From the general solutions in terms of the y parameters,

$$Y_{in2} = 0 \cdot 25 + j 0 \cdot 05 - \frac{37 \times 0 \cdot 004 \angle 270°}{0 \cdot 001 + j 0 \cdot 001 + 1} \text{ mmho}$$

Neglecting the y_{oe} term,

$$Y_{in} = 0 \cdot 25 + j 0 \cdot 05 - 0 \cdot 15 \angle 270° \text{ mmho}$$
$$= 0 \cdot 25 + j 0 \cdot 05 - 0 + j 0 \cdot 15 \text{ mmho}$$
$$= 0 \cdot 25 + j 0 \cdot 2 \text{ mmho} \qquad (9.29)$$

To find A_v, consider the original solutions, in terms of the y parameters.

$$V_1 = \frac{I_1(y_{oe} + Y_L) - 0}{\Delta} \qquad (9.30)$$

and
$$V_2 = \frac{0 - I_1 y_{te}}{\Delta} \qquad (9.31)$$

dividing 9.31 by 9.30

$$A_v = \frac{V_2}{V_1} = \frac{-y_{fe}}{y_{oe} + Y_L} \tag{9.32}$$

Substituting values

$$A_v = \frac{-37 \angle 0°}{1 + 0.001 + j0.001} \simeq -37$$

Now allowing for the effect of local feedback with a β of $\frac{1}{3}$

$$Y_{in2} = Y_{inf} = \frac{Y_{in}}{1 - \beta A} = \frac{0.25 + j0.2}{1 + 12.3} \text{ mmho}$$

$$= 0.019 + j0.015 \text{ mmho} \tag{9.33}$$

The current gain, which is unaffected by the local feedback, may be obtained from equation 9.31.

Multiplying by Y_L and dividing by I:

$$A_1 = \frac{I_2}{I_1} = \frac{V_2 Y_L}{I_1} = \frac{-y_{fe} Y_L}{y_{ie}(y_{oe} + Y_L) - y_{re}y_{fe}} \tag{9.32}$$

Substituting values and neglecting y_{oe}

$$A_{12} = \frac{-37 \angle 0°}{(0.25 + j0.05)1 - 37 \angle 0° \times 0.004 \angle 270°}$$

Comparison with the input admittance calculation shows that this denominator has the same value as result 9.29,

$$\therefore \qquad A_{12} = \frac{-37}{0.25 + j0.2} = \frac{37 \angle 180°}{0.32 \angle 38° 42'}$$

and $\qquad A_{12} = 115 \angle 141° 18' \tag{9.33}$

To calculate A_{11} we can again use result 9.32, but this will only give the current gain in terms of Tr1 collector and base currents. To find the overall current gain, current splitting factors due to bias components, Y_{L1}, Y_{in1} and Y_{in2} must be included.

First we must find the effective load on Tr1, $Y_{L1 \text{ eff}}$.

$$Y_{L1 \text{ eff}} = Y_{in2} + Y_{L1} + Y_{B3} + Y_{B4}$$

$$= 0.019 + j0.015 + 0.333 + 0.015 + 0.03 \text{ mmho}$$

$$\simeq 0.5 + j0.015 \text{ mmho}$$

From equation 9.32 neglecting y_{oe},

$$A_{11} = \frac{-37(0\cdot5 + j0\cdot015)}{(0\cdot25 + j0\cdot05)(0\cdot5 + j0\cdot15) + j0\cdot15}$$

$$A_{11} = \frac{-37(0\cdot5 + j0\cdot015)}{0\cdot125 + j0\cdot18} = \frac{37 \angle 180° \times 0\cdot5 \angle 1° 42'}{0\cdot218 \angle 55° 18'}$$

$$\therefore \quad A_{11} = 85 \angle 126° 26' \tag{9.34}$$

The interstage current splitting factor

$$= \frac{Y_{1n2}}{Y_{L1} + Y_{B3} + Y_{B4} + Y_{1n2}} = A_1''$$

$$= \frac{0\cdot019 + j0\cdot015}{0\cdot5 + j0\cdot015}$$

$$\therefore \quad A_1'' = \frac{0\cdot0232 \angle 78° 18'}{0\cdot5 \angle 1° 42'} = 0\cdot0464 \angle 36° 36' \tag{9.35}$$

To determine the input current splitting factor we require Y_{1n1} which may be obtained from the general solutions.

$$Y_{1n1} = 0\cdot25 + j0\cdot05 - \frac{37 \times 0\cdot004 \angle 270°}{0\cdot001 + j0\cdot001 + 0\cdot5 + j0\cdot015}$$

$$= 0\cdot25 + j0\cdot05 - \frac{0\cdot15 \angle 270°}{0\cdot5 \angle 1° 42'}$$

$$= 0\cdot25 + j0\cdot05 + 0\cdot009 + j0\cdot3$$

$$\therefore \quad Y_{1n1} = 0\cdot26 + j0\cdot35 = 0\cdot435 \angle 53° 30'$$

$$\therefore \quad A_1' = \frac{Y_{1n1}}{Y_{B1} + Y_{B2} + Y_{1n1}} = \frac{0\cdot435 \angle 53° 30}{0\cdot021 + 0\cdot122 + 0\cdot26 + j0\cdot35}$$

$$= \frac{0\cdot435 \angle 53° 30'}{0\cdot4 + j0\cdot35} = \frac{0\cdot435 \angle 53° 30'}{0\cdot53 \angle 41° 12'} = 0\cdot82 \angle 12° 18' \tag{9.36}$$

The overall current gain may now be obtained from equations 9.33, 9.34, 9.35 and 9.36.

$$A_1 = A_1' \times A_{11} \times A_1'' \times A_{12}$$

$$= 0\cdot82 \angle 12° 18' \times 85 \angle 126° 26'$$
$$\times 0\cdot0464 \angle 36° 36' \times 115 \angle 141° 18'$$

$$= 374 \angle 316° 18' \tag{9.37}$$

Finally we come to the design of the overall feedback circuit: the connection shown in *Figure 9.7* provides simple negative feedback if βA has zero phase angle.

$$\therefore \qquad A_{1f} = \frac{A_1}{1 + \beta A_1}$$

In this case the required gain is 100, and β will be real.

$$\therefore \qquad 100 = \left| \frac{374 \angle 316° \ 18'}{1 + 374\beta \angle 316° \ 18'} \right|$$

$$\therefore \qquad |1 + 374\beta \angle 316° \ 18'| = 3.74$$

$$|1 + 270\beta - j258\beta| = 3.74$$

$$(1 + 270\beta)^2 + (258\beta)^2 = 3.74^2$$

$$1 + 540\beta + 7.3 \times 10^4 \beta^2 + 6.7 \times 10^4 \beta^2 = 14$$

Re-arranging

$$14 \times 10^4 \beta^2 + 540\beta - 13 = 0$$

and

$$\beta = \frac{-540 \pm \sqrt{(540^2 + 52 \times 14 \times 10^4)}}{28 \times 10^4}$$

$$= \frac{-540 \pm \sqrt{[(29 + 728) \times 10^4]}}{28 \times 10^4}$$

$$= \frac{-540 \pm 27.5 \times 10^2}{28 \times 10^4}$$

Taking the positive sign,

$$\beta = \frac{2.21}{2.8} \times 10^{-2} = 0.79 \times 10^{-2} \qquad (9.38)$$

So remembering that $A_f \simeq 1/\beta$ result 9.38 is obviously correct. But

$$\beta = \frac{G_{F2}}{G_{F1} + G_{F2}} \quad \text{and} \quad G_{F1} = 3 \text{ mmho}$$

$$\therefore \qquad 0.79 \times 10^{-2}(G_{F2} + 3) = G_{F2}$$

$$G_{F2} = \frac{0.79 \times 10^{-2} \times 3}{1 - 0.79 \times 10^{-2}} \text{ mmho}$$

$$= 2.37 \times 10^{-2} \text{ mmho}$$

$$\therefore \qquad R_{F2} = 43 \text{ k}\Omega$$

Since this is a preferred value no modification to this result is necessary.

The remaining components to be selected are C_2 and C_4 which must have negligible reactances compared with the terminal input impedance to Tr2 and to R_{F2} respectively.

In each case a 0·01 μF capacitor would be satisfactory.

The completed circuit design is shown in *Figure 9.8*. The reader should appreciate that in practice many approximations could be

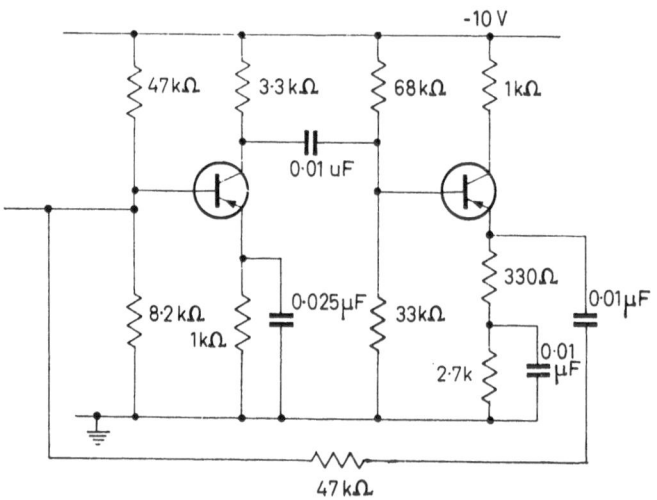

Figure 9.8. Circuit for the solution to Example 9.5

made in the above calculations. Full details have been given to demonstrate the various principles that have been discussed in this book.

In general, the equivalent circuit methods discussed in these chapters provide a very useful tool for the solution of a wide range of electronic circuits. They are not applicable to circuits involving large a.c. signals or switching circuits. Should other devices such as field effect transistors come into general use the methods will be equally useful provided suitable equivalent circuit parameters are known.

EXAMPLES

Example 9.6. Repeat Example 9.1 for an angular frequency of 10^7 rad/sec.

Ans. 587 $\angle$ 277° 42′, 6·86 $\angle$ −27° kΩ.

Example 9.7. A valve oscillator employs Miller feedback and a crystal resonator in the grid circuit. At the required frequency of 1 MHz the crystal requires an additional 30 pF to produce resonance and the loss component is equivalent to 10 kΩ in parallel with the terminals. If C_{ag} is 3 pF and g_m 8 mA/V determine the value of tuning capacitor and the Q factor for an anode coil of inductance 0·05 mH.

Ans. 420 pF, 4·5.

Example 9.8. A transistor having the hybrid π parameters given in Example 9.3 is used as a wide band amplifier over the frequency range 100 kHz to 2 MHz. If the collector load is purely resistive of value 5 kΩ, calculate the voltage gain and input impedance at these frequency extremes. Use the exact method at 100 kHz, and use valid approximations at 2 MHz.

Ans. A_v 155 $\angle$ 164° 30′, Z_{in} 390 $\angle$ −49° Ω. A_v 34 $\angle$ 100°, Z_{in} 102 $\angle$ −8° Ω.

Example 9.9. An r.f. transistor is loaded with a parallel tuned circuit, having a coil of inductance 2 μH and Q factor 25. Calculate the tuning capacitor required to tune it to 80 MHz and the voltage gain if the transistor y parameters at this frequency are y_{ie} $(0·4 + j20\omega \times 10^{-12})$ mho, y_{re} 350×10^{-6} $\angle$ 300° mho, y_{fe} 0·02 $\angle$ 320° mho, y_{oe} $(150 \times 10^{-6} + j\omega \times 10^{-12})$ mho.

Ans. 2 pF, 37·5 $\angle$ 70° 48′.

APPENDIX 1

USE OF LOGARITHMIC UNITS

In many electronic systems, the signal level may vary from kilowatts to microwatts in different locations. It is convenient to express such a range of levels in logarithmic units. Also, where a number of circuits are cascaded, each multiplying or dividing the signal by a factor, the corresponding logarithmic units may be simply added or subtracted.

Two sets of logarithmic units are in common use, one using log to the base 10 and the other using log to the base e. In both cases the units are based upon *power* ratios.

If P_1 is the input power, and P_2 the output power, then by definition:

$$\text{Number of Nepers} = \log_e \frac{P_2}{P_1} \qquad (A1.1)$$

and

$$\text{Number of Bels} = \log_{10} \frac{P_2}{P_1} \qquad (A1.2)$$

The Neper is mainly used in transmission line problems and will not be discussed further in this book.

The Bel is an inconveniently large unit and the Decibel (db) is more convenient. This may be defined as:

$$\text{Number of Decibels} = 10 \log_{10} \frac{P_2}{P_1} \qquad (A1.3)$$

If the circuit in question is an amplifier, P_2 will be greater than P_1 and the result will be a positive number. If it is an attenuator, P_1 will be greater than P_2 and expression A1.3 will involve the determination of a negative logarithm. It is simpler under these circumstances to write:

$$\text{Number of db of attenuation} = -10 \log_{10} \frac{P_1}{P_2} \qquad (A1.4)$$

In practice, voltage or current ratios are commonly used and under certain conditions the db scale can be applied. If the input resistance is equal to the load resistance, expression A1.3 may be rewritten as:

$$\text{Number of db} = 10 \log_{10} \frac{I_2^2 R}{I_1^2 R} = 10 \log_{10} \frac{V_2^2/R}{V_1^2/R}$$

$$= 10 \log_{10} \left(\frac{I_2}{I_1}\right)^2 = 10 \log_{10} \left(\frac{V_2}{V_1}\right)^2$$

$$= 20 \log_{10} \frac{I_2}{I_1} = 20 \log_{10} \frac{V_2}{V_1} \qquad (A1.5)$$

Strictly the definition in equation A1.5 should only be used with equal load and input resistance. In practice, the frequency response of amplifiers is frequently quoted in db units, even where the condition is not maintained.

Thus at ω_h and ω_L for the RC coupled amplifiers discussed in Chapters 4 and 5,

$$A_{vh} = \frac{A_{vm}}{1 + j}$$

$$|A_v| = \frac{A_{vm}}{\sqrt{2}}$$

$$\text{Number of db} = 20 \log_{10} \frac{A_{vm}}{A_{vm}\sqrt{2}} = 20 \log_{10}\sqrt{2}$$

$$= 10 \log_{10}2 = 3{\cdot}010 \text{ db}$$

Thus these frequencies, ω_h and ω_L, are referred to as the 3 db frequencies and are quoted simply as they are most convenient to calculate.

WORKS FOR FURTHER READING

Electronic Fundamentals and Applications. J. D. Ryder. 3rd edn., 1964. Pitman Technical Books, London.

Electronics. P. Parker. 1050. Edward Arnold Technical Books, London.

Feedback Circuit Analysis. S. S. Hakim. 1966. Iliffe Books Ltd., London.

Field Effect Transistors. L. J. Sevin. 1966. McGraw-Hill Book Co., New York.

Semiconductor Junctions and Devices. W. B. Burford and H. G. Verner. 1965. McGraw-Hill Book Co., New York.

Transistor Electronics. D. De Witt and A. L. Rossoff. 1957. McGraw-Hill Book Co., New York.

Transistors. D. Le Croissette. 1962. Prentice-Hall, Inc., Englewood Cliffs, New Jersey.

Transistors: Theory and Circuitry. K. J. Dean. 1964. McGraw-Hill Book Co., New York.

INDEX